DEVELOPMENTS
IN
APPLIED
SPECTROSCOPY
Volume 6

DEVELOPMENTS IN
APPLIED SPECTROSCOPY

Selected papers from the Annual Mid-America Spectroscopy Symposia

A Publication of the Chicago Section of the Society for Applied Spectroscopy

DEVELOPMENTS IN APPLIED SPECTROSCOPY

Volume 6

edited by

William K. Baer

Nalco Chemical Company
Chicago, Illinois

Alfred J. Perkins

University of Illinois
College of Pharmacy
Chicago, Illinois

E. L. Grove

Illinois Institute of Technology
Research Institute
Chicago, Illinois

Selected papers from the
Eighteenth Annual Mid-America Spectroscopy Symposium
Held in Chicago, Illinois
May 15-18, 1967

ℚ PLENUM PRESS · NEW YORK · 1968

ISBN 978-1-4684-8699-5   ISBN 978-1-4684-8697-1 (eBook)
DOI 10.1007/978-1-4684-8697-1

Library of Congress Catalog Card No. 61-17720

Plenum Press
A Division of Plenum Publishing Corporation
227 West 17 Street, New York, N. Y. 10011

Preface

Volume 6 of Developments in Applied Spectroscopy presents a collection of twenty-eight selected papers from those that were presented at the Eighteenth Mid-America Symposium on Spectroscopy held in Chicago, May 15 to 18, 1967. In general, the papers selected by the editors are those of the symposium type and not those papers pertaining to a specific research topic that one expects to be submitted to a journal. Not all of the submitted papers were included. Some revisions could not meet the deadline and others were not accepted based on the advice of the reviewers. It is the opinion of the committee that this type of publication has an important place in the literature.

The Mid-America Symposium is sponsored annually by the Chicago Section in cooperation with the Cincinnati, Detroit, Indianapolis, Milwaukee, Niagara Frontier, and St. Louis Sections of the Society of Applied Spectroscopy, and the Chicago Gas Chromatography Group. Although the Mid-America is often thought of as a regional meeting, its attendees and authors generally come from all parts of the United States and Canada. Both applied and theoretical principles were provided in sessions on X-ray, emission, atomic-absorption, nuclear magnetic resonance, infrared, Raman, nuclear-particle, and gamma-ray spectroscopy; activation analysis; and gas chromatography. In addition, there were symposia on absorption spectra of biologically significant molecules; the structure of ice, water, and aqueous solutions; air and water pollution analyses; and the practical application of statistics.

The various chairmen in the Symposium committee, Dr. William K. Baer, Dr. Bruce Murray, Dr. Alfred J. Perkins, Dr. M. S. Wang, Dr. Herman Szymanski, Mr. Leonard Afremow, Dr. Mary J. Oestmann, Mr. James Hickey, Dr. George Walrafen, Mr. Elwin Davis, Mr. Jim McGinness, Mr. Charles Reagan, Mr. Dick Terry, Mr. L. R. Pearson, Dr. L. S. Gray, Mr. Tod Engelskirchen, Mr. Lew Malter, Miss Grace Marsh, Mr. Dick Scott, and Miss Martha Beeler, and the other committee members should be commended for the program.

Thanks should also be extended to the exhibitors for their part in the Symposium and the exhibitor seminars.

The editors wish to express their appreciation to the authors and to those who helped with the reviewing and editing. These include Dr. M. S. Wang, Dr. Elma Lanterman, Mr. Elwin Davis, Mr. Dick Terry, Dr. Robert Scholz, Dr. N. Ashford, Dr. Gordon A. Noble, Mr. Hugh O'Neill, Mr. John F. Kopp, Mr. Morris E. Gales, Mr. James J. Lichtenberg, Mr. Robert Booth, Mr. Robert C. Kroner, Mr. W. Revkin, Mrs. Ethel L. Grove, Mr. James A. Gibbs, Dr. B. Jaselskis, Dr. Mary J. Oestman, Dr. J. C. Hindman, Dr. C. L. Bell, Dr. T. P. Day, Dr. K. Nakamoto, and Dr. J. Ziomek.

W. K. Baer
E. L. Grove
A. J. Perkins

Contents

X-Ray Spectroscopy

Emission–Flame–Atomic-Absorption Spectroscopy

Statistical Applications

Nuclear Applications

Spectroscopy of Biologically Significant Molecules

Structure Studies of Ice, Water, and Aqueous Solutions

Pollution Studies

Gas Chromatography

X-Ray Spectroscopy

Comparison of Methods of Standardization of X-Ray Data

F. Bernstein

General Electric Company
Milwaukee, Wisconsin

Various methods of standardizing X-ray emission calibration data for day-to-day variations have been studied. The method of using an external standard either as a ratio or slope correction is clearly the best approach. The use of internal standard lines can be as effective, but this method can be susceptible to large errors over extended periods of time. The parallel-curve shift correction method has limited application. Finally, data are presented which show that the analytical precision of a fully automated X-ray spectrograph approaches the statistical accuracy of the X-ray measurements.

INTRODUCTION

Improvements in techniques and sampling procedures, coupled with a better understanding of X-ray phenomena, have brought about widespread acceptance of the X-ray emission analysis method. However, X-ray spectrographic methods are subject to many errors which can be divided into two major classes. In the first are those errors which arise from causes within the sample, and these may be listed as follows:

1. Particle-size effects
2. Mineralogical effects
3. Matrix effects
4. Interelement effects

Errors which arise from causes external to the sample may be listed as follows:

1. Wet chemistry errors
2. Sampling errors

3. Precision errors
   a. Instrument
   b. Operator

It is beyond the scope of the present paper to discuss the above effects. There are, however, many papers available in the literature which go into greater detail regarding these factors. A few references are cited which deal with matrix [1], particle size, and mineralogical effects [2–5], interelement effects [6], and a general treatment of X-ray emission problems [7]. It is the intent of this paper to discuss the last items listed above, namely, those under "Precision," which can be attributed to the operator and the instrument.

Once a suitable analytical method has been worked out and refined, the problem of applying the method is primarily that of keeping the instrument in calibration. In order to do this, several methods are commonly resorted to:

1. Ratio to an internal standard
2. Ratio to an external standard (slope-intercept method)
3. Parallel shift

It is the object of this report to study the effectiveness of these methods of standardization in compensating for day-to-day instrumental errors which are encountered in X-ray emission analysis. Such errors may be listed as:

1. Variations in excitation (kV, mA)
2. Tuning errors (crystal or 2θ)
3. Improper pulse-height selector settings

EQUIPMENT

A General Electric XRD-6 manual spectrograph and an XRD-410 fully automated spectrograph were used for this investigation. The latter unit performs virtually all operator functions automatically, including standardization, and prints the output directly in composition per cent.

PROCEDURE AND RESULTS

In order to evaluate the effects of the instrumental errors on working curves, conditions were deliberately changed in analytical systems and the results of these changes were studied. It was found that changes both in excitation and in tuning errors produced the same type of effect on working curves. This could be best described by a change in slope of the working curve with a constant intercept on the composition axis. This is illustrated in Fig. 1, where several working curves are drawn, which show the effect of change of kilovoltage or milliamperage, or 2θ positioning.

The effect of improper pulse-height selector setting is a more complex one for a gas-filled proportional counter, and the results of these types of errors are shown in Fig. 2. The straight line (curve 1) is the working curve obtained with proper pulse-height selector setting. The upper portion of curve 3 shows the effect of setting the lower limit of the pulse-height selector window incorrectly, i.e., at such a level as to cut through the family which is being measured. Curve 2, which bears to the right below the straight working curve, shows the effect of setting the upper limit of the pulse-height selector window so as to cut through a family which is being measured. The curves can be readily explained as follows: When the lower limit of the pulse-

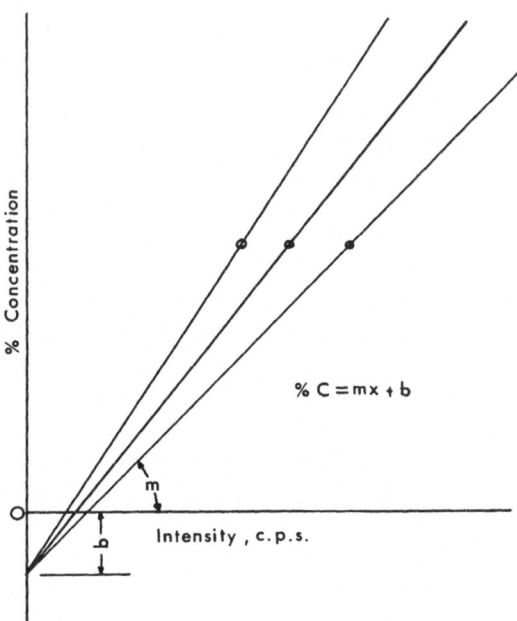

Fig. 1. Excitation changes, crystal tuning, or 2θ errors result in calibration of different slopes.

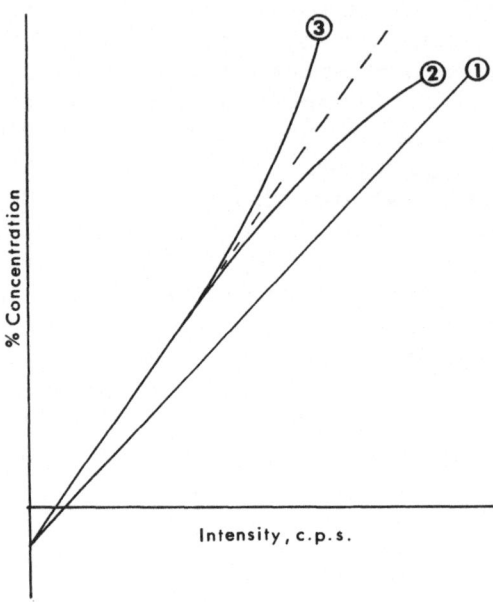

Fig. 2. Effect of improper pulse-height selector settings on calibration curves.

height selector is set too high, then the fraction of pulses which are rejected varies with the counting rate. As the counting rate increases, the relative number of pulses rejected will increase owing to the downspreading of the pulse family. Similarly, when the upper limit of the pulse-height selector window cuts through the pulse family, then an increased counting rate which produces downspreading will permit proportionately more pulses to be accepted in the window. Obviously, the magnitude of these effects depends upon instrumental factors, including the counting rate covered by the analytical system and the gain, voltage, and design of the counter tube which is being used in the system. It is quite apparent from the curves in Fig. 2 that none of the methods of standardization listed above can adequately correct for improper setting of the pulse-height selector.

Experimental data are shown in Table I for five elements in a nickel-base alloy with one standard and nine unknowns, the first sample being the standard. These data were taken on the XRD-410, which automatically standardizes by the slope-intercept method. To utilize this method, the following steps were employed: Standards were run in the conventional manner, and the equation of the regression curve was determined. The intercept of this line on the composition axis was determined, as well as the value from the regression curve of the standard to be used for subsequent standardizations. Whenever possible, this standard was chosen from the upper portion of the working curve to minimize statistical errors. When standardizing, the slope of the curve was redetermined from the count rate on the standard and combined with the fixed intercept determined above. It can be shown that this method of standardization is equivalent to utilizing ratios of the unknowns to the standard. However, the slope-intercept method permits more flexibility of operation since it is not necessary to run the standard with every set of unknowns.

It should be noted that the analytical results on the standard (sample 1, Table I) are always printed out uncorrected. Thus, the values for the elements in the standards are indicative of any changes which have occurred in the analytical system. In the data shown, the kilovoltage between the first and second runs was reduced by approximately 10%, which produced significant changes in the values for the standard. However, it is evident that the corrected values for unknowns 2 through 10 are independent of the kilovolt setting.

One other feature of the data shown in Table I is worthy of note, namely, the results obtained on the element columbium. The K_α line

TABLE I

Effect of Change of Kilovoltage on Analytical Data for Five Elements with Automatic Standardization

Sample	1	2	3	4	5	6	7	8	9	10
				Run 1:	60 kV					
Cr	S 7.96	8.44	7.61	7.09	5.93	7.80	9.54	7.90	7.70	7.66
Fe	S 0.06	0.15	0.34	0.79	0.05	0.05	0.05	0.04	0.04	0.04
Co	S 9.54	10.47	9.94	9.59	10.42	9.88	9.36	8.13	10.08	11.98
Cb	S 0.11	0.02	0.17	0.36	0.02	0.03	0.02	0.02	0.01	0.02
Mo	S 5.89	3.52	6.39	8.17	5.86	5.79	5.66	5.84	5.68	5.65
				Run 2:	55 kV					
Cr	S 4.84	8.42	7.59	7.07	5.90	7.77	9.51	7.87	7.64	7.60
Fe	S 0.05	0.16	0.35	0.81	0.05	0.05	0.05	0.04	0.04	0.04
Co	S 8.25	10.49	9.93	9.56	10.43	9.90	9.38	8.14	10.09	11.99
Cb	S 0.02	0.02	0.17	0.36	0.02	0.03	0.02	0.02	0.01	0.02
Mo	S 4.86	3.53	6.41	8.20	5.88	5.81	5.66	5.85	5.70	5.65

for columbium occurs in a region of relatively high background. Thus, the signal-to-noise ratio for low columbium concentrations is not very good. It was felt that the slope-intercept method of correction might not be applicable to this analytical case. The net signal-to-noise ratio for the 0.11% columbium was 0.63, while the first sample at 0.02% columbium was 0.36. As the data clearly show, in spite of the drastic change in the columbium standard from 0.11 to 0.02%, the unknown results are almost identical with those obtained before the kilovoltage was adjusted. Thus, the effectiveness of the slope-intercept correction for standardization even in the case of unfavorable background situations is confirmed.

Table II shows data taken on aluminum and titanium in nickel-base alloys. The first set shows a normal run on four samples and a standard for three elements using pentaerythritol (PET) and ethylene-diamine d-tartrate (EDT) crystals for aluminum. The second set shows the effect of changing two parameters: The kilovoltage was reduced to approximately half of its original value, and the PET crystal was detuned to the side of the aluminum peak. The data for runs 2, 3, and 4 were taken over an elapsed time of approximately 2 hr. The PET crystal is quite susceptible to drift due to temperature change, and this can be seen by the variation in percentages obtained on the standard. However, the ability of the slope-intercept method to compensate adequately for the changes which have been introduced into the system is clearly shown by the data on the unknown samples.

TABLE II

Effect of Change of Kilovoltage and Detuning Crystal on
Analytical Data with Automatic Standardization

	Sample	1	2	3	4	5	
Run 1	Al (PET)	S 6.15	5.73	5.65	5.87	5.45	60 kV
	Al (EDT)	S 6.18	5.65	5.71	5.82	5.55	
	Ti	S 1.07	0.98	0.97	0.97	0.97	
Run 2	Al (PET)	S 0.28	5.39	5.52	5.87	5.34	35 kV
	Al (EDT)	S 1.68	5.70	5.82	5.70	5.57	
	Ti	S 0.58	0.96	0.97	0.98	0.98	
Run 3	Al (PET)	S 0.63	5.69	5.82	5.90	5.36	35 kV
	Al (EDT)	S 1.76	5.61	5.66	5.90	5.47	
	Ti	S 0.57	0.96	0.97	0.98	0.98	
Run 4	Al (PET)	S 0.94	5.73	5.72	5.94	5.39	35 kV
	Al (EDT)	S 1.76	5.86	5.62	5.81	5.56	
	Ti	S 0.57	0.96	0.97	0.98	0.98	

TABLE III

Comparison of Standardization Methods
(Cobalt in nickel-base alloys)

No. of observations	Average determina- tion	Std. dev.	Relative std. dev., obs., %	Relative std. dev., calc., %
25 *	11.78	0.014	0.12	0.11
31 *	8.00	0.023	0.29	0.22
31 *	11.81	0.020	0.17	0.19
31 †	8.09	0.027	0.33	0.22
31 †	11.72	0.026	0.22	0.19

*Slope intercept.
†Parallel shift.

In order to determine the effectiveness of the standardization methods under actual operating conditions, nine samples and a standard of nickel-base alloys were run over a period of 3 days. Analyses were made for nine elements in the samples with automatic standardization immediately preceding each set of runs. Typical values for cobalt obtained by automatic standardization with the slope-intercept method are shown in Table III. It is apparent that the observed relative standard deviation is virtually the same as the calculated standard deviation. The latter was determined by combining the statistics of the standard sample with the statistics of the unknown sample. The same set of samples was rerun 31 times over a 2-day period. In addition to automatic standardization, intensity data were printed out for each run so that percentages could be calculated by means of the parallel-shift correction. The standard used for standardization was at the middle of the calibration curve. The average values for the high and low samples and the standard deviations obtained are shown in Table III. The average value for the low sample by the parallel-shift method is above the value obtained with the slope-intercept method, while the reverse is true for the high sample. These deviations would be expected since the changes in the working curve are primarily those of slope changes. Consequently, the use of a midrange standard would overcorrect the sample at the low end of the curve and under-correct the sample at the high end of the curve. The magnitude of the correction errors would be a function of the range which is covered by the working curve and the instrumental changes. An increase in either of these factors would increase the errors in samples at the ends of the range. The standard deviations observed in the samples by the parallel-shift method were reasonably good but slightly higher in both cases than those obtained by using the slope-intercept method. If the changes in intensity of the standard, which were of the order of ±1.5% over the period in question, were considerably larger, then the standard deviations observed with the parallel-shift correction method would also be considerably larger.

In order to verify the effectiveness of the standardization procedure applied to working curves for high concentration levels, a series of slag samples were run over a 4-day period during which 50 sets of analyses were made. Typical results are shown in Table IV for calcium oxide in the low and high samples with automatic standardization. It is apparent, as was true for the nickel-base alloys, that the observed standard deviations are very close to those of the calculated values.

As a basis of comparison, Table IV also shows data taken on a manual unit measuring calcium oxide in raw mix samples. The analytical working curve was established by determining the linear regression curve for a set of standards and by the running of three samples with a midrange standard once per day for 10 days. Data for the regression curve and sample runs were based upon ratios of their intensities to that of the external standard. The observed relative standard deviations are not as consistent as those obtained with the automatic unit, but they represent fewer data points. They are presented primarily to show the type of precision attainable with standardization on a manual unit.

The use of an internal standard element as a means of standardization was also investigated. If the material under study does not already contain such a reference line, it is possible when working

TABLE IV

Comparison of Precisions on Manual and Automated Units
(CaO in slag and raw mix)

No. of observations	Average determination	Std. dev.	Relative std. dev., obs., %	Relative std. dev., calc., %
50 *	30.28	0.081	0.27	0.24
50 *	52.35	0.125	0.24	0.20
10 †	40.23	0.051	0.12	0.13
10 †	43.50	0.103	0.23	0.13
10 †	43.12	0.130	0.30	0.13

*Automated unit.
†Manual unit.

with powders or solutions to add a known amount of an element not present in the system. A variation on this approach is to use a line from the X-ray tube scattered by the sample. Results obtained by the internal standard method were comparable to those with the external standard in some cases and much worse in others. It is believed that the nature of the internal standard method, which utilizes an intensity measurement on a second element, must be subject to errors which are independent of the first element. This is not to be construed as a reflection on the use of internal standards to compensate for matrix and interelement effects; the above remarks pertain only to internal standards for standardization of X-ray data.

CONCLUSIONS

1. The best method for day-to-day standardization of X-ray emission data is the use of an external standard, which can be used to adjust the slope of the regression curve. The intercept on the composition axis is maintained constant. An alternative approach is to ratio the unknowns to the external standard. These approaches will correct for errors in turning and those due to changes in excitation. They can produce data, the precision of which is close to predictions based upon statistical considerations.

2. An internal standard element can be used for standardization, but this method is susceptible to significant errors and is therefore not recommended.

3. No practical evidence of parallel shifts in working curves was found in this investigation. The method is simple and rapid and can give reasonably good results if elemental ranges and instrumental variations are small.

REFERENCES

1. E.L. Gunn, Advances in X-Ray Analysis, Vol. 6, Plenum Press, New York (1962), p. 403.
2. F. Bernstein, Advances in X-Ray Analysis, Vol. 6, Plenum Press, New York (1962), p. 436.
3. F. Bernstein, Advances in X-Ray Analysis, Vol. 7, Plenum Press, New York (1963), p. 555.
4. F. Claisse, Quebec Dept. Mines Prelim. Repts. No. 327 (1956).
5. W.J. Campbell, M. Leon, and J.W. Thatcher, U.S. Bur. Mines Rept. Invest. No. 5497 (1959).
6. J. Lucas-Tooth and C. Pyne, Advances in X-Ray Analysis, Vol. 7, Plenum Press, New York (1963), p. 523.
7. H.A. Liebhafsky, H.G. Pfeiffer, E.H. Winslow, and P.D. Zemany, X-Ray Absorption and Emission in Analytical Chemistry, John Wiley and Sons, Inc., New York (1960).

X-Ray Emission Quantitation
of Trace Elements in
Biomedical Research*

Marvin Goldman
and E. D. Beckman

Radiobiology Laboratory
University of California
Davis, California

Determination of trace-element concentrations in biologic materials is often limited by matrix effects and interelement interferences. X-ray spectrometry can provide rapid, nondestructive quantitation of trace metals at the parts per million level. Derivation of empirical absorption coefficients by utilizing standardized specimen preparation can optimize the precision and accuracy of determinations. The specimen is used as an absorber of the characteristic wavelength of the element to be quantitated and provides an index of the excitation efficiency essential to the quantitation of emission intensity. Food and tissue specimens were analyzed and the results compared with those obtained by other analytical methods. Calibration techniques using internal and reference standards are illustrated. Determinations of Ca, Sr, Zn, Tl, Se, Fe, I, and Ti were applied to the study of specific problems in metabolic, geriatric, and cancer research.

INTRODUCTION

X-ray emission spectrometry is utilized to good effect in several problems at the Radiobiology Laboratory in a program sponsored by the Atomic Energy Commission. The mission of this program is to assist in the evaluation of the hazards of certain radionuclides. During these studies, the metabolism of several radioelements and their distribution within biologic compartments is essential to a clear understanding of some of the pathways involved. Many of the examples cited were derived from experiments involving over 1000

*This work was supported by the U.S. Atomic Energy Commission, Contract No. AT(04-3)-472.

13

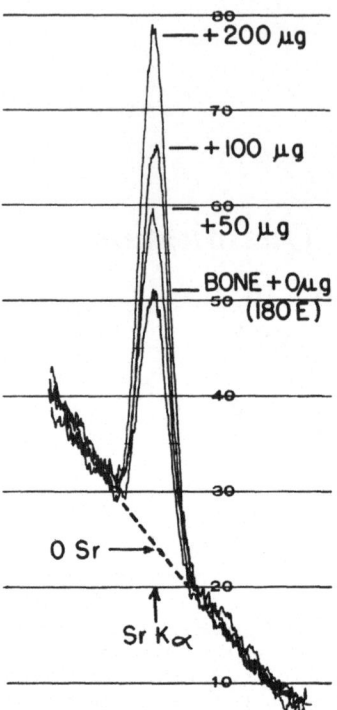

Fig. 1. Spectral scan of bone ash showing additions of known amounts of strontium.

beagles which are studied from embryonic time until senility. One of the major aspects of the program is an evaluation of the hazards of radioactive strontium 90 and radium 226, alkaline earths that concentrate in bone. In addition, several nutritional parameters relative to alkaline earth concentrations in the diet have been evaluated by means of X-ray spectrometry. Part of the plant operation involves decontamination of wastes containing radioactivity. Again, X-ray spectrometry has been of use in evaluating the efficiency of ion exchange. Certain of the more volatile elements, such as selenium and iodine, have also been investigated for sensitivity of quantitation and distribution within tissues.

Common to all of the above studies is the analyst's concern with accurate quantitation of trace concentrations of elements within light matrices such as food and tissue specimens and bone and waste materials. The purpose of this paper is to illustrate some empirical solutions to the problems of matrix effects using experimental material at hand and to indicate some areas in which further application of X-ray spectrometry may be of use to the biomedical scientist.

EXPERIMENTAL PROCEDURES

Internal Standards

The simplest empirical approach to circumventing the matrix effect is the use of an internal standard. There is, however, a practical limitation related to the number of samples necessary for a single quantitation. Figure 1 is a determination of elemental strontium in samples of beagle bone ash [1]. By the internal standards method, i.e., by addition of known quantities of elemental strontium to the unknown, a calibration for the strontium content was achieved. Notice that the intensity of the peak does not alter the appearance of the base line and that the slope of the base line is such that it is necessary to evaluate not only the Bragg peak but at least two "off angles." In this case, the bone labeled 180 E contained 212 μg Sr/g ash. Figure 2 shows the results of a calibration to determine the elemental strontium concentration in the feed on which the dogs are raised. By internal standardization again, the value for the food was estimated at 350 μg Sr/g ash. Notice, however, that eight specimens were prepared for the single determination.

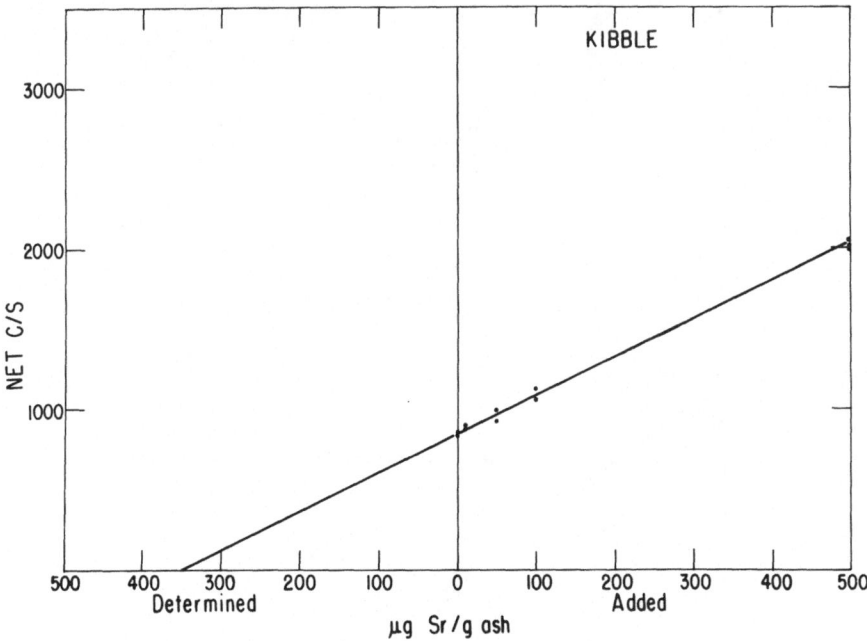

Fig. 2. Dog-food strontium content determined by the method of internal standards.

Matrix Transmittance Corrections

In biologic materials the presence of traces of strontium represents that of an element of intermediate atomic number within a matrix of relatively low effective atomic number [2]. Thus X-ray absorption within the matrix increases in proportion to increasing elemental concentrations. Therefore, the degree of internal absorption is related to the mass absorption coefficients weighted for the percentage of each of the elements within the matrix. If the matrix elemental composition is known, one may calculate the absorption coefficient from the sums of the coefficients of each element present. However, an empirical measure of the mass-absorption coefficient can be obtained from the determination of the transmittance of a mono-chromatic beam of X rays by the sample under investigation. The particular wavelength of the element being analyzed was used. This

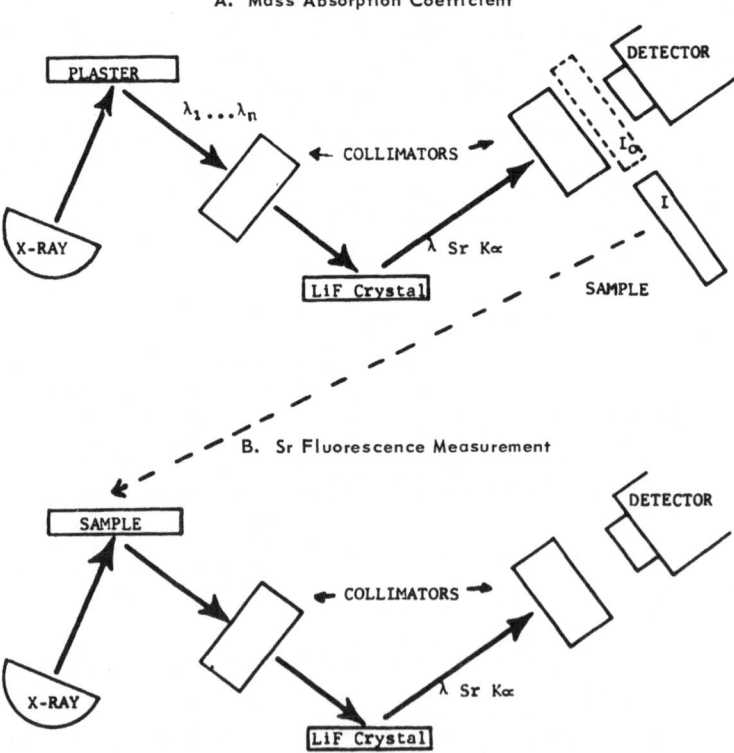

Fig. 3. Schematic arrangement for X-ray transmittance and fluorescence measurements of strontium.

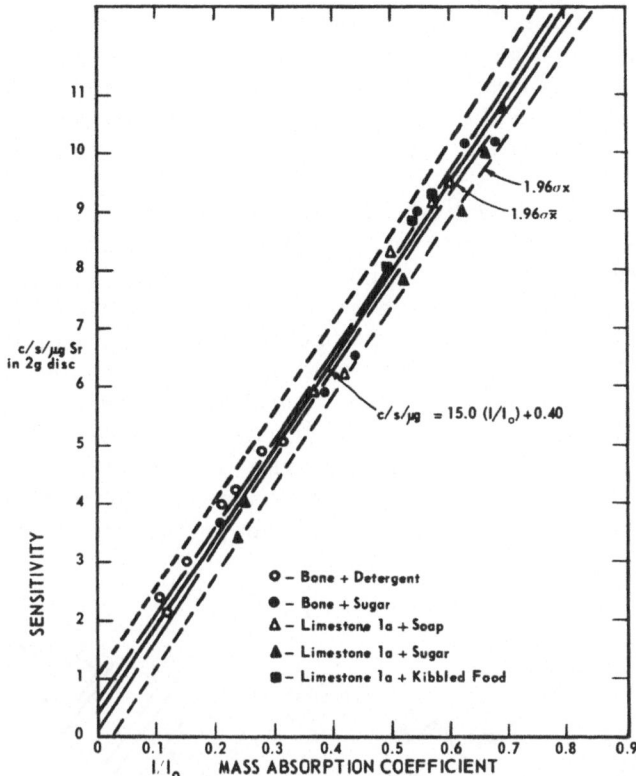

Fig. 4. Relation of Sr sensitivity (the count per second per microgram of Sr in the briquet) to transmittance (I/I_0) in several prepared matrices; (a)-- --, 95% error interval for a single measurement; (b) — —, 95% error interval for all data.

selection provides a practical solution of the matrix problem by quantitating interelement–absorption effects at the wavelength of the emergent analytical line. If the sample and the spectral energy of the exciting X-ray beam are kept constant, an integrated mass–absorption coefficient for any matrix not infinitely thick may be established for the selected set of conditions by using the appropriate attenuation of the energy of the analytical line. Thus, for strontium, the transmission at 14.1 keV provided a practical measure of the mass–absorption coefficient and permitted interelement-absorption corrections to be calculated without further knowledge of the elemental composition of the specimen.

In briquets of equal weight, the excitation of fluorescent radiation by the incident beam follows Beer's law in relation to sample density if it is assumed that an effective wavelength energy is incident on the briquet. Therefore, in a given specimen, a total absorption coefficient, which is the sum of the incident polychromatic beam and the subsequent fluorescent coefficients, influences the fluorescent intensity and, hence, sensitivity. The ratio of the total absorption coefficient to the fluorescent absorption coefficient in a sense characterizes the specimen.

With the X-ray spectrometer utilized, the absorption and intensity measurements were made as demonstrated schematically in Fig. 3. The strontium contaminant in plaster was used as a strontium source of radiation which collimated when excited by the polychromatic X-ray beam and then, diffracted by the crystal, provided a source of mono-chromatic 14.1-keV strontium K_α X rays. The unknown briquet was placed to intercept the emergent monochromatic beam. The ratio of the transmittance seen by the collimated detector with the beam unattenuated by the sample to that with the sample in the beam path was used to compute the effective mass-absorption coefficient. Subsequently, the same briquet was put in the spectrometer in the

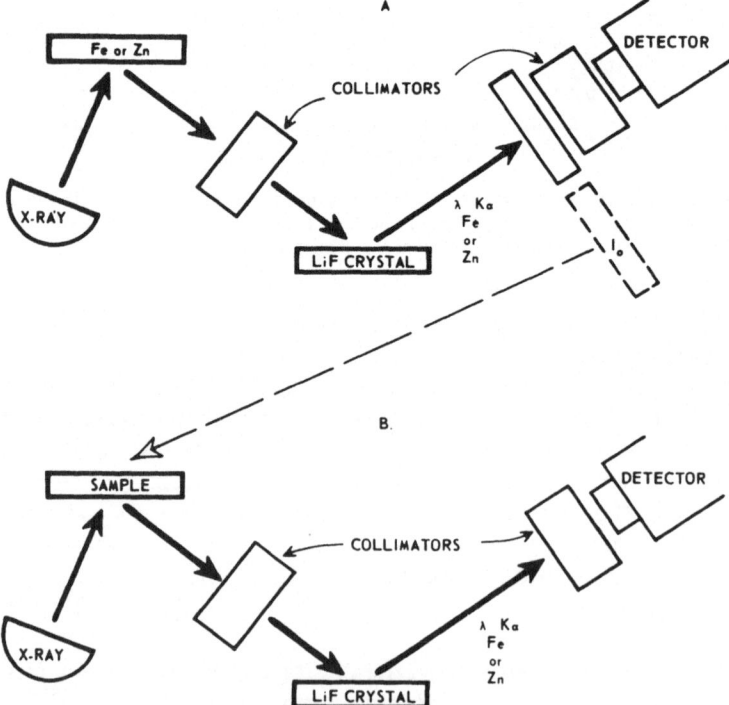

Fig. 5. Schematic arrangement for X-ray transmittance and fluorescence measurements of iron and zinc.

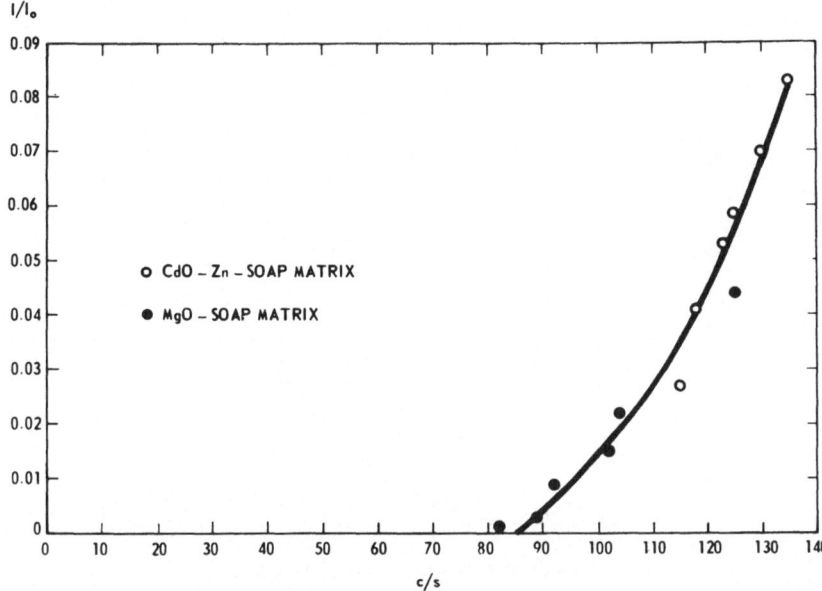

Fig. 6. Transmittance and background scattering at Fe K_α wavelength in nonferrous samples.

conventional position, and the strontium K_α net intensity was recorded. A series of briquets containing strontium was examined and a trans-mittance–sensitivity curve computed as shown in Fig. 4. The outer set of lines indicates the 95% error interval for a single measurement; the inner set brackets the same interval for all of the data. Regardless of the quality of the matrix in the 2-g briquets used, the values fit fairly linearly to the regression between sensitivity as counts per second per microgram and the fractional transmittance of the primary wavelength.

The effects of matrix scattering and attenuation are more marked as the energy of the characteristic radiation decreases [3]. Thus, in determinations for iron, whose K_α radiation is only 6.4 keV, the 2-g briquets used for strontium were too dense even when the lightest of the binders were used. Therefore, 1.25-in.-diameter, 5-ton-pressed briquets were prepared with a soap binder (Ivory Snow). A calibrated source of iron was the National Bureau of Standards' limestone 1a, which contains 1.63% of ferric oxide. When mixed with varying fractions of soap to form briquets, varied amounts of mag-nesium oxide or of cadmium oxide in zinc oxalate were used to vary the effective density and thus to alter the transmittance and scattering characteristics of these briquets. The positioning of the fluorescence source and sample are again shown in Fig. 5. Figure 6 illustrates the

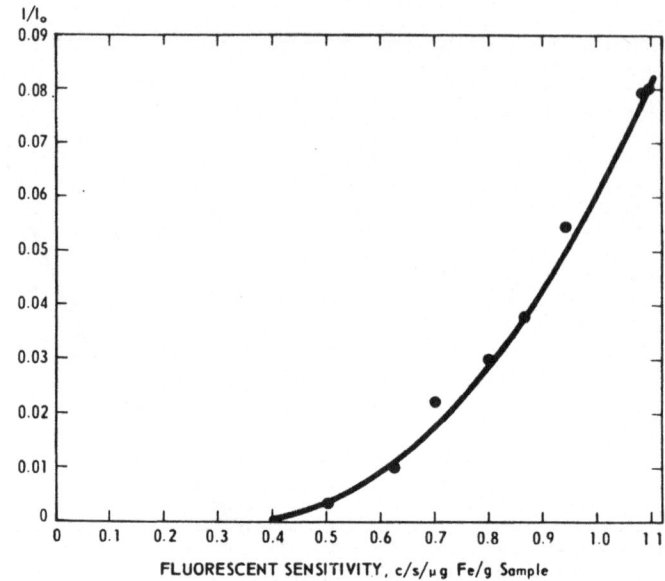

Fig. 7. Transmittance and Fe K_α fluorescent sensitivity of National Bureau of Standards' limestone standards.

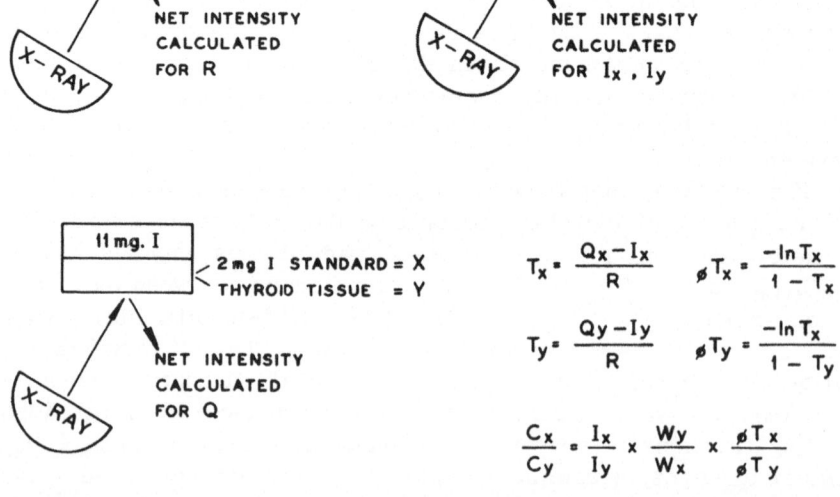

Fig. 8. Positioning of iodine standards and sample for iodine K_α transmittance and intensity measurements, and (lower right) the formulas used to calculate the weight fraction of iodine in the sample.

relationship between the background-scattering count rate of the iron K_α characteristic energy and the transmittance for the two matrices used. From these transmittance measurements, a sensitivity curve for iron in the limestone standards of varying transmittance was constructed as shown in Fig. 7. By use of these techniques, it was possible to determine the concentration of iron in such diverse specimens as ashed dog hair (8 mg/g), the dog food fed the experimental animals (4.5 mg/g of ash), and a synthetic milk product used in nutritional experiments on bone development (0.5 mg/g of ash). A similar procedure was utilized in the determination of zinc in biologic specimens. Concentrations of 0.2 mg zinc/g bone ash were found, while the food on which the animals were raised contained 0.6 mg/g of ash. The dog-hair ash contained 20 mg zinc/g of ash.

The previous figures showed a transmittance method for evaluating matrix effects in which two independent measurements were performed on the specimen to be analyzed. Leroux and Mahmud [4] recently published a similar method in which a sandwich is formed of the

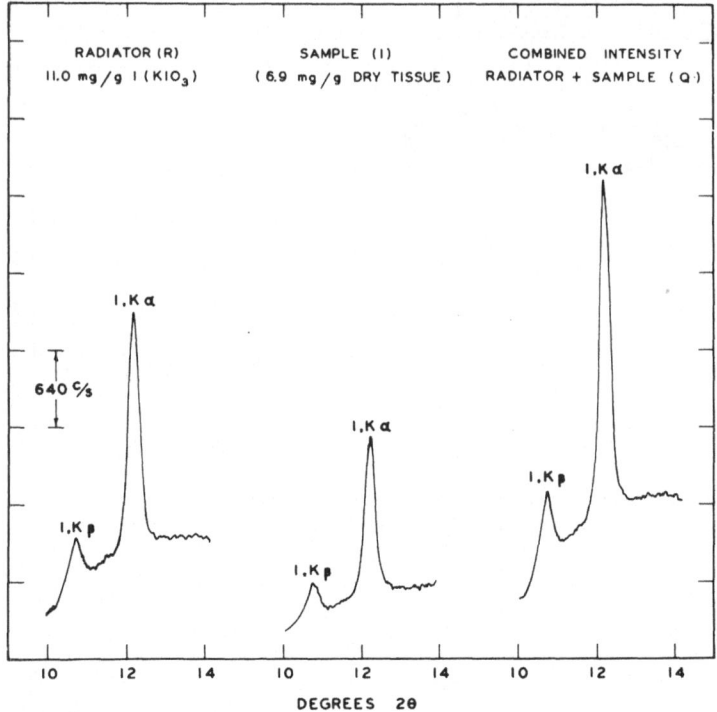

Fig. 9. Spectral scans of iodine standards and sample of swine thyroid (1.6 mg/g of wet tissue).

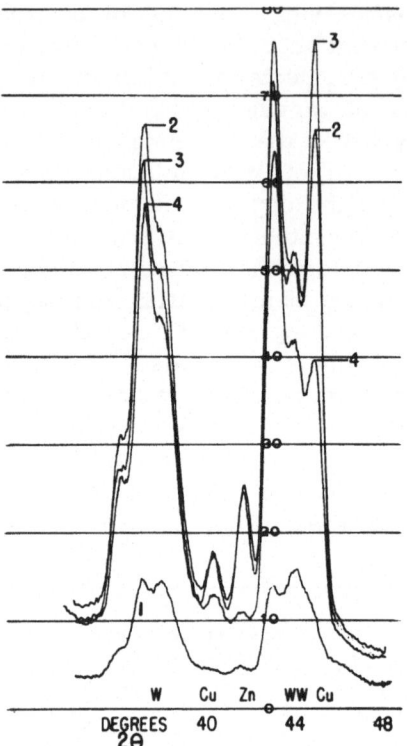

W   Cu   Zn   WW Cu

DEGREES   40   44   48
2θ

Fig. 10. Spectral scan demonstrating the removal of copper and zinc from waste samples by ion exchange resins.

Fig. 11. Calcium removal from raw and chemically clarified waste by ion exchange resins.

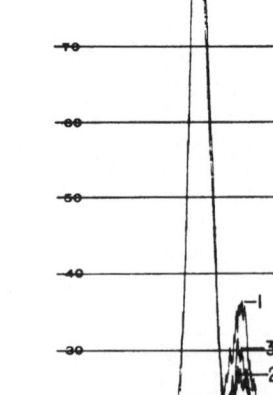

Ca   K

DEGREES 2θ   37   39

specimen to be analyzed and a source of characteristic energy.
Figure 8, a schematic of the determination of iodine, shows the
positioning of the briquets and the formulas used to quantitate the
element according to the principles discussed in their report. At the
upper left, the first of the series of five measurements provides a
value for I_0 in the analogy presented earlier. At the upper right is
the sample and a standard specimen which are alternately measured
for their iodine fluorescent radiation. These are denoted as I_x and I_y.
In this case, the iodine standard contains 2 mg I/g in a cellulose
matrix, and the sample is 1 g of lyophilized swine thyroid tissue.
Finally, the radiator, which contains 11 mg I/g, is placed above the
tissue specimen and the iodine standard alternately, and the net
intensity of the combined fluorescent radiation is measured. A
computation summarized at the lower right provides the necessary

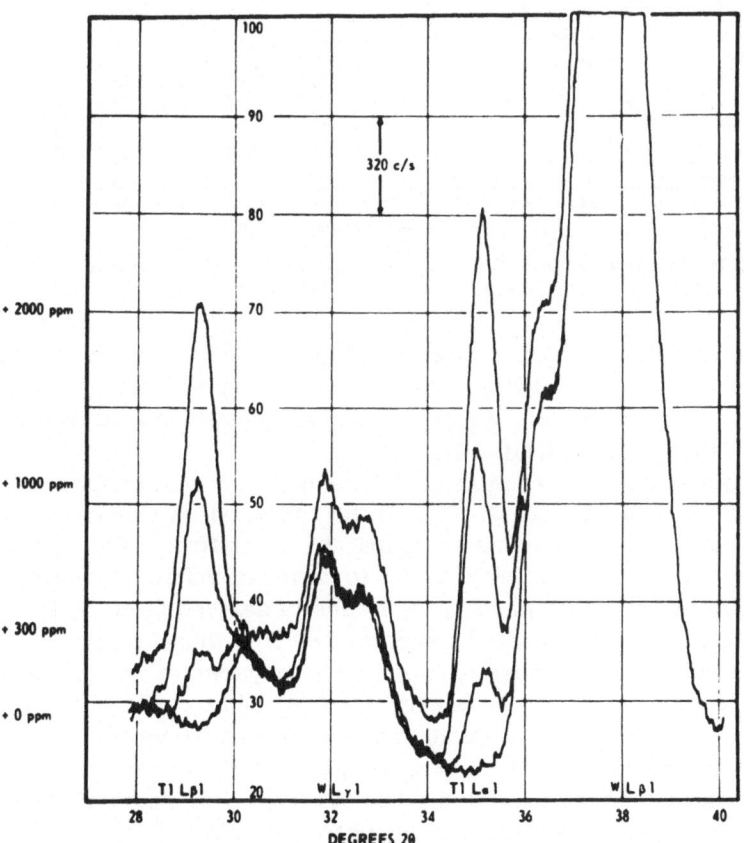

Fig. 12. Spectral scan of dried ground bird tissue with additions of thallium.

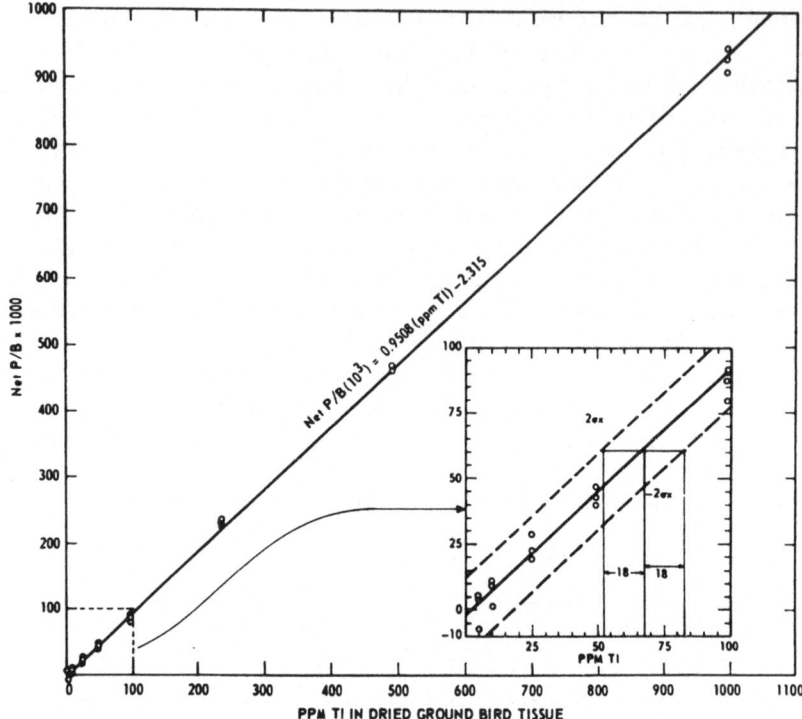

Fig. 13. Thallium L_β net peak-to-background ratio in dried ground bird tissue containing known increments of thallium.

factors to compute the concentration of iodine in the specimen, usually in parts per million. In this case of a 1-g briquet of lyophilized tissue, W refers to the weight of the briquet being analyzed and the weight of the standard (also 1 g).

A graphical representation of results for a specimen of swine thyroid is seen in Fig. 9. Notice the radiation intensity from the 11 mg/g iodine (KIO_3) source of iodine K_α's relative to the center scan of the thyroid (which was found to have an iodine concentration of 6.9 mg/g of dry tissue) and to the intensity derived from both of these irradiated simultaneously in sandwiches like those shown in Fig. 8. This method was generally applicable to most of the elements for which a characteristic wavelength could be detected. However, the sample itself must be sufficiently radiolucent to permit the characteristic wavelength of interest to be detected after transmission from the radiator through the sample or standard. For some of the lighter elements, 1.25-in.-diameter briquets weighing 0.5 g were used successfully. Mechanical stability limits the amount of specimen analyzed since a considerable fraction must be devoted to an adequate binder.

Other Applications

Another application of X-ray spectrometry is seen in Fig. 10. The results shown are from an attempt to evaluate the efficiency of an ion exchange system in removing cations from an effluent waste stream generated by a radionuclide treatment facility [5]. The lowest scan is the characteristic background in the energy region shown. The tungsten lines are characteristic of the tube target. Notice in this case that the Dowex 50 succeeded in removing much of the copper and zinc contaminant (resulting from the plumbing system). Curves 2 and 3 represent various stages prior to ion exchange; curve 4, the same sample following resin treatment. Strontium-90 wastes are also decontaminated in this fashion, and, since strontium is an alkaline earth, this process is closely paralleled by calcium removal. It was therefore of interest to determine the efficiency with which the alkaline earth cations were removed by this procedure. Figure 11 shows two of the chemical-treatment steps prior to ion exchange; curve 3 shows the same material following calcium removal by ion exchange.

One collaborative study relates to an interest in a heavy metal poison, thallium, which has been used in an attempt to control nuisance birds [6]. In Fig. 12, notice the qualitative nature of the thallium spectrum in which the thallium added to a light biologic matrix altered the background continuum. In this figure are shown the scans of 1 g of dried bird tissue with thallium increments added. Notice the relationship of the X-ray tube tungsten L lines to the line of the thallium L_β energy at 29° 2θ. Matrix effects in this instance were satisfactorily handled by computing the net peak-to-background ratio for each of the internal standards. The peak-to-background ratio as a function of thallium concentration is shown in Fig. 13. The inset indicates that the sensitivity of the method is approximately ± 18 ppm. Corrected for the desiccation value, at least 5 ppm of thallium in a 10-g sample of fresh tissue appears to be required for an adequate thallium quantitation.

Recently, interest has increased in the role of selenium in biology [7]. Selenium is considered an essential nutrient, but, like iodine, it is sometimes volatile and difficult to analyze. X-ray spectrometry was recently used to determine selenium in human and swine tissues. Figure 14 shows a typical scan of a gram of desiccated swine kidney, which was found to contain 10.5 ppm of selenium. Notice the proximity

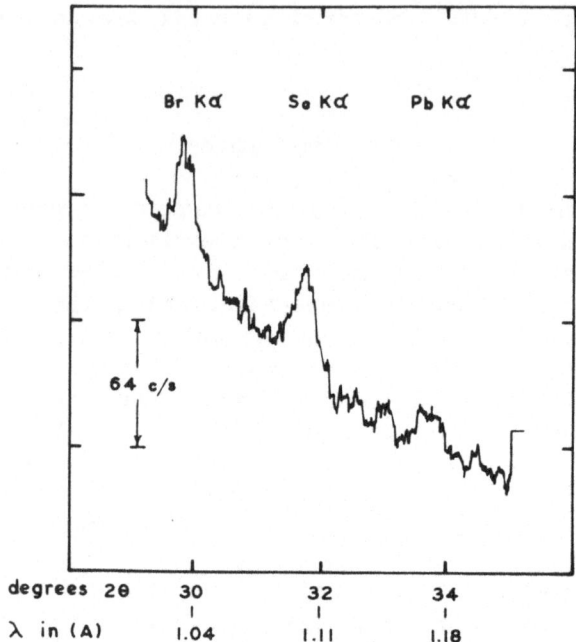

Fig. 14. Spectral scan of a sample of lyophilized kidney from Se-injected swine in the region
of the Se K_α peak.

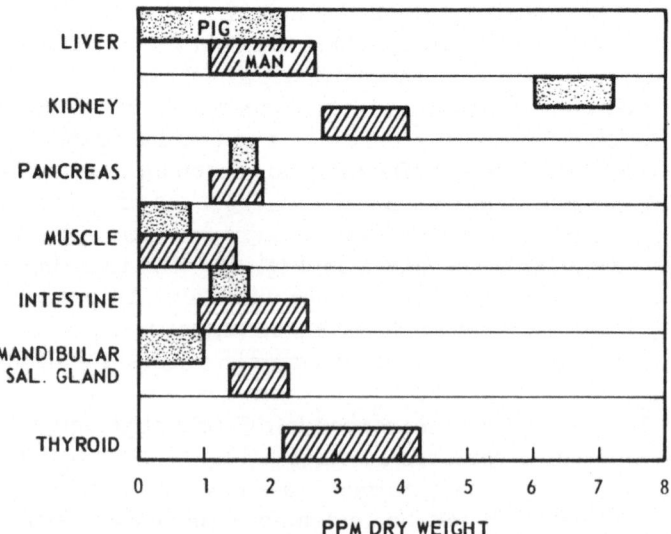

Fig. 15. Selenium content of lyophilized human and swine tissues as determined by X-ray
fluorescence spectrometry.

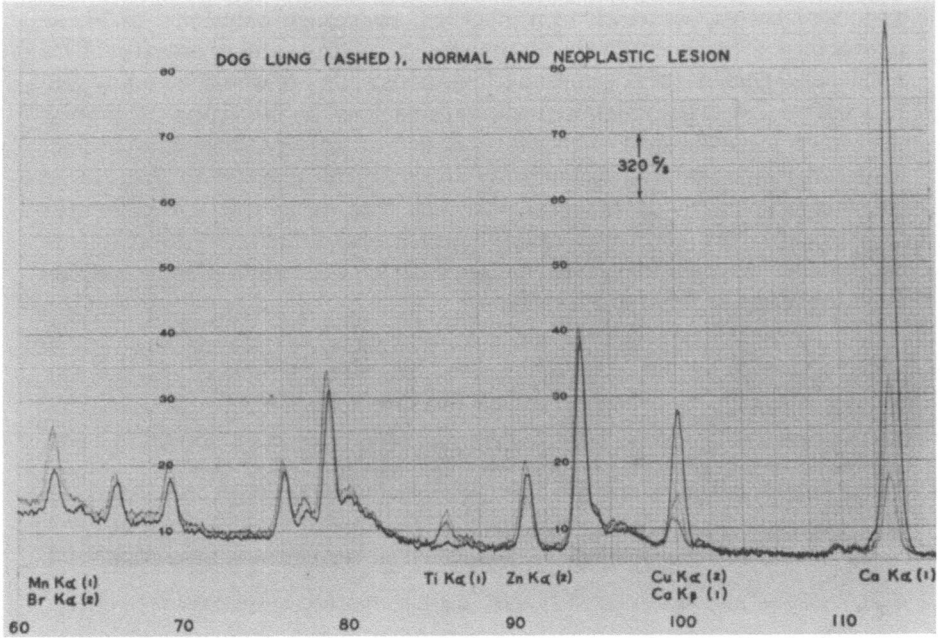

Fig. 16. Spectral scans of 700–mg samples of ashed dog lungs comparing normal and neoplastic tissues.

of the bromine peak. The lead is part of the background continuum in
the particular molybdenum target tube utilized. Selenium was
quantitated by using the matrix–absorption method of Leroux described
above. The results, in Fig. 15, demonstrate that the selenium
distribution is comparable in these samples from the two species
and indicate a 0 to 10 ppm range of concentration in these lyophilized
tissues.

Figure 16 illustrates another area of research in which it may be
possible to apply quantitative X–ray spectrometry. The three scans
are of 700 mg of ashed lung tissue from aged dogs. The high calcium
values are associated with an animal that died of a metastatic mammary
carcinoma. These neoplasms have been reported to be high in calcium.
In this case of metastatic invasion, the titanium, which is usually found
in lung, appears to have been displaced by the neoplastic tissue. It is
quite possible that several qualitative changes may be associated
with different types and degrees of neoplasia. Some of these changes
may be difficult to distinguish morphologically; in this case, X–ray
spectrometry may serve as an adjunct to quantitative pathologic
diagnosis.

To conclude, this has been a summary of some of the problems in
sample preparation and evaluation encountered in biomedical research

and illustrates several examples of proposed solutions to these problems. An additional advantage of X-ray spectrometry is its nondestructive nature. In many instances, e.g., it would be desirable to perform a radiochemical determination on the same aliquot or specimen for which the elemental concentrations have been determined. With the improvement in intensity and stability of a newer generation of spectrometers, it is quite possible that the X-ray spectrometer will equal or exceed many of the conventional laboratory instruments in the determination of trace quantities of elements of intermediate atomic weight in biologic tissues.

REFERENCES

1. M. Goldman, R.P. Anderson, and W. Gee, U.S. At. Energy Comm. Rept. UCD 108:75 (1963).
2. M. Goldman and R.P. Anderson, Anal. Chem. 37:718 (1965).
3. M. Goldman, C.K. Hui, and R.P. Anderson, U.S. At. Energy Comm. Rept. UCD 112:59 (1965).
4. J. Leroux and M. Mahmud, Anal. Chem. 38:76 (1966).
5. M. Goldman, R.P. Anderson, and F. Jassar, U.S. At. Energy Comm. Rept. UCD 108:123 (1963).
6. M. Goldman, R.P. Anderson, J.P. Henry, and S.A. Peoples, J. Agr. Food Chem. 14:367 (1966).
7. J.M. Fuller, E.D. Beckman, M. Goldman, and L.K. Bustad, "Selenium Determination in Human and Swine Tissues by X-Ray Emission Spectrometry," in: Symposium: Selenium and Biomedicine, O.H. Muth, ed., The Avi Publishing Co., Inc., Westport, Conn. (1967), pp. 119–124.

Determination of Element Localization in Plant Tissue with the Microprobe

H. P. Rasmussen

Michigan State University
East Lansing, Michigan

and

V. E. Shull and H. T. Dryer

Applied Research Laboratories
Dearborn, Michigan

Elemental distribution of aluminum, calcium, phosphorus, and potassium in corn seedling root and shoot tissues was determined with the microprobe. The seedlings had been grown in quartz sand for two weeks and transferred to a nutrient culture containing 10^{-3} M $AlCl_3$ or $CaCl_2$ for four days. The method of sample preparation was found to be critical to the localization of these elements. Aluminum and phosphorus were found in high concentrations at the root surface and within the root cap. Calcium and potassium, however, were distributed through the root and shoot tissues. Aluminum appeared to enter the plant conductive tissue through the break in the root surface caused by secondary root growth.

INTRODUCTION

Studies on element distribution and localization in plants has in the past been to a large extent dependent on the use of radioisotopes and autoradiography. Gross distribution of elements in whole plants or plant parts were obtainable by the use of X-ray sheet film in contact with the isotopically treated plant. More recently, micro-autoradiography has been employed to determine element distribution within tissues and to localize the elements in cells and subcellular organelles. This technique, however, has been limited since only those isotopes which emit low-energy beta radiation could be used on the micro-scale. A further unresolved problem in the use of isotopes is whether or not the plant responds to and utilizes an isotope as it does the unlabeled ion or compound.

The development of botanical histochemistry has made additional determinations of element distribution and localization by staining possible with a number of elements essential to plant growth and development.

The above techniques combined, however, have not been sufficient to study adequately either the essential or the toxic elements which influence the growth of plants. Electron microprobe X-ray analysis seems to present itself as a means by which these gaps can be filled. Not only will this allow analysis of elements heretofore impossible, but it also is a method for determining the accuracy of previously used techniques such as isotope tracer techniques. Microprobe analysis potentially allows for detection of elements with an atomic number from 5 on up. Detection limits of an element have been found to be about 10^{-15} g, as shown by Ingram and Hogben [1].

Fig. 1. A comparison of 2-week-old corn plants grown in nutrient culture: left, plant grown on full-strength Hoagland's solution plus 10^{-3} M calcium chloride at pH 5.5; right, plant grown on full-strength Hoagland's solution plus 10^{-3} M $AlCl_3$ at pH 3.2.

Fig. 2. Roots of a corn plant grown in nutrient
culture with high aluminum at pH 3.2. Note the
secondary root formation almost to the root tip,
also the nodule formation occurring throughout
the root system.

Use of the microprobe for plant-tissue analysis has been limited.
Läuchli and Schwander [2] in Switzerland have been able to detect
potassium, phosphorus, calcium, strontium, iron, silica, and sulfur
in the midvein of Zea mays (corn).

The element of interest to us in this study is aluminum. It has
been known for many years [3] that some plants grown on a soil with
a low pH are subject to injury by aluminum. McClean and Gilbert [4]
indicated that the greatest accumulation of aluminum was in the roots
of corn compared with that in the leaves. They also found that the
aluminum was localized in the cells of the cortex of those roots in
contact with the solution. This localization appeared to be in the
protoplasm and nucleus with little, if any, localized in the cell walls
or vacuoles. These results were obtained by the hematoxylin test
and should be interpreted with care since other polyvalent cations, such
as iron, will also give a positive test with hematoxylin. The above

TABLE I

Solution Culture Composition,
Concentrations, and Conditions
Used in the Aluminum-Toxicity
Experiments

Nutrient culture	
Major elements	g/liter
$Ca(NO_3)_2 \cdot 4H_2O$	1.18
KNO_3	0.51
KH_2PO_4	0.14
$MgSO_4 \cdot 7H_2O$	0.49
Iron sequestrene	0.0115
Minor elements	g/liter (1 ml/liter soln)
H_3BO_3	0.6
$MnCl_2 \cdot 4H_2O$	0.4
$ZnSO_4$	0.05
$CuSO_4 \cdot 5H_2O$	0.05
Treatments	
$AlCl_3$	0.917
$CaCl_2$	0.566
pH	
Aluminum treatment	3.2
Calcium treatment	5.5

results establish the theory that aluminum freely penetrates the epidermis and cortex but is blocked there by the endodermis from entrance into the xylem stream, which prevents movement of aluminum to the shoot. Furthermore, it is suggested that the primary action of aluminum is on the roots causing the precipitation of phosphorus and thus a phosphorus deficiency in the shoot [5].

Since aluminum toxicity is restricted to the root tissue, it may be postulated that the transition zone would block aluminum movement to the shoot. This would result in a gradient of aluminum from the blockage point down through the root system.

The effect of low pH on plants has been demonstrated by Rorison [6] and by McClean and Gilbert [4]. The root system of corn plants grown in a nutrient culture that contains $10^{-3}M$ of aluminum as $AlCl_3$

at pH 3.2 is shown in Fig. 1. The roots are severely stunted with severe abnormalities and secondary root formation occurring almost to the root cap as seen in Fig. 2. These protrusions, according to Rorison [6], are root initials which failed to penetrate the cortex. The shoots of treated plants, Fig. 1, exhibit reduced growth, and the older leaves develop a red color characteristic of phosphorus deficiency.

The nutrient culture compositions used with their corresponding pH are given in Table I. Control plants were grown on excess $CaCl_2$ to eliminate any possible effect of excess chlorine.

MATERIALS AND METHODS

Corn plants (Springold) for microprobe analysis were germinated in white quartz sand under approximately 600 ft-c of fluorescent light for 2 weeks. The plants were then transferred to nutrient cultures, Table I, for 96 hr. Two types of tissue preparation were used, the paraffin and cryostat methods. All samples were mounted on polished carbon discs, 1 in. in dameter.

Samples for paraffin section were fixed in formalin—acetic acid—alcohol (FAA). The tissue was then carried through the normal dehydration and infiltration steps [7]. Sections were cut at 12 μ on a rotary microtome and affixed to the carbon disc with Haupt's adhesive [7], and the paraffin removed with xylene.

Tissue for cryostat sectioning was cut with a razor blade and immediately mounted and frozen in Optimum Cutting-Temperature Compound (OCT −15 to −30°C, Fisher Scientific Company). Sections 16 μ thick were cut at −16°C and placed on the room-temperature polished carbon discs. The OCT cutting compound serves also as a stable mounting medium. The samples were allowed to air-dry before microprobe analysis.

Instrument conditions used for all analyses were 25 kV accelerating voltage and 0.05 μA sample current. No conductive coating was necessary with sections of these thicknesses.

RESULTS AND DISCUSSION

With the use of paraffin sections of the root tip, the cellular detail was preserved extremely well, as shown in the insert of Fig. 3. The delineation between the root cap and the epidermal layer of the root is very marked. Analyzing for aluminum from point A to B in Fig. 3 with a line scan indicated that aluminum was highest at the root surface and diminished toward the center of the root. The X-ray oscillogram, Fig. 4, for aluminum gave similar results.

In contrast to the paraffin preparation, oscillograms of cryostat sections of the root tip, Fig. 5, indicated that the aluminum is uniformly distributed within the root cap but that very little aluminum is present after traversing the epidermal layer of the root. The corresponding oscillogram for phosphorus indicated that its localization was

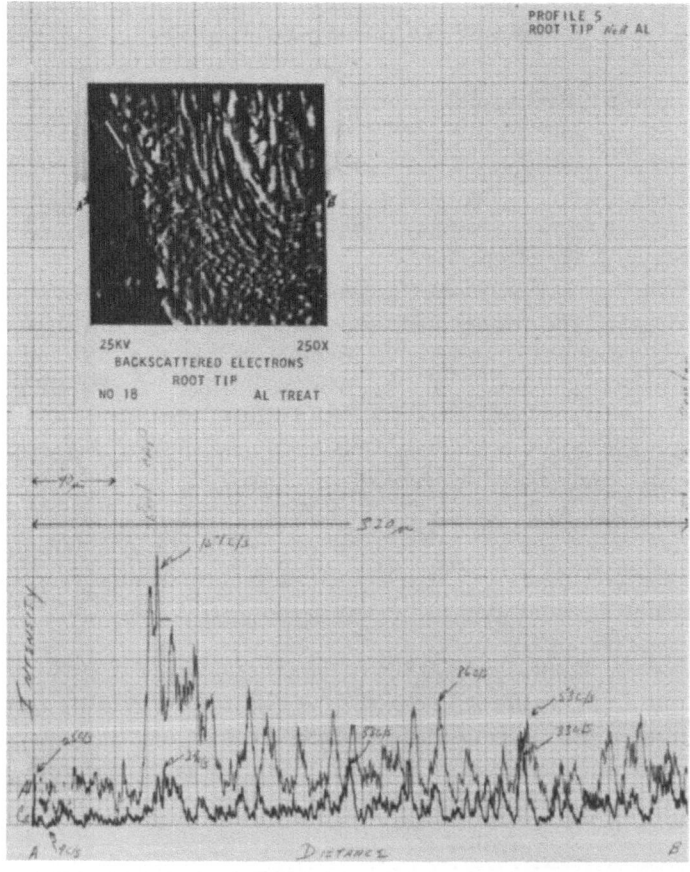

Fig. 3. Line profile analysis of the root-tip area of a corn plant trated with a high-aluminum nutrient solution. This tissue was prepared by the paraffin method. The line scan progressed from point A to B as indicated on the inserted backscattered-electron oscillogram.

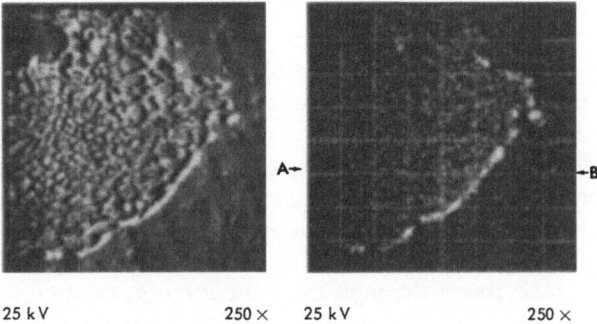

25 kV   250 ×   25 kV   250 ×

Fig. 4. Oscillograms of the root tip of an aluminum-treated
plant which was prepared by the paraffin technique: left,
backscattered electron oscillogram; right, aluminum X-ray
oscillogram.

the same as for aluminum. This was confirmed by the line scan shown
in Fig. 6. Concomitantly, Fig. 6 also confirms the fact that aluminum
readily penetrates the root cap but is blocked almost completely by
the epidermal layer of the root. The line scan for calcium does not
follow this pattern. These data confirm the hypothesis that aluminum
phosphate is precipitated within the root cap and at the epidermal layer.

The fact that the paraffin preparation resulted in a decreasing
amount of aluminum from the root surface toward the center of the
root tip indicated that aluminum phosphate must be bound to different
degrees within the tissue.

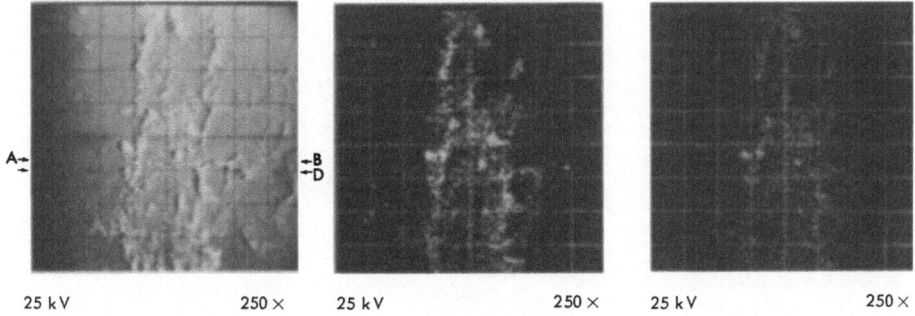

25 kV   250 ×   25 kV   250 ×   25 kV   250 ×

Fig. 5. Root tip of a corn grown in high-aluminum nutrient culture and sectioned by using
the cryostat procedure: left, backscattered electron oscillogram; center, aluminum X-ray
oscillogram; right, phosphorus X-ray oscillogram.

Analysis of a cryostat section just above the root cap also indicated that aluminum was not penetrating the epidermal layer but was accumulating at the surface. This aluminum was very tightly bound, as shown by paraffin sections which gave results similar to those of cryostat sections.

Longitudinal cryostat sections taken at the point of the emergence of a secondary root provided an answer to the mode of entry of aluminum into the root. Aluminum and phosphorus oscillograms, Fig. 7, revealed that the surface of the secondary root, which had physically forced itself through the endodermis, cortex, and epidermis

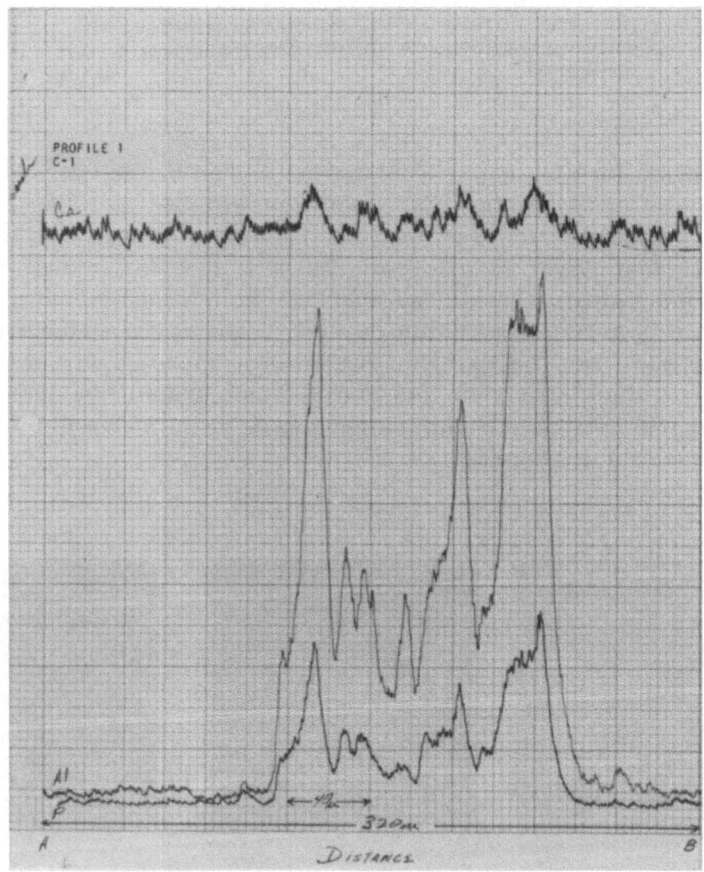

Fig. 6. Line profile analysis of Fig. 5 demonstrating the distribution of aluminum Al, phosphorus P, and calcium Ca. One 40-μ square on the graph equals a similar and corresponding square on the oscillogram.

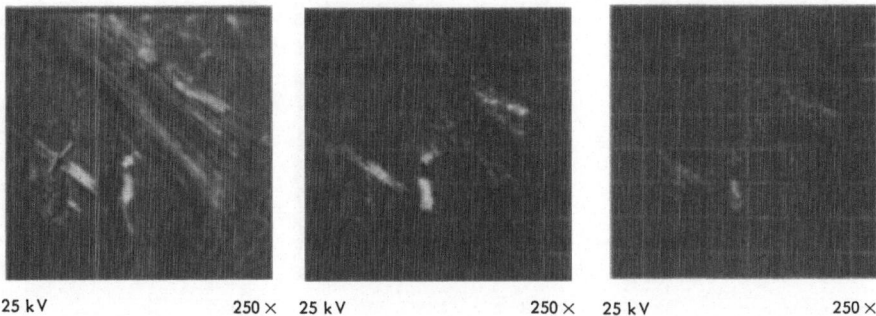

25 kV   250×   25 kV   250×   25 kV   250×

Fig. 7. Oscillograms of an aluminum-treated secondary root (longitudinal cryostat section) which has penetrated the cortex (at arrow) and epidermis of the main root: left, backscattered electrons; center, aluminum X-ray distribution; right, phosphorus X-ray distribution.

of the main root, had aluminum phosphate present in high concentrations. Correspondingly, as the aluminum entered the root along this surface, it moved into the cortex of the main root and on into the vascular tissue of the secondary root, Fig. 7. Once, however, that aluminum has penetrated the root epidermis, the endodermis does partially block movement of aluminum into the vascular tissue. This is indicated in Fig. 8. In these oscillograms, as before, the root surface is shown to be high in aluminum phosphate (lower right of the aluminum and phosphorus oscillograms of Fig. 8 at arrow), as is the endodermis (middle of center and right oscillograms of Fig. 8). Thus, the mode of entry of aluminum is through the injury made by the emergence of a secondary root from the main root. Once aluminum has entered the root, it appears to diffuse throughout the root system where a major portion is precipitated out as aluminum phosphate.

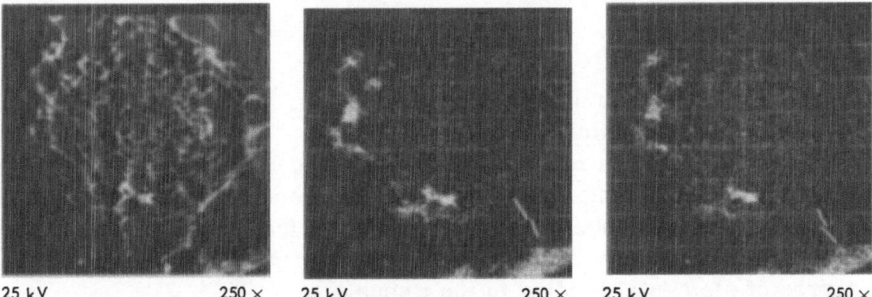

25 kV   250×   25 kV   250×   25 kV   250×

Fig. 8. Secondary root (cross section) of an aluminum-treated secondary root, with root surface indicated by arrow: left, backscattered electrons; center, aluminum X-ray distribution; right, phosphorus X-ray distribution.

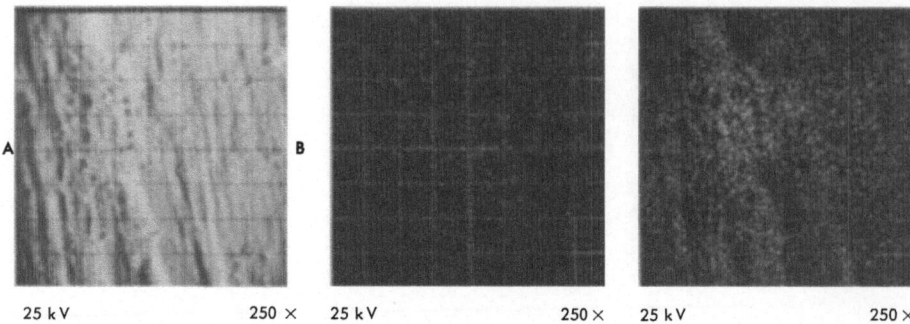

25 k V   250 ×   25 k V   250 ×   25 k V   250 ×

Fig. 9. Cryostat section of the transition zone of an aluminum-treated plant, focusing particularly on conducting tissue: left, backscattered electrons; center, aluminum X-ray distribution; right, potassium X-ray distribution.

Since aluminum phosphate is insoluble in water, very little is available for translocation to the transition zone, Fig. 9, or the leaf. The transition zone does not form a barrier to movement of aluminum. An analysis of the transition zone for potassium confirmed the validity of our sample preparation technique (cryostat) in that the potassium seemed to be localized within the vascular tissue of the transition region, Fig. 10. Paraffin techniques, on the other hand, resulted in complete loss of potassium from all tissues, as is shown in Fig. 11. The small localized areas of calcium are artifacts from the paraffin. In this particular wax (Fisher Tissue mat), we have found calcium carbonate crystals by X-ray diffraction analysis; therefore, for calcium or carbon analysis, this material proves unsatisfactory.

Analysis of sheath and leaf tissue revealed only small quantities of aluminum in the aboveground parts. Potassium, on the other hand, was found to be very high in the leaves, especially the leaf veins, as illustrated in the control section Fig. 12.

SUMMARY

This paper is summarized in two parts, (1) sample preparation and (2) aluminum toxicity.

Paraffin preparation, while giving good cellular detail, causes considerable leaching and movement of elements within the tissue and may also introduce artifacts which could lead to false conclusions. This technique may be useful, however, in giving some idea of the degree of elemental binding in the tissue.

The cryostat technique, with the use of OCT cutting medium, while not yielding the quality of cellular detail of paraffin, retains the elements within the sections. The retention, however, does not extend to the cellular level. In a single xylem element of a transition zone, Fig. 13, potassium has moved from the middle of the element to the wall and spread as well somewhat beyond the wall owing to knife penetration of the tissue and the subsequent air drying of the sample on the carbon disc. This technique on a tissue level proves to be adequate as well as time saving.

A moving-gas freeze-drying apparatus [7] is being constructed to prevent such element movement.

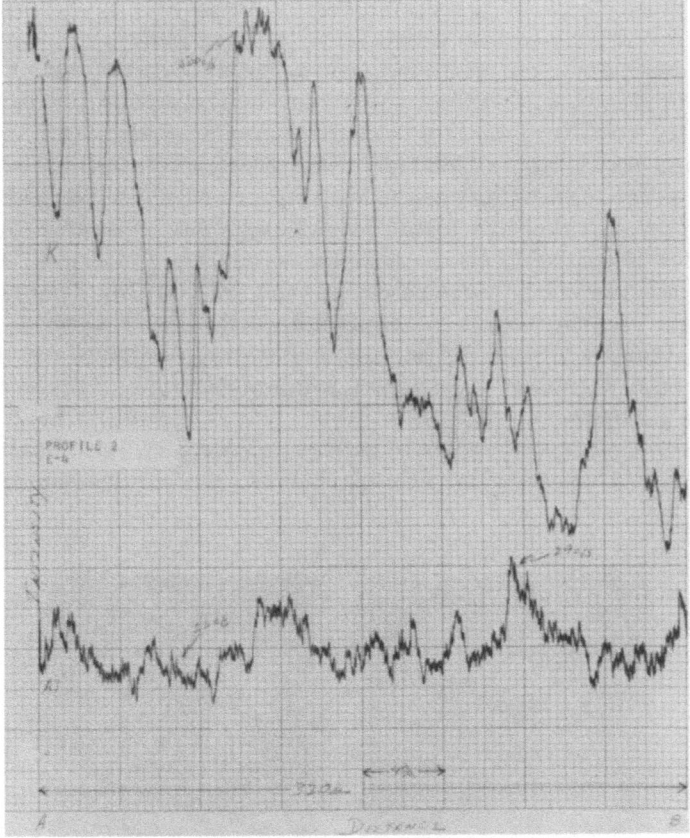

Fig. 10. Line profile analysis of the transition zone of an aluminum-treated corn plant, indicating the extremely low aluminum Al content: potassium K remained high in this cryostat section (line profile from point A to B of Fig. 9).

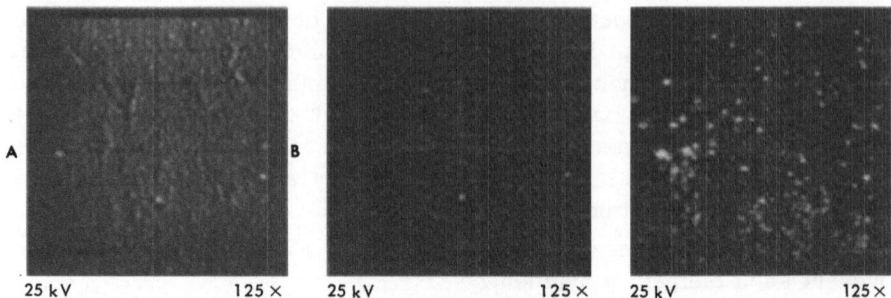

25 kV   125 ×   25 kV   125 ×   25 kV   125×

Fig. 11. Oscillograms of the transition zone of an aluminum-treated plant prepared for analysis by the paraffin technique: left, backscattered electrons; center, potassium X-ray distribution; right, calcium X-ray distribution showing contaminants present in the paraffin.

To summarize our aluminum work, the root epidermis of the corn plant appears to be an effective barrier to aluminum penetration even though the root cap is freely permeable to aluminum. This epidermal layer and also the outer surface of the root cap bind aluminum tightly enough to retain it during washing with either organic solvents or water. This factor explains why many investigators have advanced the theory of cortical aluminum buildup and blockage by the epidermis. Although the roots may be thoroughly washed in distilled water, enough aluminum remains to give a strong stain with hemotoxylin.

The transition zone did not accumulate any aluminum and therefore could not be a means by which aluminum is prevented from entering the tops of plants.

From our data, we propose that the root effectively selects against aluminum as long as the root surface is intact. The penetration of

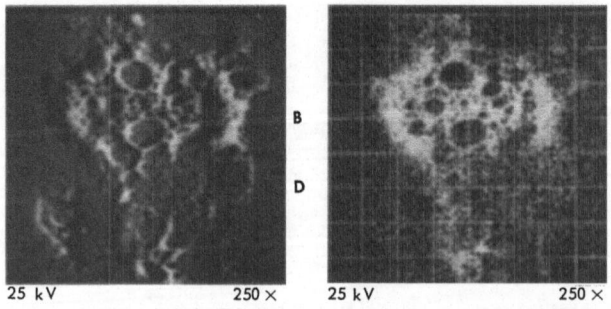

25 kV   250 ×   25 kV   250 ×

Fig. 12. Oscillograms of a corn leaf blade grown in high $CaCl_2$: left, backscattered electrons; right, potassium X-ray distribution showing good localization in the vein of the leaf.

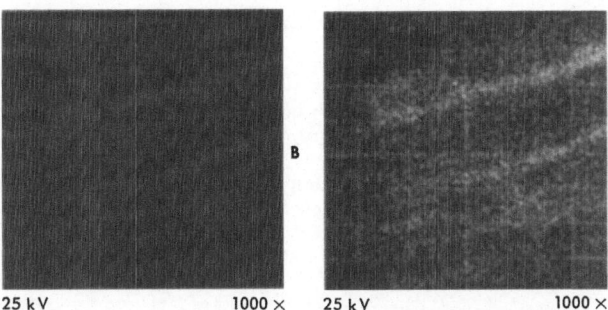

25 kV   1000 ×   25 kV   1000 ×

Fig. 13. Analysis of a single xylem element from the transition zone of the corn plant, the oscillograms indicating that the cryostat technique, although adequate on a tissue level, is not sufficient for analysis on a cellular or subcellular level: left, backscattered electrons; right, potassium X-ray distribution indicating movement away from the xylem element.

aluminum into the root occurs when a secondary root forces its way through the endodermis, cortex, and epidermis of the main root. The aluminum moves in with the soil solution along the surface of the secondary root from which it diffuses readily into the cortex of the main root and the xylem and cortex of the secondary root. Most of the aluminum is precipitated out of the solution as aluminum phosphate, which causes a severe phosphorus deficiency, especially in the top of the plant. That aluminum which remains in solution causes injury and death of the meristematic cells, which results in stunting the roots. Those meristematic cells which are not killed form the nodules which appear on roots treated with aluminum.

The stunting of growth of the root system, coupled with the precipitation of phosphorus, causes the top to be stunted or killed by nutrient starvation. Several investigators—Rorison [6], Jones [8], Wright [5], Clymo [9], and Bollard and Butler [10]—have hypothesized that aluminum is acting on the root primarily and secondarily on the shoot. We have provided evidence for the first time to substantiate this theory.

REFERENCES

1. M.J. Ingram and C.A.M. Hogben, "Electrolyte Analysis of Biological Fluids with the Electron Microprobe," Anal. Biochem. 18: 54–57 (1967).
2. A. Läuchli and H. Schwander, "X-Ray Microanalyzer Study on the Localization of Minerals in Native Plant Tissue Sections," Experientia 22: 503–505 (1966).
3. G.E. Hutchinson, "The Biogeochemistry of Aluminum and Certain Related Elements," Quart. Rev. Biol. 18: 1–29, 129–153, 242–262, and 331–363 (1943).

4. F.T. McClean and B.E. Gilbert, "The Relative Aluminum Tolerance of Crop Plants," Soil Sci. 24: 163–175 (1927).
5. K.E. Wright, "Internal Precipitation of Phosphorus in Relation to Aluminum Toxicity," Plant Physiol. 708–712 (1943).
6. I.H. Rorison, "The Calcicole–Calcifuge Problem II. The Effects of Mineral Nutrition on Seedling Growth," Ecology 48:679–688 (1960).
7. W.A. Jensen, Botanical Histochemistry, W.H. Freeman, San Francisco, Calif. (1962).
8. L.H. Jones, "Aluminum Uptake and Toxicity in Plants," Plant Soil 13: 297–310 (1961).
9. R.S. Clymo, "An Experimental Approach to Part of the Calicole Problem," Ecology 50: 707–731 (1962).
10. E.G. Bollard and G.W. Butler, "Mineral Nutrition of Plants," Ann. Rev. Plant. Physiol. 17: 77–112 (1966).

Procedures for the Study
of Biological Soft Tissue
with the Electron Microprobe*

Mary Jo Ingram and C. Adrian M. Hogben

University of Iowa
Iowa City, Iowa

Advances in electron microprobe X-ray analysis have opened a new frontier in biology. To be useful for the biologist, a low accelerating voltage is required to achieve an X-ray spot size of 1 μ or less.

When operating conditions are compatible with this spatial resolution and there is a sufficient concentration of element to be analyzed, special problems of sample preparation must be solved. If the constituent can diffuse readily, rigorous precautions are necessary. We shall describe steps in sample preparation permitting the display of K, Cl, and Na. Small blocks of tissues are rapidly frozen in liquid propane and subsequently dehydrated under vacuum. The dried tissue is embedded in epoxy. Sections about 2 μ are mounted on quartz.

The results of sample preparations will be described and correlated with morphological structure.

Standards for quantitative analysis have been prepared from solutions of albumin and salts.

INTRODUCTION

The functions of living systems depend importantly on the maintenance of unique concentrations of salts within and without the cells of an organism. Consequently, biologists would like to be able to display the distribution of ions in tissues. To appreciate some of the problems encountered in achieving such a display, it may be helpful to know that the soft tissues of animals are about 80% water with the common ions being present in concentrations of up to 0.5%. Characteristically, the concentration of K^+ is relatively high, and the concentrations of Cl^- and Na^+ are low within the cell. The reverse is true for the solution between cells. Individual cells are often 5 to 10 μ in diameter, although

*Supported by NSF Grant GB 1597.

there are larger cells such as the fibers of skeletal muscle that may be as large as 100 μ in cross section.   Manipulation of tissues may damage the barriers that maintain considerable concentration differences.   Since the ions are in solution, injury can result in rapid diffusion and redistribution unless special precautions are adopted.

The electron microprobe will permit analysis of the principal ions K^+, Cl^-, and Na^+, while achieving a necessary spatial resolution. In applying the electron microprobe to this problem, the primary concern has been to develop a sample preparation that will minimize ionic movement.   Other problems of concern include the attainment of a minimum X-ray spot, the development of suitable standards, and the reduction of background. This paper presents the progress made by the authors during experiments in the handling of these problems.

SPECIAL EQUIPMENT

In the work described subsequently, a 1963-model Applied Research Laboratories EMX electron microprobe was used.   The following 4-in. crystals with corresponding half-slit widths were used for the detection of characteristic X rays: potassium $K\alpha$ LiF, 7×10^{-3} Å; chlorine $K\alpha$ NaCl, 9×10^{-3} Å; and sodium $K\alpha$ KAP with pulse-height analysis.

Freeze drying of tissues has been accomplished with a system originally manufactured by Canalco.   However, this has since been substantially modified in our shop, and it is understood to be no longer marketed.

EXPERIMENTAL PROCEDURES AND DISCUSSION

Sample Preparation

To minimize ion movement as much as possible a technique involving freeze drying was adopted. Small bits (1 mm^3) of fresh tissue are plunged into liquid propane, cooled to the temperature of liquid nitrogen, and placed immediately in the vacuum system.   Liquid nitrogen is placed around the drying chamber.   In the early studies, temperature was maintained with a controlled heating element at -30°C for 12 hr and then brought to room temperature.   The dry tissue was then dropped in liquid Epon 826, to which the accelerator had been added previously. After 2 hr of infiltration, the system was brought to atmospheric pressure and the tissue samples removed and embedded. The plastic was polymerized for 3 days with a temperature sequence of 45, 52, and 60°C.

The Epon 826 mixture selected for optimum cutting consistency consists of 43 ml resin and 41 ml dodecenyl succinic anhydride with the addition of 1.5% dimethylaminomethyl phenol accelerator.

The blocks were cut by a steel knife on a Leitz rotary microtome, to a thickness of 4 to 5 μ. The sections were individually transferred to a carbon-cast quartz slide and flattened on a light coat of hot paraffin. In the early procedure and subsequently, a light coat of carbon has been evaporated over the section to make it electrically and thermally conductive.

The choice of Epon 826 as an embedding medium has evolved from the trial of a number of different polymers, some of which may yet prove to be suitable embedding media. Epon exhibits fairly good handling qualities and seems to allow the retention of the original ionic distribution. However, it presents a problem in the study of chloride in that it contains 30 mM/kg of organically bound chlorine. This amount is substantially less than in other Epons.

The use of an embedding media is controversial. Some of those critical of the procedure have placed their faith in a method of cutting frozen sections on a cryostat, followed by freeze drying the sections [1]. The authors hope to demonstrate in the near future that plastic embedding is not only compatible with retaining the original distribution of ions but is to be preferred for analyses with the microprobe since variations in density are largely avoided.

On the advice of Drs. Stirling and Kinter [2], tissue has more recently been dried at lower temperatures to prevent damage by ice-crystal formation and also to minimize the possibility of element redistribution. Starting at -65°C, the temperature is now increased by 5° every 12 hr. After the specimens have been brought to 30°C, they are gradually brought to room temperature. The tissue is subjected to vapor fixation. The dried tissue is placed in a vacuum desiccator with a desiccant and the selected fixative, which is followed by moderate evacuation of the desiccator. The infiltrating and embedding procedures have not been changed. The results from this approach are good, as we are now able to cut excellent sections of a thickness in the range of 1 to 3 μ. Without fixation, it was difficult to cut sections much below 5 μ.

Two fixatives that have been used, glutaraldehyde and osmium tetroxide, work equally well. Several milliliters of glutaraldehyde are introduced into the desiccator in a flat container. Fixation seems to be complete in about 24 hr. The fixed tissue, after embedding, has superior cutting qualities, appears normal to microscopic examination, and will accept common stains.

Osmic acid is placed in the desiccator in a crystalline form, from which it sublimes. Fixation occurs in 12 hr. The fixed samples are darker, and, under the microscope, the sections appear lightly stained. These samples do not accept the common aqueous stains.

The handling of thinner sections of 1 or 2 μ has proved to be more difficult. More curling is observed during cutting, but it has been found that effective flattening can be accomplished by heating the sections on a thin layer of paraffin and then gently pulling out the wrinkles.

Among the several methods of mounting the biological specimen, the following have been tried:

1. Sections can be mounted on a carbon-cast quartz slide, flattened as mentioned above, then recast with carbon. Quartz, however, may make a contribution to background because the section may not be infinitely thick. However, the transparency of the quartz slide is an asset for examination by transmitted light.

2. Certain types of grids, when covered with a thin film of carbon-cast colloidin will support sections nicely. The grid-supported sections are less thermally conductive, and visible distortion occurs under the beam.

3. A very smooth, flat surface is presented to the probe by facing the plastic block with the microtome and mounting the entire block in the specimen holder. Carbon casting provides conduction, but heat conduction does not seem to be as efficient as with sections on carbon-cast quartz. In addition, microscopic examination of the specimen is limited to reflected light.

Display of Tissue Electrolytes

Figure 1 is an example of one of the early successes. Frog skeletal muscle cut in cross section was advanced mechanically in steps of 3 μ under a static electron beam. Potassium and chlorine were measured simultaneously. The abscissa is the distance traveled; the location of the cells is indicated in the upper part of the figure by bars. The ordinate is the X-ray intensity of K K_α and Cl K_α. The black line traces the variations in the signal registered for potassium. If you relate this curve to the bars indicating when the beam was over a cell, you will note that potassium increases as a step function when a cell passes under the beam and decreases abruptly when the beam strikes interstitial space.   With the acceleration voltage employed in this instance, spatial resolution was not better than $6\,\mu$.

Two conclusions are drawn. The method of tissue preparation is compatible with the retention of the normal marked difference between the intracellular and extracellular concentration of potassium. Second, a sensitivity adequate to demonstrate this marked difference has been realized.   No conclusion is warranted about the homogeneity or hetero-geneity of intracellular potassium.

The circumstance for the analysis of tissue chloride is clearly less satisfactory in that one is faced with the disadvantage of an inherently high matrix background owing to the organic chlorine of the

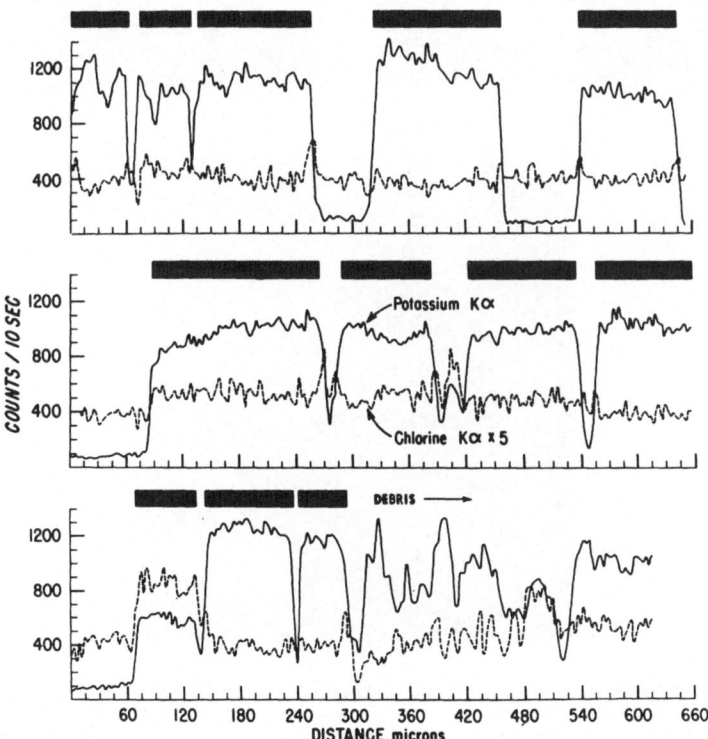

Fig. 1. Mechanical linear scans of frog skeletal muscle cut in cross section. The sample was advanced by steps of $3\,\mu$ under a stationary electron spot. The abscissa distance of sample advance is in microns; the ordinate detector output is for $K\,K_\alpha$ and $Cl\,K_\alpha$. The dark bars above each scan indicate that the electron spot was over a muscle cell. (20 kV, 0.05 μ A, 1 sp.)

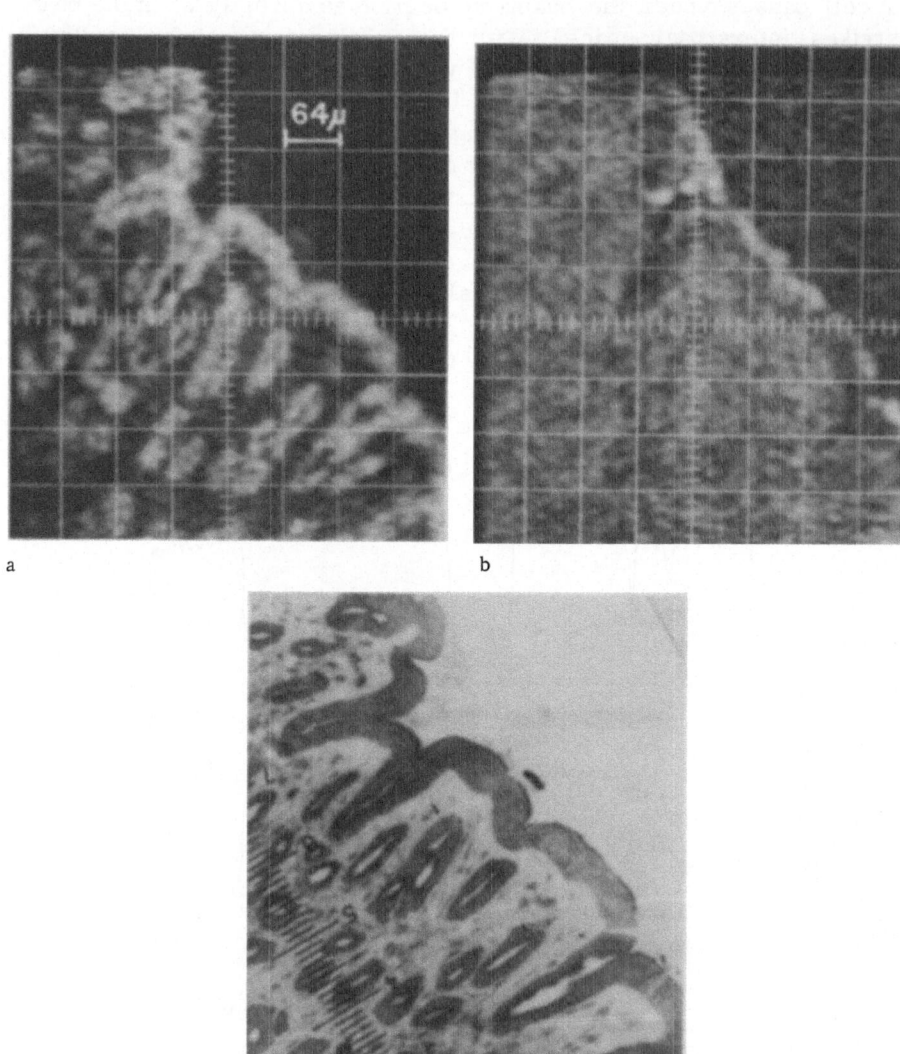

Fig. 2. Analysis of frog gastric mucosa. (a) Oscilloscopic display of the potassium K K_α ; lighter areas indicating regions of higher potassium content; (b) the same region scanned for Cl K_α ; and (c) light micrograph at the same magnification of the adjacent section stained with crystal violet.

plastic. It is clear that there is an additional problem. One would expect extracellular chloride to be displayed as a step function. Given the less favorable circumstances for chloride analysis, this might not be as apparent as the step increase of intracellular potassium. However, scrutiny of this figure and other similar studies indicates that there are real increases of chloride, varying in magnitude, at the cell edges. It is suspected, but not yet established, that the difficulty arises at least in part from the relatively low protein concentration of amphibian interstitial fluid. As a consequence, one or a combination of factors comes into play: a "salting out" during freezing; a "settling" of salt during freeze drying in the absence of a "sponge framework" of organic material; or a displacement by the embedding material.

The mechanical linear advance under a static spot has been preferred for more precise display. However, while modifying sample preparation had been the dominant concern, attention has turned to oscilloscopic display. Figure 2 is an example of the frog gastric mucosa shown with a light micrograph of an adjacent section. Gastrointestinal mucosa has the advantage of having rather clearly identifiable cell masses (epithelium) separated by a considerable interstitial space. The analysis was conducted at 22 kV with $0.15 \mu A$ on 5μ sections of Epon 826.

After osmium fixation, the resultant variation in sample current serves a useful function. The more conductive osmium modulates the sample current image of the section with excellent contrast, which allows a precise view of the area scanned and solves the problem of correlation of X-ray signal with tissue structure. Figure 3 is an example.

X-Ray Spot Size

It has become increasingly clear that, to be of great value to the study of biological tissues, a spatial resolution of at least 1μ must be attained. Several factors are involved.

1. The electron-beam size must be less than 1μ at the point of impact. Beam size is dependent upon the condition of many components as well as the design of the electron probe. The difficulty of achieving a submicron electron spot becomes accentuated at low acceleration voltages.

2. Independent of beam size, the spatial resolution is still a function of the energy of the electrons.

TABLE I

Approximations from Andersen's Graphs of
the Dependence of Electron Range on Acceleration Voltage, Computed for Matrix Density of
1.0; Si K_α in SiO_2 [3]

kV	Range, μ	50% X-ray production, μ	Width of X-ray volume, μ
20	9.50	2.50	5.8
15	7.00	1.60	2.8
10	3.00	0.80	1.0
5	0.95	0.25	0.4

TABLE II

Decrease in Signal-to-Noise
Ratio with Acceleration Voltage
KCl Crystals versus Quartz
Slide

kV	Potassium K_α	Chlorine K_α
20.0	1290	1960
17.5	1400	2620
15.0	1170	2060
12.5	980	2120
10.0	740	1770
7.5	470	1160
5.0	130	500

According to the findings of C. A. Andersen [3], the resolution can
be predicted for a given acceleration energy, critical excitation energy,
and matrix sample density. From his work, the information in Table I,
has been computed.   On the basis of this, work at 10 kV has been
elected for the best resolution without excessively compromising
sensitivity.

At this acceleration voltage, it is possible to work with thinner
sections in the knowledge that they will be infinitely thick, or nearly
so, to the beam's penetration. The reduced beam penetration and the
ability to use thinner sections is advantageous for the biologist:
The microstructure below the surface of the tissue is often different
from the surface structure (unless the tissue consists of fibers cut
in cross section).

Operation at a lower energy than 10 kV is a problem since electron backscattering does not provide adequate information for focusing the beam. The approach has been to observe the sample current images of 1000-mesh silver screen. Focus by this method seems to be comparable to backscatter methods of focusing; an example is Fig. 4. Fluorescent crystals such as benitoite are helpful for observing the beam.

Decreasing the accelerating voltage is accompanied by the expected decrease in sensitivity, Table II. However, the background also decreases so that the signal-to-noise ratio is not much less with 10 kV than with 20 kV. To reduce the background, pulse-height analysis is being employed. To attain a better signal-to-noise ratio, further steps will have to be considered, such as the use of a backing material of low atomic number.

Because the signal is reduced with a lower accelerating voltage, electronic storage has been used to ensure that the difference between the signal and background signal becomes significant. With the probe operated in the line-scan mode, detector output is stored in a 1024-channel analyzer called an "Enhancetron." (The number of channels utilized in the Enhancetron is excessive, and a modification is anticipated to allow a reduction to 100 storage channels.) The data are

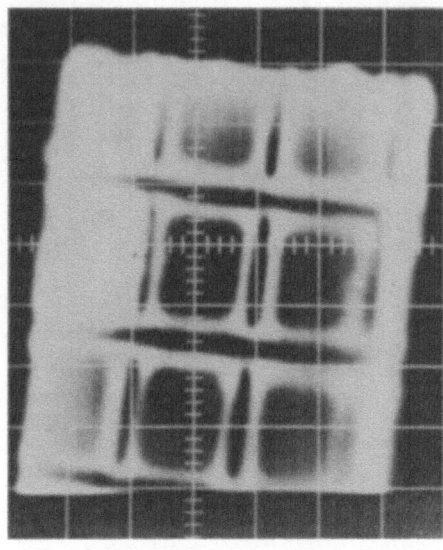

Fig. 3. Sample current image of silver grid at 10 kV (25 μ/grid).

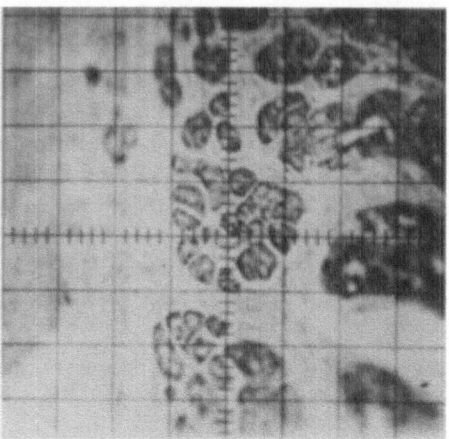

Fig. 4. Sample current image of frog gastric mucosa at 10 kV (about 500 by 500 μ). A 1- to 2-μ-thick section of tissue fixed with osmium.

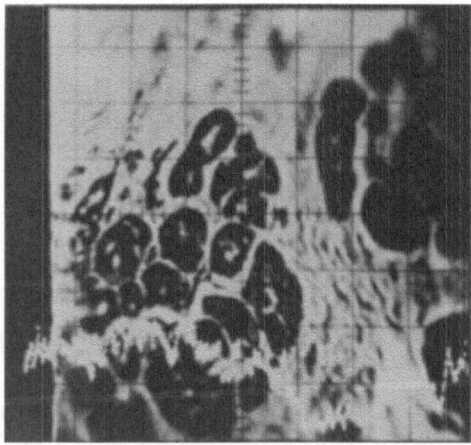

Fig. 5. Stored line scan of potassium counting rates from a section of frog gastric mucosa, superimposed on the sample current image at 10 kV. The horizontal line scan traverses the center of the image, and the length of scan is approximately 250 μ. The tissue was fixed with osmium.

accumulated in the form of an integrated counting rate until it is obvious that the differences represented are significant. The resulting curve of detector output can be photographed and superimposed upon a sample current image displaying morphological detail. Figure 5 is an example of the combined information.

Standards

Thus far, the results described in this paper present only a semi-quantitative analysis of tissue since quantitative analysis requires reliable standards for comparison. Efforts to develop standards that would be both similar to tissue and homogeneous at the micron level were frustrating for a long time. The first breakthrough was with the use of Epon 812 and 826 in a mixture of 1400 mM/kg and 30 mM/kg of chlorine, respectively. When mixed in various proportions, a set of standards for chlorine was obtained.

An alternative approach, which is also suitable for potassium and sodium, is to prepare standards from a 20% solution of bovine albumin. This has been adapted from methods employed by Dr. W.B. Kinter [3]. Solutions containing different but known concentrations of potassium or sodium chloride are ejected into cold liquid propane. The frozen drops are then treated like tissue specimens. The result is a matrix which is smooth in appearance and homogeneous by probe analysis. Though not ideal, it is felt that these standards are proving a prom-ising avenue for the quantitative analyses of ions in tissue. Over

TABLE III

Analysis of Electrolyte Standards*

Ionic concentration, mM/kg	KK_α cts/sec	K, cts/sec per mM/kg	$Cl\,K_\alpha$ cts/sec	Cl, cts/sec per mM/kg	$Na\,K_\alpha$ cts/sec	Na, cts/sec per mM/kg
0	1.9		12		5.5	
8	3.0	1.4	13	1.7	6.3	0.97
21	5.0	1.6	15	1.7	6.7	0.57
41	8.0	1.5	21	2.2	8.8	0.80
62	11.8	1.6	24	2.0	9.4	0.63
83	16.2	1.7	32	2.4	12.0	0.78
124	23.9	1.8	38	2.1	13.9	0.67

*Analyses of KNO_3 + NaCl solutions in 20% albumin. Drops were frozen, dried, fixed in osmium vapor, and embedded in Epon 826, 10 kV, 0.05 μA, and minimal spot size. Data given for O concentration consist of ions present in the albumin and chlorine in the plastic, in addition to background.

a period of time, several modifications have been employed. Typical analyses are given in Table III.

A few additional references of interest [4-8] are listed in this still very new field of study.

CONCLUSION

There has long been a need in the biological field for a means of analysis which is sensitive, microscopic, and versatile. It is felt that electron-microprobe analysis is rapidly moving forward to fill this need as the difficulties involved are overcome. As a result, possibilities for research can be seen that previously were not feasible. It is certain that electron-microprobe analysis can make a substantial contribution to physiology.

The electron probe has been operated with an X-ray spot of $1\,\mu$ and an ability to discriminate differences of concentration of about 10 meq/liter. A method of sample preparation has evolved that is compatible with retaining the normal marked concentration difference for potassium between the inside and outside of the cell. Though wholly satisfactory standards for quantitative analysis are not available, adequate standards for appropriate concentrations of K, Cl, and Na have been prepared in a matrix very similar to that of tissue which has been embedded in plastic.

REFERENCES

1. W. E. Stumpf and L. J. Roth, "Frozen sectioning below -60°C with a refrigerated microtome," Cryobiology 1:227–232 (1965).
2. C. Stirling, Ph.D. Thesis, State University of New York, Upstate Medical Center, Syracuse, N. Y. (1966).
3. C. A. Andersen and M. F. Hasler, "Extension of electron microprobe techniques to biochemistry by the use of long wavelength x-rays," presented at the Congres International—L'Optique des rayons X et la microanalyse, Applied Research Laboratories, Orsay, Seine et Oise, France (Sept., 1965).
4. R. Castaing, Ph.D. Thesis, Univ. Paris Publ. O.N.E.R.A. No. 55 (1951).
5. P. Galle, "Nouvelle méthode d'analyse des incrustations minérales du rein," Actualités Néphrologiques de L'Hôpital Necker, pp. 193–206 (1965).
6. R. C. Mellors, "Electron microprobe analysis of human trabecular bone," Clin. Orthopaedics and Related Res. 45:157–167 (1966).
7. R. W. G. Wyckoff, R. A. Laidley, and V. J. Hoffmann, "The probe analysis of nonconducting samples," Norelco Reptr. 10: (1963).
8. M. J. Ingram and C. A. M. Hogben, "Electrolyte analysis of biological fluids with the electron microprobe," Analyt. Biochem. 18:54–57 (1967).

Emission—Flame—Atomic-Absorption Spectroscopy

Recent Advances in Analytical Emission Spectroscopy*

Anna M. Yoakum

Oak Ridge National Laboratory
Analytical Chemistry Division
Oak Ridge, Tennessee

This review covers the significant developments in analytical emission spectroscopy which have occurred in the past year.

INTRODUCTION

An article entitled "Work Ahead in Emission Spectroscopy" appeared in the March, 1961, issue of The Spex Speaker [1]. This article reported on the findings of a survey of 30 recent articles on emission spectroscopy. Only four papers dealt with theoretical aspects of the subject, one discussed a minor new accessory, another the use of inert gases for spark analysis, but the vast majority told of the specific problems faced by the authors. The basic question asked at that time was: "Should there not be more papers on subjects bordering on the unknown aspects of the science—art of spectroscopy?" Apparently this same question was asked by many spectroscopists, and the response has been strictly in the affirmative, as is evidenced by current literature. In preparing for this review, only those articles were surveyed which have appeared since this meeting last year, an 11-month period. Of the 57 papers reviewed, 28 dealt with theoretical aspects and fundamental problems of emission spectroscopy, 7 with instrumental developments, 2 with new applications using standard spectrographic equipment, and the remaining 20 described applications and improved techniques.

*Research sponsored by the U.S. Atomic Energy Commission under contract with the Union Carbide Corporation.

EXCITATION SOURCES

The excitation source contributes in a most significant manner to the ultimate sensitivity, precision, and accuracy of spectrochemical analysis. Because of this important role, there is a never-ending search for new excitation sources and for improvements in old ones.

An excellent review article by Scribner [2] has appeared recently, covering advances in excitation sources for spectrochemical analysis. Several types of excitation sources, including flames, the high-frequency torch, the plasma jet, arcs, sparks, and the laser are considered.

Laser Excitation

Since the announcement, in early 1963, of the commercial avail-ability of a ruby laser excitation source for spectrochemical analysis, many significant applications and studies have been described.

Rasberry, Scribner, and Margoshes [3, 4] have published their findings on laser-probe excitation in spectrochemical analysis. The first article deals with characteristics of the source, and typical applications are discussed. The second article describes an investigation of quantitative aspects. They found that random errors come largely from variations in laser energy and from photometric errors. Correlations have been established between the energy of the laser beam, the size of the pit formed, and spectral intensities. One serious problem relating to the quantitative analysis of microspecimens is finding or synthesizing suitable standards.

Runge and associates [5] have described their efforts to test the quantitative nature of pure laser excitation of major constituents in molten-metal samples located within a small, laboratory-size induction furnace. They found that background continuum from the surface of the molten metal does not prevent the measurement of lines of major constituents of the melt. A major problem in analyzing molten metal remotely by a laser would appear to be trace-element detectability in the presence of background due to the intense hot spot produced by the laser discharge itself, rather than background due to the molten condition of the bulk metal.

An improved laser microprobe with a neodymium tube replacing the ruby has been announced by the manufacturer, the Jarrell–Ash Company. This neodymium tube is reported to have more dependable firing characteristics. The laser as an excitation source is gaining recognition in the ceramics, metals, glass, biological, geological, and pharmaceutical fields.

In addition to the use described in the preceding paper [6], the laser has been employed to determine trace metallics in living tissue, without harm to the subject [7]. At least, the subject, an anesthetized mouse, raised no serious objection.

Another interesting application of laser excitation involved the analysis of the hub of the rotor of a jet engine from a commercial aircraft [7]. Spectra obtained showed the presence of boron, which is not a component of the alloy nor was it present at any other location. The presence of boron confirmed the suspicion of the chief engineer that a structural defect had been repaired by welding.

High-Frequency Torch

The high-frequency torch as an excitation source for the spectro-chemical analysis of solutions has generated considerable interest. The torch applies the well-known technique of induction heating to heat-conducting gases. Although the idea behind the torch is not new, the commercial availability of such an excitation source has just been announced [8]. Several important advantages are reported. Detection limits are increased for many elements; e.g., phosphorus is detectable to 0.1 ppm with the use of the 2535-Å line. Depressive interference is reduced. The excitation atmosphere can be neutral, reducing, or oxidizing, depending on the gases used in the torch. This feature also reduces background and banding problems. The unit is readily adaptable for use on most conventional spectrographs and spectrometers.

Plasma Jet

Since 1959, when Margoshes and Scribner [9] reported the successful application of the plasma jet to quantitative spectrochemical analysis, widespread acceptance and usage of the source has occurred.

A comparative study involving dc-arc and plasma-arc excitation in uranium isotope analysis by optical emission spectrometry was conducted by Leys and Perkins [10]. They found the plasma arc to be less intense than the dc arc and the line-to-background ratio was less favorable. With both sources, background corrections substantially improved the precision.

Schrenk, Ho, and Lehman [11] also report a comparison of dc-arc and plasma arc excitation. Their problem dealt with the determination of rhenium in extractions from molybdenite. They found the plasma arc to be twice as precise as the dc arc, but the dc arc was considerably more sensitive than the plasma arc.

It is of interest to note that the use of the plasma jet as an excitation source is not limited to optical emission spectroscopy. It has been used successfully in the determination of refractory oxide elements by atomic absorption spectrometry [12].

Other Excitation Sources

Other excitation sources which have been reported as suitable for use in spectrochemical analysis include the disk-stabilized arc [13] and a unipolar arc [14].

Schroll and Sauer [15] have described a new excitation and heating source called a "mixed source." It consists of an exciting spark or ac source and a separately controlled heating dc arc. The purpose of the dc arc is to heat and vaporize the sample, while the spark or ac source is used to excite the vapors. In conjunction with this double-arc technique, large graphite beakers, which can be filled with more than 10 g of sample, have been developed. The larger sample size gives the advantage of better detection limits (in the parts per billion range). A detection limit for Hg of 1 ppb and for Mo of 10 ppb is reported.

The mathematical theories which are applicable to the excitation and ionization of atoms in plasmas at thermal equilibrium have been described by Margoshes [16]. He has shown that these theories can be applied to practical problems which arise in spectrochemical analysis.

FUNDAMENTAL STUDIES

During the past 3 years, there has been a heartening and significant increase in the number of papers dealing with fundamental studies relating to emission spectroscopy. Yamamoto [17] has studied the profile of ionized-calcium lines in an arc-plasma jet. The lines were found to exhibit Stark-effect broadening which was accompanied by small violet shifts. This result can be interpreted by taking into account the effect of both strong and weak collisions of electrons disturbing the radiating process of the ion. The broadening is attributed to strong collisions, while the shift can be explained to be due to weak collisions.

The effect of diffusion of components in the sparking zone on the results of spectral analysis of alloys has been examined by Buravlev [18]. He has also reported on the effect of sparking and the mechanism of the influence of elements on the results of spectral analysis [19].

Vukanović [20] reports his observations on the effect of mass separation perpendicular to the current in the plasma of a dc arc under usual spectrochemical conditions. The dependence of the effect on the degree of ionization is also given.

A number of articles relating to fundamental studies of various plasmas have appeared, including a study on the population of hydrogen levels in the argon–hydrogen plasma jet [21], elementary processes in the plasma of a dc gas discharge in helium [22], electron–ion recombination in a plasma [23], and, finally, a mathematical treatment which allows a graphic representation of ionization temperature of a plasma [24].

De Galan [25] has investigated the possibility of a truly absolute method of spectrographic analysis. He has derived a quantitative expression for the relation between the concentration of an element in a sample and the line intensity. A low-current dc carbon arc with anode excitation is employed. By use of his equation, the concentration of an element can be determined without previous establishment of a working curve.

A recent paper by Boumans and Maessen [26] considers the evaluation of detection limits in photographic emission spectroscopy. Another fundamental investigation dealing with integrated spectral-line intensities is reported by Chaney [27]. He found that the integrated intensity and the integrated photographic density of the entire spectral line are essentially linear functions of one another from a density of 0.05 to above 1.5. Analytical curves from these integrated measurements are unaffected by self-absorption, are more linear, and extend over a wider concentration range than curves involving peak measurements.

Lowenthal, Rank, and Wiggins [28] have reported on their study involving resolution and efficiency of single- and double-pass spectrographs. The results of their study indicate that double-pass operation can be very efficient for high-resolution spectrographs. Svoboda [29] has described his study concerning the relation of optimum density of developed photographic plates to characteristic curve constants.

Prince, Ellgren, and De Glopper have studied the chemical composition of solid-metal buttons produced by the dc-arc fusion of wire and chips under argon shielding. Materials studied included low-alloy steel, stainless steel, high-temperature alloys, and copper-base and aluminum-base alloys. A correlation was found to exist between boiling points and element losses. With the exception of zirconium, no significant changes in the composition of metallic elements having boiling points greater than manganese (2150°C) were noted. Consistent and significant losses of zinc were noted. The boiling-point correlation does not apply to nonmetallics, such as sulfur and phosphorus, which are retained in chemical combination.

Slavin [31] has reported on a stable source for plate calibration. The source, a commercial, low-pressure, quartz, mercury lamp, was found to be stable at both high and low voltage loads. The intensity of the lamp is high enough to permit short exposures with slow spectrographs and slow emulsions. The main objection to this source is its lack of a sufficient number of lines to cover adequately all wavelength ranges.

INSTRUMENTAL DEVELOPMENT

Another area closely related to fundamental studies is that of instrumental development. Cremers and Winter [32] have reported on an image-rotating device for a spectrograph illumination system.

An inexpensive, semiconductor-controlled regulator for the dc spectrographic arc is described by Conover, Peters, and Lalevic [33]. The described model offers the advantages of compactness, easier installation, and economy over the conventional multipurpose power source with current regulators.

An article by Eberhardt [34] should be of interest to those using direct-reading equipment. Threshold sensitivity and noise ratings of multiplier phototubes are discussed.

Another instrumental development is an automatic line-centering device for direct-reading emission spectrometers. Piepin, Schroeder, and Jacobs [35] describe a system by which the entrance slit is automatically set at the optimum position by an electronic measuring and control system.

A device for the continuous introduction of powdered samples into spark and arc sources is presented by Kántor and Erdey [36]. The sample is pushed by a motor-driven spindle into the electrode gap through a tubular electrode.

TRACE ANALYSIS

Micro and trace analysis still presents the greatest single challenge to emission spectroscopy. (This has been clearly demonstrated in this meeting by the papers dealing with this topic.) In this area, it is often necessary to employ a technique for the separation and enrichment of trace components prior to the spectrochemical determination. Minczewski [37] has reviewed the enrichment and separation methods most commonly employed. A preliminary extraction was employed by Karpenko and co-workers [38] in the quantitative spectral determination of yttrium and rare-earth elements of the yttrium subgroup in rocks and minerals.

Barton [39] has reported on the spectrographic determination of plutonium, thorium, and the rare earths in americium. By utilizing anionic complexing properties in conjunction with an anion exchange resin, he was able to separate quantitatively the impurities from americium prior to spectrographic determination. Detection limits in the range from 10 to 100 ppm were achieved for the rare earths with an initial sample of 22.6 mg of AmO_2.

Balfour and co-workers [40] have utilized preconcentration and separation prior to emission analysis in the determination of trace impurities in metallurgical materials. The trace elements are coprecipitated with copper as mixed sulfides which provide an excellent matrix for stable excitation in the dc arc. Another chemical procedure for concentrating the rare-earth elements from silicate rocks is described by Herrmann and Wedepohl [41]. An oxalate precipitation of the rare-earth elements was used.

A number of direct spectrographic methods for trace analysis have also been reported [42–47]. Bevege and Gallion [48] have studied the dc-arc excitation of aqueous hydrochloric acid solutions by using the vacuum-cup technique. The method proved to be simple, fast, and sensitive, but the precision suffers owing to arc wander.

Using a 3.4-m plane grating spectrograph and a red filter to remove second- and third-order interferences, Gurney and Erlank [49] have developed a direct-emission dc-arc technique for trace amounts of rubidium, cesium, and lithium in silicate rocks. Determination of the optimum operating conditions has resulted in a detection limit of 0.1 ppm or less for each element.

Goodfriend and his associates [50] have developed a unique technique for the determination of trace impurities in semiconductor filaments. By discharging through the filament a 2-μf capacitor charged to 6000 V, an entire section of the filament is electrically exploded. Sufficient light results from the explosion to produce an emission spectrum.

UNIQUE APPLICATIONS AND IMPROVED TECHNIQUES

There is, in general, a tendency to become somewhat stereotyped in the applications of emission spectroscopy. Sommer and Kick [51] have described a new application that uses normal spectrographic equipment. They have developed a method for determining N^{15} in the atom-concentration range of 0.38 to 20.0%. The relative isotopic abundance is determined by a calibration curve.

Sawatzky and Kay [52] have used emission spectroscopic techniques to study sputtered particles emitted into a high-vacuum environment by means of the combination of a high-current-density ion beam and electron beam. Intense atomic copper lines were easily observed when a copper target was bombarded with 5000-eV Ar^+ ions. The intensities of these lines were determined to obtain the sputtering yield and relative excitation functions for copper.

Grove and associates [53] have demonstrated the response of H_2, N_2, and O_2 in an inert gas atmosphere with the use of a Stallwood Jet and dc-arc excitation. Their work illustrates the feasibility of exciting the spectra of these gases, as well as chlorine, at atmospheric pressure and in other relatively simple conditions with a minimum of special facilities. Mellichamp [54] has developed a technique for stabilizing the dc arc. Stability of the dc arc can be increased by using a cathode that is cored with a material containing an element with an ionization potential lower than carbon. He found an effective core material to be a 1:2 mixture of $BaCO_3$, or Li_2CO_3, and graphite powder. In the arc, the added element forms a stationary positive-ion cloud at the tip of the cathode and acts as a ballast to electron flow. Arc temperature is not lowered. Current and voltage fluctuations and arc wandering are reduced. The desired tip shape is maintained while the added compound is slowly distilled into the arc during sample consumption. An improvement in the resulting reproducibility could be seen, but no percentage evaluation was given because of incomplete control over the other factors that are involved with reproducibility.

TIME-RESOLVED SPECTROSCOPY

In conclusion, the literature relating to time-resolved spectroscopy should be considered. An article discussing apparatus and techniques in this field has just appeared by Bardócz [55]. Several details of time-resolved spectra of high- and low-voltage sparks and ac arcs are demonstrated in the paper.

Increased interest in transient spectroscopic phenomena has led to new developments in rapid-scan spectrometers. Dolin and associates [56] have converted a grating spectrometer to rapid scanning without loss of optical quality. This is accomplished by sweeping a sequence of corner mirrors through an intermediate focal plane. Performance is limited only by the detector signal-to-noise ratio, determined by the scan time and the electrical bandwidth required and not by the optical scanning technique.

A high-resolution rapid-scanning spectrometer is described by Liberman, Church, and Asars [57]. By placing a rotating mirror near the exit slit, existing commercial instruments can be converted for rapid-scan use without extensive modification.

A very useful time-resolving spectroscopic technique for the detection of self-reversed spectrum lines is described by Bardócz and Vanyek [58]. A direct-reading, time-resolving technique in emission spectroscopy is reported by Gotô and associates [59]. The application of the technique to the analysis of iron, steel, and iron ores is discussed.

REFERENCES

1. A.J. Mitteldorf, Spex Speaker 6:1 (1961).
2. B.F. Scribner, Pure Appl. Chem. 10:579 (1965).
3. S.D. Rasberry, B.F. Scribner, and M. Margoshes, Appl. Opt. 6:81 (1967).
4. S.D. Rasberry, B.F. Scribner, and M. Margoshes, Appl. Opt. 6:87 (1967).
5. E.F. Runge, S. Bonfiglio, and F.R. Bryan, Spectrochim. Acta 22:1678 (1966).
6. I. Harding-Barlow, E.S. Beatrice, and D. Glick, Eighteenth Annual Mid-America Symposium on Spectroscopy, Paper No. 45, Chicago (May 15–18, 1967).
7. Spectrum Scanner 21:4 (1966).
8. Spectrum Scanner 22:9 (1967).
9. M. Margoshes and B.F. Scribner, Spectrochim. Acta 15:138 (1959).
10. J.A. Leys and R.E. Perkins, Anal. Chem. 38:1099 (1966).
11. W.G. Schrenk, Show-jy Ho, and D.A. Lehman, Appl. Spectry. 20:241 (1966).
12. K.E. Friend and A.J. Diefenderfer, Anal. Chem. 38:1763 (1966).
13. K. Doerffel and J. Lichtner, Spectrochim. Acta 22:1245 (1966).
14. P.M. Shvartsberg and P.D. Korzh, Zavodsk. Lab. 32:631 (1966).
15. E. Schroll and D. Sauer, Appl. Spectry. 20:404 (1966).
16. M. Margoshes, Appl. Spectry. 21:92 (1967).
17. M. Yamamoto, Phys. Rev. 146:137 (1966).
18. Y.M. Buravlev, Zavodsk. Lab. 31:1341 (1965).
19. Y.M. Buravlev, Zavodsk. Lab. 32:554 (1966).
20. D.D. Vukanović, Spectrochim. Acta 22:815 (1966).
21. V.M. Goldfarb, E.V. Ilyina, I.E. Kostygova, G.A. Lukyanov, and V.A. Silantyev, Opt. i Spektroskopiya 20:1085 (1966).
22. Y.A. Tolmachev, Opt. i Spektroskopiya 21:397 (1966).
23. V.A. Abramov and B.M. Smirnov, Opt. i Spektroskopiya 21:19 (1966).
24. C.A. Berthelot, Spectrochim. Acta 22:829 (1966).
25. L. de Galan, Anal. Chim. Acta 34:2 (1966).
26. P.W.J.M. Boumans and F.J.M.J. Maessen, Z. Anal. Chem. 220:241 (1966).
27. C.L. Chaney, Spectrochim. Acta 23A:1 (1967).
28. J.A. Lowenthal, D.H. Rank, and T.A. Wiggins, J. Opt. Soc. Am. 56:1473 (1966).
29. V. Svoboda, Appl. Spectry. 20:219 (1966).
30. L.A. Prince, A.J. Ellgren, and T.J. De Glopper, Appl. Spectry. 20:372 (1966).
31. M. Slavin, Appl. Spectry. 20:333 (1966).
32. C.J. Cremers and E.R.F. Winter, Appl. Spectry. 20:421 (1966).
33. H.H. Conover, J.T. Peters, and M. Lalevic, Appl. Spectry. 20:334 (1966).

34. E.H. Eberhardt, Appl. Opt. 6:251 (1967).
35. H. van der Piepen, W.W. Schroeder, and P.P.J. Jacobs, J. Sci. Instr. 43:597 (1966).
36. T. Kántor and L. Erdey, Talanta 13:1289 (1966).
37. J. Minczewski, Pure Appl. Chem. 10:567 (1965).
38. L.I. Karpenko, L.A. Fadeeva, and S.V. Bel'tyukova, Zavodsk. Lab. 32:424 (1966).
39. H.N. Barton, Anal. Chem. 38:1077 (1966).
40. B.E. Balfour, D. Jukes, and K. Thornton, Appl. Spectry. 20:168 (1966).
41. A.G. Herrmann and K.H. Wedepohl, Z. Anal. Chem. 225:1 (1967).
42. D.L. Nash, Appl. Spectry. 20:392 (1966).
43. R. Gerbatsch and G. Artus, Z. Anal. Chem. 223:81 (1966).
44. B. Strzyzewska, Z. Radwan, and J. Minczewski, Appl. Spectry. 20:236 (1966).
45. A.S. Sambueva and S.A. Shipitsyn, Zavodsk. Lab. 31:1087 (1965).
46. F. Fehér, H.D. Lutz, and K. Obst, Z. Anal. Chem. 224:407 (1967).
47. B. Podobnik and M. Špenko, Anal. Chim. Acta 34:294 (1966).
48. E.E. Bevege and R.E. Gallion, Appl. Spectry. 21:20 (1967).
49. J.J. Gurney and A.J. Erlank, Anal. Chem. 38:1836 (1966).
50. P.L. Goodfriend, H.P. Woods, and L.J. Parcell, Anal. Chem. 38:1433 (1966).
51. K. Sommer and H. Kick, Z. Anal. Chem. 220:21 (1966).
52. E. Sawatzky and E. Kay, Rev. Sci. Instr. 37:1324 (1966).
53. V. Raziunas, W.A. Loseke, and E.L. Grove, Appl. Spectry. 20:395 (1966).
54. J.W. Mellichamp, Appl. Spectry. 21:23 (1967).
55. Á. Bardócz, Appl. Spectry. 21:23 (1967).
56. S.A. Dolin, H.A. Kruegle, and G.J. Penzias, Appl. Opt. 6:267 (1967).
57. I. Liberman, C.H. Church, and J.A. Asars, Appl. Opt. 6:279 (1967).
58. Á. Bardócz and U.M. Vanyek, J. Opt. Soc. Am. 56:756 (1966).
59. H. Gotô, S. Ikeda, A. Saitô, and M. Suzuki, Z. Anal. Chem. 220:95 (1966).

Recent Developments in
Atomic Absorption and
Flame Emission Spectroscopy

S. R. Koirtyohann

University of Missouri
Columbia, Missouri

This review will cover the significant developments in atomic absorption and flame emission spectroscopy which were published during 1966 and early 1967. The applications of these two methods are now too numerous to be covered completely in a review of this type and therefore only those applications which are novel or which contribute to the general development of the methods will be considered.

BOOKS AND REVIEWS

Two books on atomic absorption were published during 1966: one by Robinson [1] and the other a rather complete revision of an earlier edition by Elwell and Gidley [2]. The book by Mavrodineanu and Boiteux [3], which was published in 1965 but did not become generally available until 1966, is probably the most complete treatment of analytical flame spectroscopy to appear to date. It is a large book which covers the theory and practice of emission and absorption flame methods. The high cost ($50) will limit its usefulness, however.

Although other reviews have been published [4–6], those by Slavin [7, 8] and by Margoshes and Scribner [9] will be most useful to workers in the United States. Slavin's reviews and his 1965 bibliography [10] are particularly valuable for those interested in applications of atomic absorption. A general article by Kahn [11] presents the essentials of atomic absorption instrumentation in an easily understandable way.

INSTRUMENTATION

Flames

Gilbert [12] has described a novel flame which he used for the determination of chlorine. The air—hydrogen flame is split into two separate combustion zones. The first portion of the flame converts any chlorine present in the sample to HCl, which reacts with an indium-coated tube as the hot gases pass to the second flame, where the remainder of the hydrogen is burned with air. The InCl-band emission is observed in the second flame. Gilbert used the method only with volatile chlorine-containing organic compounds but indicated that it should be quite sensitive for chloride in solution as well.

Zacha and Winefordner [13] describe an argon—hydrogen-entrained air flame for emission analysis. The system was similar to the one used earlier by Veillon et al. [14] for atomic fluorescence. They found reduced background emission and improved detection limits compared with the conventional hydrogen—oxygen flame for 9 of the 14 elements which they investigated. A large, well-defined inner cone seems to contribute to the success of the flame. In the section on fundamental developments, reduced quenching of excited atoms will be considered as an additional contributing factor.

The premixed nitrous oxide—acetylene flame, which was first introduced by Willis [15], has continued to be very successful in atomic absorption. All major manufacturers now provide burners for this flame, and its ability to decompose stable compounds is well established [16—21]. It is still not clear if the difference in behavior is due only to the higher temperature of the flame or if basic differences in flame chemistry also contribute. Ionization, which is more severe at the higher temperature, was studied by Manning and Capacho-Delgado [22] after being mentioned by several other authors. They observed the change in absorption of ionic and atomic lines of barium and calcium as ionization was suppressed by the addition of potassium and were able to calculate that in the absence of potassium (or some other easily ionized element), 91% of the barium and 36% of the calcium were ionized in the nitrous oxide—acetylene flame. Addition of a large excess of potassium reduces the ionization quite effectively.

Pickett and Koirtyohann [23] found that, in spite of the rather intense background radiation, the nitrous oxide-acetylene flame holds great promise as a source for emission analysis. They used the same slot burner and flame configuration that are normally recommended for atomic absorption and obtained detection limits for a number of elements that were significantly better than those previously reported by any flame method (see Table II).

For atomic absorption, flame adapters designed to increase the number of atoms in the light path have received continued attention, principally by Pulido, Fuwa, and Vallee [24], Rubeska and Stupar [25], Stupar [26], Robinson and co-workers [27, 28], and Thilliez [29]. Stupar was able to measure the temperature and atomic population along the length of a flame confined in a quartz tube [26]. The extended flames are quite sensitive, but optical alignment can be critical, interferences are probably more severe [30, 31], and contamination of the tube walls can be troublesome [32].

Fassel and Golightly [33] have continued their work with fuel-rich oxyacetylene flames [33], and Skogerboe, Heybey, and Morrison [34] have found that oxide-forming elements can also be excited in a fuel-rich oxyhydrogen flame if alcoholic solutions are nebulized. The published description of the three-slot Boling burner for atomic absorption [35] was almost anticlimactic since it has been commercially available for almost 2 years.

The rather specific advantage of an air—hydrogen flame for the determination of tin was observed in atomic absorption [36, 37], just as it was for emission several years ago [38]. No explanation for the higher atomic population of tin in this flame has been presented. Experiments with a universal sampling method for flames have been described [39], and a significant step toward a convenient, practical, ultrasonic nebulizer system was made by Kirsten and Bertilsson [40]. The high cost of such nebulizers will discourage their use, however.

Several devices have been suggested to replace the conventional flames as sources of atoms for some sample types. A slow-burning propellant mixture containing both fuel and oxidant was used to atomize solid samples for atomic-absorption measurements [41], and graphite furnaces can be used for the same purpose [42, 43]. It is too early to judge the general utility of either of these methods. Induction-coupled plasmas have found application in both flame-emission and atomic-absorption method [44, 45]. A commercial plasma generator for solution analysis is now available from the Jarrell—Ash Company. The plasmas should, because of the high temperature attained, provide a source which is quite free of interferences, but the sensitivities reported so far are better than those from combustion flames for only a very few elements. Atomic absorption can also be performed with a plasma jet, a magnetically pinched, flamelike arc [46]. This, too, is a very high temperature source, but it has yet to provide sensitivities for any element that are equal to those from the best combustion flames.

Sources

High-brightness cathode lamps, which were first described by Sullivan and Walsh [47], are now commercially available for a number of elements [48, 49]. A method to increase the emission intensity from standard cathode lamps was described by Dawson and Ellis [50]. They operated the lamp at a low direct current and superimposed 15-μsec pulses of several hundred milliamperes at a frequency of about 300 cps. Several hundredfold gains in intensity were obtained with no deterioration in absorption signal. Rapid warm-up, longer lamp life, and even rejuvenation of old lamps are claimed as fringe benefits from this method of operation. These lamps may be more valuable in atomic fluorescence than in atomic absorption because the added intensity is usually not needed in the latter method.

Bowman, Sullivan, and Walsh [51] have described a method for selective modulation of resonance lines. A high-brightness lamp is operated on direct current and another cathode containing the element to be determined and powered by alternating current is placed in the light path. The sputtered atoms in the second cathode absorb the resonance lines. The atomic population is modulated at the selected frequency, which causes only the resonance intensities to be modulated. All other lines are steady and are not measured by the ac detector system. Some sort of monochromator is still required to avoid the intense dc signal, although simple filters may often suffice. The fact that the lamp requires three separate power supplies would seem to be a rather severe limitation.

Vollmer [52] has described a tin hollow-cathode lamp in which the metal is molten as the temperature of operation, and Goodfellow [53] described a demountable lamp. Frank, Schrenk, and Meloan [54] have investigated the use of an iron hollow-cathode lamp as the light source for the determination of other elements. The fact that they were partially successful indicates that there may be important exceptions to the high degree of specificity normally found in atomic absorption methods. Electrodeless discharges have been used instead of hollow-cathode lamps as the light source for the determination of indium, gallium, bismuth, antimony, thallium, lead, magnesium, calcium, and copper [55].

Continuous sources were investigated by a number of workers [56, 57]. Sensitivities are still roughly an order of magnitude poorer than with cathode-lamp sources. Lang has described a method for measuring the difference between the incident and transmitted intensity, rather than their ratio [58], and has used the method in conjunction with a continuous light source [59]. An alternating absorption signal is produced by pulsating the nebulizer air supply; and the ac component of the transmitted signal, which is relatively independent of the spectrometer bandpass, is measured. The sensitivities reported so far are not as good as can be obtained with line sources.

Resonance Monochromators

Resonance monochromators, which were first described by Sullivan and Walsh [60] and which operate on an atomic fluorescence principle, have been developed into practical analytical tools. Applications to the determination of calcium, magnesium, sodium, potassium, lead [61, 62], and lithium [63] have been described, and a commercial instrument with this type of monochromator is now available [64].

Complete Instruments

Elwell and Gidley [2] have placed a foldout sheet in the back of their book which describes 16 instruments from 10 different manufacturers, most of which can be used for atomic absorption or flame emission. At least three new instruments have been described since the book was published [64, 65, 66]. There should still be a place for simple, low-cost, special-purpose instruments such as the one which was described by Haggen-Smit and Ramírez-Muñoz [67] for the simultaneous determination of sodium, potassium, and calcium in biological materials.

Automation and Computer Applications

Automatic flame-emission determinations of alkali metals has been standard practice in clinical chemistry for several years, based primarily on the Technicon equipment. This equipment has recently been applied to potassium in fertilizers [68, 69], and portions of it are used in some automated atomic-absorption methods. Slavin et al. [70] have described an automated atomic-absorption method applicable to wear metals in lubricating oils and to various trace and major constituents in blood serum. The instrument automatically dilutes the samples, measures the absorption, and provides a paper-tape output.

Operator errors associated with these steps are avoided. The method gave precision within about 3% and required about 25 sec/determination on a large number of samples. This included sample logging, instrument calibration, data reporting, and presumably coffee breaks as well.

Klein et al. [71] were able to determine calcium in serum automatically at the rate of 40 samples/h with a precision of within 2%. Butler et al. [72] have designed an instrument for the automatic extraction and atomic-absorption determination of gold in reduction plant tailings.   The method is sensitive to about 5 ppb in the sample and gives precision within about 3.5%.   The instrument is sufficiently rugged to operate satisfactorily in a gold reduction plant.

Most automatic methods can be adapted to read out in a form which can be used as a computer input. Computers can be helpful in other phases of flame analysis, such as optimizing operating parameters, as shown by Ramírez-Muñoz and co-workers [73–75].

INTERFERENCES

It is now apparent that interferences of various sorts are more common in atomic absorption than they were believed to be a few years ago.   Some types that are well understood from emission-flame spectrophotometry are being discovered in atomic absorption.   An excellent discussion of the various types of chemical interferences is presented by Alkemade [76], while the effects of flame temperature on interferences is discussed by de Galan and Winefordner [77]. Interferences by aluminum or silicon on the alkaline earth elements are commonly reported [78–81], but others such as iron on chromium [82, 83], aluminum on chromium [84], and iron on calcium in emission [85], also occur. These interferences can usually be removed by the addition of lanthanum, ammonium chloride, or potassium peroxy-disulfate $(K_2S_2O_8)$. Generally speaking, the nitrous oxide–acetylene flame suffers less from chemical suppression interferences in either emission or atomic absorption than does the air-acetylene flame [16, 23]. Interelemental enhancements are reported, however, between oxide-forming elements. Sachdev et al. [86] report enhanced vanadium absorption in the presence of oxide-forming elements. The enhancement reaches a constant level, after which the addition of more of the same element or a second interfering element has no effect. A constant amount of aluminum was added to all samples and standards to avoid errors.   A similar enhancement by iron on titanium absorption was noted by Headridge and Hubbard [87]. In this case, the presence of aluminum with iron caused a slightly greater enhancement than iron alone. The effects of ionization in this flame were discussed above.

Premixed flames appear to suffer less from several types of interferences than those of equal temperature formed with total-consumption nebulizer burners [23, 88, 89]. This is probably because the flame zones are less well defined in the turbulent flame and because larger droplets reach the flame.

A rather subtle effect due to the solvent used for dilution of gasoline prior to the determination of lead was reported by Wilson [90]. His results were high compared with those of other methods when the samples were diluted with isooctane but were comparable when a 1:1 mixture of acetone and isooctane were used as the diluent. He does not offer an explanation for the effect.

One frequently encounters statements indicating that atomic absorption is completely free from spectral interferences. These statements are usually made without experimental evidence, theoretical justification, or even a precise definition of what the author means by a spectral interference. Koirtyohann and Pickett [91] observed molecular absorption by alkali halides in an oxygen–hydrogen flame which was passed through a vycor tube. These spectra were not observed in the air–acetylene premixed flame, but in the air–natural gas flame the alkali halides gave absorption spectra similar to those reported by Willis [92], who attributed the losses to scattering by particles in the flame. Koirtyohann and Pickett also observed absorption spectra closely resembling the emission for alkaline earth oxide and hydroxide molecules [91]. The most serious interference was by CaOH absorption in the vicinity of the Ba 5535.5 Å line. The absorption from a solution containing 1% calcium was about the amount expected from 75 ppm of barium. Less serious interferences were noted on Li 6708 due to SrO bands, on Na 5890 due to CaO, and on Cr 3579 by MgOH. In each case, the molecular absorption band was identical in shape with the corresponding emission band, although much weaker. Later results [32] indicate that an unknown magnesium compound absorbs in the region of Mn 2795 and some calcium species absorb at all wavelengths below about 3000 Å. A correction for this background can easily be made by measuring the absorbance at wavelengths near the resonance line with the use of a continuous source and subtracting it from the absorbance at the resonance wavelength. A lamp mount which allows convenient switching from cathode lamp to continuous source was described [31]. The calcium absorption band in the vicinity of the Ba 5535.5 line was also studied by Capacho-Delgado and Sprague with similar results [93].   Spectral interferences are comparatively rare in atomic absorption, but those working with small absorbances on concentrated solutions of complex materials must keep them in mind.   Scale-expansion devices multiply the potential error from background absorption just as much as they increase the sensitivity.

Light scattering by particles in the flame had been used to explain small absorption losses which were observed in atomic absorption [92].   Some of the scattering data were examined by Koirtyohann and Pickett [94] and found to be inconsistent with the appropriate light-scattering theory.   They concluded that many of the observed losses were probably due to molecular absorption.

A spectral interference of a different type was reported by Jaworowski and Weberling [95].   They reported some absorption by iron, chromium, and manganese at Ni 2320 when using a multi-element, hollow-cathode lamp.   The interference disappeared when the combination lamp was replaced with one containing nickel only. Evidently, some lines which were active in absorption for the other elements were passed by the monochromator.   They point out the need for a high-quality monochromator, especially for working with some of the multielement lamps.

CHEMICAL SEPARATIONS

Chemical separations can be used to avoid interferences and to increase sensitivity.   In atomic absorption, extraction into methyl isobutyl ketone (MIBK) with ammonium pyrrolidine dithiocarbamate (APDC)* is the most popular method of separation.   Recent work by Mulford [96] extends the list of elements extractable with this reagent to arsenic, indium, palladium, platinum, selenium, tellurium, and thallium.   Other reagents are also used for separations.   Takeuchi et al. [97] tested the appropriate combinations of diethyldithiocar-bamate, dithizone, iodide, cupferron, oxine, thenoyltrifluoroacetone, and salicylaldoxine with cadmium, silver, iron, palladium, manganese, copper, nickel, and lead. Kirkbright, Peters, and West [98] extracted

---

*It might be pointed out that pyrrolidine dithiocarbamate is a bad chemical name for this compound because two nitrogen atoms are implied in the name and only one is present in the compound.   A more correct name would be ammonium pyrrolidine carbodithioate; APCD could then serve as a short name. Because of its popularity, however, APDC will be used in this review.

copper into ethyl acetate with oxine, and West [99] presented a more general article on chemical separations. Butler and Mathews [100] found that oxine extracted molybdenum more efficiently than APDC did, while Kirkbright et al. [20] separated the same element from niobium and tantalum by extracting it into butanol as the oxine complex. Fluoride and EDTA (ethylenediaminetetraacetic acid) served as masking agents. Podobnik, Dular, and Korosin [101] used APDC for the extraction of iron, cobalt, copper, and nickel from waters. Aluminum was extracted as the cupferrate, and chromium as $HCrO_3Cl$. Joyner and Finley [102] combined coprecipitation with diethyldithiocarbamate extraction for the determination of manganese in sea water. Sizonenko et al. [103] described a method for the extraction of calcium, and Akaza [104] extracted alkali metals as their polyiodides. Anion exchange resin was used by Calkins [105] to separate zinc from aluminous materials. He also extracted manganese into chloroform with 8-hydroxyquinaldine. The complex was decomposed with acid prior to the determination.

Nearly any chemical separation method which concentrates the desired constituents and rejects the bulk of the matrix can be used for flame methods. It is usually necessary to avoid solvents such as chloroform because they do not burn well.

FUNDAMENTAL STUDIES

Zeegers and Alkemade [106] studied recombinations of H, OH, and O radicals in air–acetylene flames. They found that the decay of radical concentrations is governed mainly by the H + OH combination and, to a lesser extent by the CO + O. They also obtained the dissociation energy of LiOH as a side result. Alkemade and Hooymayers [107] considered the role of electrons in excitation of metals in flames and concluded that their effect was not important. These authors also studied the quenching of excited alkali atoms in flames [108, 109]. They found by measuring fluorescence yields that the deexcitation of sodium and potassium atoms was much less efficient in flames formed by H_2-O_2-Ar mixtures than when nitrogen or carbon dioxide was used in place of the argon. They concluded that collisions between water molecules and argon atoms are quite inefficient in removing energy from an excited alkali metal atom, while, in the presence of nitrogen or CO_2, most of the atoms which do become excited are quenched before they can radiate. This observation probably helps to explain the advantages of the hydrogen–argon-entrained air flame which Winefordner and co-workers described [13, 14]. The quenching of excited atoms would have no effect, of course, on atomic-absorption measurements.

Winefordner and co-workers are presently the most prolific authors on flame methods. They have examined the causes for the bending of working curves in emission methods [110], the intensity of thermal radiation from flames [111], and the rates of solvent evaporation in turbulent flames [112]. The solvent evaporation was 50 to 90% complete, depending on flame conditions and the individual nebulizer, in the area of the flame normally viewed. The influence of flame temperature in emission and absorption methods was studied [77], and the approximate half-intensity line widths for 136 lines of 60 elements in 5 different flames were calculated [113]. Partition functions were calculated and tabulated in their laboratory [114], and the free-atom fraction for 22 elements in the air–acetylene flame was estimated from atomic-absorption spectra [115]. The fraction of free atoms ranged from near 1 for copper and sodium to less than 10^{-5} for aluminum. The values are given in Table I.

Dean and Adkins [116] reported on excitation gradients in oxy-acetylene flames and pointed out the need for careful adjustment of the burner position when working with elements where chemilumines-cence plays a part in the excitation. Lang [117] developed a method to compensate for variations in flame-emission intensity. Andersen

TABLE I

Free-Atom Fractions in the Air–Acetylene Flame*

Element	Atom fraction	Element	Atom fraction
Ag	0.66	Li	0.20
Al	$<10^{-5}$	Mg	0.59
Au	$<10^{-3}$	Mn	0.45
Ba	0.0011	Na	0.50
Ca	0.14	Na + Cs	1.00
Cd	0.50	Pb	0.44
Co	0.052	Sn	$<10^{-4}$
Cr	0.064	Sr	0.13
Cu	0.98	Tl	0.36
Fe	0.66	Zn	0.45
Ga	0.16		
In	0.67		
K	0.25		

*From de Galan and Winefordner [115].

and Hume [118] studied barium and strontium emission as a function of fuel composition in $H_2-O_2-N_2$ flames. They found maximum emission with 40% O_2 and 60% N_2 as the nebulizing gas. Rains and Menis [119] described a special form of scale expansion for flame-emission and atomic-absorption spectroscopy. Precision within 0.5% and better was obtained.

APPLICATIONS

As stated earlier, only those applications which are unique or which contribute to the development of the method as a whole will be considered here. Robinson and Smith [120] observed emission from molecular fragments when organic compounds were aspirated into a flame. The observed spectra were characteristic of the compound aspirated, but the intensities were not proportional to the concentration. More work is needed to get a useful analytical method, but the possibilities are interesting.

Goleb [121 was able to determine uranium isotope ratios by using a cathode lamp as the absorption source. He also observed visible and ultraviolet atomic-absorption lines of the noble gases in the same type of discharge [122] but failed in an attempt to differentiate between boron isotopes [123].

Indirect methods in which the component to be determined reacts with a metal which can be determined by atomic absorption or flame emission continue to appear. Kumamaru et al. [124-126] described such methods for the determination of nitrate, phthalic acid, and inorganic phosphate. Zaugg and Knox [127] also determined inorganic phosphate. In these cases, the extraction of a metal (Cu and Mo) depends on the presence of the nonmetal in a quantitative manner. Similar determinations have been suggested for APDC and oxine, while masking of these extractions can be used to determine EDTA [89]. Precipitation followed by the determination of the excess metal in solution or separation and dissolution of the precipitate were used to determine chloride (Ag) and sulfate (Ba) [128, 129]. Sulfate can also be determined by its suppression on calcium emission or absorption [89].

These indirect methods can be quite valuable in some cases. However, they must be used most cautiously because the specificity of the determination depends on the chemistry of the reactions, not on the final determinations. Methods that depend on an interference, for example, assume that no other interference is present. This may be far from true in practical samples.

TABLE II

Flame-Emission and Atomic-Absorption Sensitivities

Element	Atomic absorption det. limit, μg/liter	Ref	Flame emission * det. limit, μg/liter	Ref
Ag	5	130	20	23
Al	100	7	10	23
As	500	7	6,000	131
Au	10	130	500	23
B	6,000	7	300	130
Ba	100	7	2	23
Be	2	7	200	23
Bi	20	7	9,000	131
Ca	2	7	0.1	23
Cd	1	130	900	132
Ce			30,000	33
Co	7	7	30	130
Cr	5	7	5	23
Cs	50	7	2	130
Cu	5	7	10	23
Dy	200	7	300	130
Er	100	7	300	130
Eu	200	7	9	33
Fe	10	7	30	130
Ga	70	7	10	23
Gd	4,000	7	300	130
Ge	1,000	133	500	23
Hf	15,000	133		
Hg	200	7	120,000	33
Ho	300	7	300	33, 130
In	50	7	5	23
Ir	4,000	7	300,000	33
K	5	7	0.2	130
La	8,000	133	100	23
Li	5	7	0.003	33
Lu	50,000	7	150	130
Mg	0.3	7	4	23
Mn	5	7	5	23
Mo	100	7	90	33
Na	5	7	0.1	33, 130
Nb	20,000	7	600	130
Nd	2,000	7	3,000	33
Ni	10	7	30	23, 130
Os			30,000	33
P			3,000	132
Pb	10	7	300	130
Pd	500	7	50	23
Pr	6,000	7	6,000	33

TABLE II (Continued)

Element	Atomic absorption det. limit, μg/liter	Ref	Flame emission* det. limit, μg/liter	Ref
Pt	500	7	2,000	23
Rb	5	7	1	130
Re	1,500	7	3,000	33
Rh	30	7	900	33
Ru	300	7	200	130
Sb	200	7	1,500	131
Sc	200	7	30	23, 130
Se	500	7		
Si	100	133	15,000	33
Sm	5,000	7	2,000	33
Sn	100	7	300	23
Sr	10	7	0.2	23
Ta	6,000	7	60,000	33
Tb	2,000	7	300	130
Te	300	7	30,000	132
Th			500,000	33
Ti	100	7	200	23
T1	200	7	2	130
Tm	100	7	300	130
U	12,000	7	30,000	33
V	20	133	10	23
W	3,000	133	500	23
Y	300	133	60	130
Yb	40	133	30	130
Zn	2	7	100,000	23
Zr	5,000	7	3,000	23

*Flame-emission results obtained with organic solvents were multiplied by 3.

COMPARISON OF FLAME EMISSION AND ATOMIC ABSORPTION

All too often these methods are compared by persons who have worked primarily with only one of them, always the one which looks best in their comparison, and the results frequently include prejudice or commercial considerations as well as scientific facts. The present comparison, based on literature reports and the author's experience with both methods, will consider the following analytically important factors:

Instrument cost   Operator training
Flexibility   Sensitivity
Elements covered   Interferences
Probable future developments

The instruments for the two methods cost about the same; in fact, they are frequently the same basic unit. The best commercial models sell for $6000 to $10,000 depending on accessories. Atomic-absorption instrument prices do not normally include the price of the hollow-cathode lamps. If 30 elements are to be determined (about half the maximum number), these would cost roughly an additional $2500 above the basic instrument cost. A complete library of lamps is usually not purchased at one time, however, and the cost may be spread over several years. Today's commercial lamps have a long life, which makes lamp-replacement costs only a minor item.

The flame-emission method is more flexible because no light source needs to be changed between elements and, more importantly, no light source needs to be on hand prior to running an analysis. There is no need to anticipate a given analytical requirement several weeks in advance. Demountable hollow-cathode lamps [53, 31] over-come this limitation of the atomic-absorption method, but at the cost of operator training.

Atomic absorption is favored somewhat for elemental coverage because elements with resonance lines at very short wavelengths are not excited appreciably in flames. Any element which can be deter-mined by atomic emission can, at least in principle, be determined by atomic absorption. The converse is not true, but molecular emis-sion can be used for a few elements which cannot be determined by atomic absorption.

A greater number of variables affect the observed signal in flame-emission methods. This is almost sure to mean that more operator training will be required, but, except in rare cases, the difference will be small. If demountable hollow-cathode lamps are used in atomic absorption, considerably more training is needed.

The detection limits which have been reported in the literature for atomic-absorption and flame-emission methods are presented in Table II. In all cases, the limit is the concentration of the element giving a signal equal to about twice the fluctuation in background. All of the data for atomic absorption are for aqueous solutions; therefore, the flame-emission detection limits which were measured with organic solvents were multiplied by 3 before being entered in the table.

Data for 69 elements are presented. Of the 56 elements which have a detection limit of 1 μg/ml or less by one of the methods, 15 yield about equal sensitivity ($\pm 3 \times$), 17 are significantly more sensitive by atomic absorption, and 24 are significantly more sensitive by flame emission. The elements in each group, as presented in Table III, show that neither of the methods is clearly superior to the other in over-all sensitivity; for individual elements, however, one or the other

TABLE III

Comparison of Sensitivities by Flame Emission and Atomic Absorption

More sensitive by atomic absorption		About equal sensitivity		More sensitive by flame emission		
17 Elements		15 Elements		24 Elements		
Ag	Mg	Cr	Ni	Al	In	Sc
As	Pb	Cu	Ru	B	K	Sr
Au	Pt	Dy	Sn	Ba	La	Tb
Be	Rh	Er	Ti	Ca	Li	Tl
Bi	Sb	Fe	Tm	Cs	Lu	W
Cd	Se	Ho	Yb	Eu	Na	Y
Co	Si	Mn	V	Ga	Nb	
Hg	Te	Mo		Gd	Pd	
	Zn			Ge	Rb	

may be much superior. A similar conclusion was reached by Fassel and Golightly [33].

Any interference which affects the population of free atoms in the flame will be equally serious in the two methods. Some confusion seems to exist on this point because the turbulent-flame, total-consumption nebulizer burners are more popular in emission methods and premixed flames are usually used in absorption. The interference behavior is different for the two flame types, but, since each type can be used for either purpose, this should not be considered as a difference between the methods. In addition to the free-atom population, flame-emission results depend on all factors which affect the excitation and deexcitation processes. The most obvious of these is flame temperature, but flame composition and the presence of extraneous elements undoubtedly have their own influence.

Spectral interferences are more serious in emission methods because the bandpass of most monochromators is much wider than the atomic line, but they are easier to correct for, too. The facilities for making continuous-source background correction, as was suggested for atomic absorption [31], are not now available on commercial instruments, and it does not appear that they will be soon. The advertising literature from the leading manufacturers still denies the existence of spectral interferences in atomic absorption. When the interferences do occur, they are likely to go undetected.

The total effect of interferences undoubtedly favors atomic absorption, and the ideal detection limits given in Table II will deteriorate more rapidly in emission in the presence of extraneous elements.

Concerning likely future developments, the technology is now available to bring about, roughly, tenfold improvement in emission-flame results; only money is needed. Better spectrometers, cooled detectors, very-high-gain lock-in amplifiers, and signal-averaging devices are available to anyone with the necessary cash. Putting the components together into a workable instrument should present no problems.

So long as only the present flames are considered, it is much less clear where further improvements in atomic absorption are to be found. Larger-scale-expansion factors are not likely to be very helpful, even if electronic stability will permit it. Flame stability probably will not. The picture changes drastically, however, if absorption by static atomic vapors is considered rather than by flames. Available information for mercury permits an estimate of the improvements that may be expected. Roughly, 5 μg of mercury are required to produce 1% of absorption in the air–acetylene flame. With the use of a cell cross section of 0.25 cm^2, the same amount of absorption is produced by 0.00005 μg of mercury in the static vapor [32] for a gain of 10^5. Perhaps more work should be done on static-vapor methods in atomic absorption.

REFERENCES

1. J.W. Robinson, Atomic Absorption Spectroscopy, Marcel Dekker, Inc., New York (1966).
2. W.T. Elwell and J.A.F. Gidley, Atomic-Absorption Spectrophotometry, Pergamon Press, New York (1966).
3. R.Mavrodineanu and H.Boiteux, Flame Spectroscopy, John Wiley and Sons, Inc., New York (1965).
4. C. Iida and K. Fuwa, Nippon Kagaku Zasshi 4:303 (1966).
5. M. Suzuki and T. Takeuchi, Bunseki Kagaku 15:1003 (1966).
6. A.Walsh, Zh. Prikl. Spektroskopii Akad. Nauk Beloruss. SSR 4:471 (1966).
7. W. Slavin, Appl. Spectry. 20:281 (1966).
8. W. Slavin, Perkin-Elmer Corp. Atomic Absorption Newsletter 5:42 (1966).
9. M.Margoshes and B.F. Scribner, Anal. Chem. 38:297R (1966).
10. W. Slavin, Perkin-Elmer Corp. Atomic Absorption Newsletter 5:50 (1966).
11. H. Kahn, J. Chem. Educ. 43:7A (1966).
12. P.T. Gilbert, Anal. Chem. 38:1920 (1965).
13. K. Zacha and J.D. Winefordner, Anal. Chem. 38:1537 (1966).
14. C.Veillon, J.M. Mansfield, M.E. Parsons, and J.D. Winefordner, Anal. Chem. 38:204 (1966).
15. J.B.Willis, Nature 207:715 (1965).
16. M.D. Amos and J.B. Willis, Spectrochim. Acta 22:1325 (1966).
17. W. Slavin, A. Venghiattis, and D. C. Manning, Perkin-Elmer Corp. Atomic Absorption Newsletter 5:84 (1966).
18. J.R. Deily, Perkin-Elmer Corp. Atomic Absorption Newsletter 5:119 (1966).
19. G.F. Kirkbright, A.M. Smith, and T.S. West, Analyst 91:700 (1966).
20. G.F.Kirkbright, M.K. Peters, and T.S. West, Analyst 91:705 (1966).

21. S.L. Sachdev, J.W. Robinson, and P.W. West, Anal. Chim. Acta 37:12 (1967).
22. D.C. Manning and L.Capacho-Delgado, Anal. Chim. Acta 36:312 (1966).
23. E.E. Pickett and S.R. Koirtyohann, 18th Mid-America Symposium on Spectroscopy, Chicago, Ill. (May, 1967).
24. P. Pulido, K. Fuwa, and B.L. Vallee, Anal. Biochem. 14:393 (1966).
25. I. Rubeška and J. Štupar, Perkin-Elmer Corp. Atomic Absorption Newsletter 5:69 (1966).
26. J. Štupar, Mikrochim. Acta 1966:722.
27. C.L. Chakrabarti, J.W. Robinson, and P.W. West, Anal. Chim. Acta 34:269 (1966).
28. T.V. Ramakrishna, J. W. Robinson, and P.W. West, Anal. Chim. Acta 37:20 (1967).
29. G. Thilliez, Anal. Chem. 39:427 (1967).
30. W. Slavin and D.C. Manning, Appl. Spectry. 19:65 (1965).
31. S.R. Koirtyohann and E.E. Pickett, Anal. Chem. 37:601 (1965).
32. S.R. Koirtyohann, Ph.D. Thesis, University of Missouri, Columbia, Mo. (1966).
33. V.A. Fassel and D.W. Golightly, Anal. Chem. 39:466 (1967).
34. R.K. Skogerboe, A.T. Heybey, and G. H. Morrison, Anal. Chem. 38:1821 (1966).
35. E.A. Boling, Spectrochim. Acta 22:425 (1966).
36. M.D. Amos, The Element, No. 14, Aztec Instruments Inc., Westport, Conn. (1966).
37. L.Capacho-Delgado and D.C. Manning, Spectrochim. Acta 22:1505 (1966).
38. J.H. Gibson, W.E.L. Grossman, and W.D. Cooke, Anal. Chem. 35:226 (1963).
39. E.C. Simmons, R.W. Gress, and L.G. Bockstie, Conference on Analytical Chemistry and Applied Spectroscopy, Pittsburgh, Pa. (1967).
40. W.J. Kirsten and G.O.B. Bertilsson, Anal. Chem. 38:648 (1966).
41. A.A. Venghiattis, Conference on Analytical Chemistry and Applied Spectroscopy, Pittsburgh, Pa. (1967).
42. R.Woodriff and G.Ramelow, 5th National S.A.S. Meeting, Chicago, Ill. (June, 1966).
43. H.Massman, Z. Anal. Chem. 225:203 (1967).
44. R. H. Wendt and V.A. Fassel, Anal. Chem. 37:920 (1965).
45. R.H. Wendt and V.A. Fassel, Anal. Chem. 38:337 (1966).
46. K.E. Friend and A.J. Diefenderfer, Anal. Chem. 38:1763 (1966).
47. J.V. Sullivan and A. Walsh, Spectrochim. Acta 21:721 (1965).
48. J.S. Cartwright, C. Sebens, and W. Slavin, Perkin-Elmer Corp. Atomic Absorption Newsletter 5:22 (1966).
49. J.S. Cartwright, C. Sebens, and D. C. Manning, Perkin-Elmer Corp. Atomic Absorption Newsletter 5:91 (1966).
50. J.D. Dawson and D.J. Ellis, Spectrochim. Acta 23A:565 (1967).
51. J.A. Bowman, J. V. Sullivan, and A.Walsh, Spectrochim. Acta 22:205 (1966).
52. J. Vollmer, Perkin-Elmer Corp. Atomic Absorption Newsletter 5:35 (1966).
53. G.I.Goodfellow, Appl. Spectry. 21:39 (1967).
54. C.W. Frank, W.G. Schrenk and C.W. Meloan, Anal. Chem. 38:1005 (1966).
55. N.P. Ivanov, L.V. Minervina, S.V. Boranov, and L.G. Pofralidi, Zh. Analit. Khim. 21:1129 (1966).
56. V.A. Fassel, B. G. Mossotti, W.E.L. Grossman, and R.N. Knisely, Spectrochim. Acta 22:347 (1966).
57. C.W. Frank, W.G. Schrenk, and C.E. Meloan, Anal. Chem. 39:534 (1967).
58. W. Lang, Z. Anal. Chem. 217:161 (1966).
59. W. Lang, Z. Anal. Chem. 219:321 (1966).
60. J.V. Sullivan and A. Walsh, Spectrochim. Acta 21:727 (1965).
61. J.V. Sullivan and A. Walsh, Spectrochim. Acta 22:1843 (1966).
62. P.L. Boar and J.V. Sullivan, Fuel (London) 46:47 (1967).
63. J.A. Bowman, Anal. Chim. Acta 37:465 (1967).
64. M.D. Amos, Pittsburgh Conference on Analytical Chemistry and Applied Spectroscopy, 1967.
65. J.J. Chisholm and H.J. Emmel, Pittsburgh Conference on Analytical Chemistry and Applied Spectroscopy, 1967.
66. P.C. Wildy and R.C. Rooney, Pittsburgh Conference on Analytical Chemistry and Applied Spectroscopy, 1967.
67. J.W. Haggen-Smit and J. Ramírez-Muñoz, Anal. Chim. Acta 36:469 (1966).
68. C.W. Gehrke, J.P. Ussary, and G.H. Kramer, J. Assoc. Off. Agr. Chemists 47:459 (1964).

69. J.P. Ussary and C.W. Gehrke, J. Assoc. Off. Agr. Chemists 48:865 (1965).
70. S. Slavin, M.W. Gaumer, and W. Slavin, Pittsburgh Conference on Analytical Chemistry and Applied Spectroscopy, 1967.
71. B. Klein, J.H. Kaufman, and J.F. Marten, Pittsburgh Conference on Analytical Chemistry and Applied Spectroscopy, 1967.
72. L.R.P. Butler, personal communication, 1967.
73. J. Ramírez-Muñoz, J. L. Malakoff, and C.P. Aime, Anal. Chim. Acta 36:328 (1966).
74. J. L. Malakoff, W. Z. Scott, and J. Ramírez-Muñoz, Pittsburgh Conference on Analytical Chemistry and Applied Spectroscopy, 1967.
75. J.L. Malakoff, C.P. Aime, and J. Ramírez-Muñoz, Pittsburgh Conference on Analytical Chemistry and Applied Spectroscopy, 1967.
76. C. Th. J. Alkemade, Anal. Chem. 38:1252 (1966).
77. L. de Galan and J.D. Winefordner, Anal. Chem. 38:1412 (1966).
78. P.B. Adams and W.O. Passmore, Anal. Chem. 38:630 (1966).
79. D.J. Halls and A. Townshend, Anal. Chim. Acta 36:278 (1966).
80. R.E. Dickson and C. M. Johnson, Appl. Spectry. 20:214 (1966).
81. T.V. Ramakrishna, J. W. Robinson, and P.W. West, Anal. Chim. Acta 36:57 (1966).
82. L. Barnes, Anal. Chem. 38:1083 (1966).
83. A. Giammarise, Perkin-Elmer Corp. Atomic Absorption Newsletter 5:113 (1966).
84. J.B. Ezell and B.B. Elrod, 5th National S.A.S. Meeting Chicago, Ill. (June, 1966).
85. J. Ch. van Schouwenburg and A.D. van der Wey, Anal. Chim. Acta 36:247 (1966).
86. S. L. Sachdev, J.W. Robinson, and P.W. West, Anal. Chim. Acta 37:12 (1967).
87. J.B. Headridge and D.P. Hubbard, Anal. Chim. Acta 37:151 (1967).
88. W. Slavin, Perkin-Elmer Corp. Atomic Absorption Newsletter 6:9 (1967).
89. G.D. Christian and F.J. Feldman, American Chemical Society Meeting, Miami, Fla. (April, 1967).
90. H.W. Wilson, Anal. Chem. 38:920 (1966).
91. S.R. Koirtyohann and E.E. Pickett, Anal. Chem. 38:585 (1966).
92. J.B. Willis, in David Glick, ed., Methods of Biochemical Analysis, Vol. II, Interscience, New York (1964).
93. L. Capacho-Delgado and S. Sprague, Perkin-Elmer Corp. Atomic Absorption Newsletter 4:363 (1965).
94. S.R. Koirtyohann and E.E. Pickett, Anal. Chem. 38:1087 (1966).
95. R.J. Jaworowski and R.P. Weberling, Perkin-Elmer Corp. Atomic Absorption Newsletter 5:125 (1966).
96. C.E. Mulford, Perkin-Elmer Corp. Atomic Absorption Newsletter 5:88 (1966).
97. T. Takeuchi, M. Suzuki, and M. Yanagisawa, Anal. Chim. Acta 36:258 (1966).
98. G.F. Kirkbright, M.K. Peters, and T.S. West, Analyst 91:411 (1966).
99. T.S. West, Analyst 91:69 (1966).
100. L.R.P. Butler and P.W. Mathews, Anal. Chim. Acta 36:319 (1966).
101. B. Podobnik, M. Dular, and J. Korosin, Mikrochim. Acta 713 (1966).
102. T. Joyner and J.S. Finley, Perkin-Elmer Corp. Atomic Absorption Newsletter 5:4 (1966).
103. N.T. Sizonenko, E.S. Zolotovitskaya, B.M. Fidelman, and A.M. Bulgakova, Zh. Analit. Khim. 21:264 (1966).
104. I. Akaza, Bull. Chem. Soc. Japan 39:465 (1966).
105. R.C. Calkins, Appl. Spectry. 20:146 (1966).
106. P.J. Ch. Zeegers and C.Th.J. Alkemade, Combustion Flame 9:247 (1965).
107. C.Th. J. Alkemade and H.P. Hooymayers, Combustion Flame 10:306 (1966).
108. H.P. Hooymayers and C.Th.J. Alkemade, J. Quant. Spectry. Radiative Transfer 6:501 (1966).
109. H.P. Hooymayers and C.Th.J. Alkemade, J. Quant. Spectry. Radiative Transfer 6:877 (1966).
110. T. J. Vickers, L.D. Remington, and J.D. Winefordner, Anal. Chim. Acta 36:42 (1966).
111. J.D. Winefordner, W.W. McGee, J.M. Mansfield, M.L. Parsons, and K.E. Zacha, Anal. Chim. Acta 36:25 (1966).
112. M.L. Parsons and J.D. Winefordner, Anal. Chem. 38:1593 (1966).
113. M.L. Parsons, W.J. McCarthy, and J.D. Winefordner, Appl. Spectry. 20:223 (1966).
114. L. de Galan, personal communication, 1967.

115. L. de Galan and J.D. Winefordner, J. Quant. Spectry. Radiative Transfer 7:251 (1967).
116. J. A. Dean and J.E. Adkins, Analyst 91:709 (1966).
117. W. Lang, Spectrochim. Acta 23A:471 (1967).
118. N.R. Andersen and D.N. Hume, Anal. Chim. Acta 35:441 (1966).
119. T.C. Rains and O. Menis, Conference of Analytical Chemistry and Applied Spectroscopy, Pittsburgh, Pa. (1967).
120. J. W. Robinson and V. Smith, Anal. Chim. Acta 36:489 (1966).
121. J.A. Goleb, Anal. Chim. Acta 34:135 (1966).
122. J.A. Goleb, Anal. Chem. 38:1059 (1966).
123. J.A. Goleb, Anal. Chim. Acta 36:130 (1966).
124. T. Kumamaru, E. Tao, N. Okamoto, and Y. Yamamoto, Bull. Chem. Soc. Japan 38:2204 (1965).
125. T. Kumamaru, Y. Hayashi, N. Okamoto, E. Tao, and Y. Yamamoto, Anal. Chim. Acta 35:524 (1966).
126. T. Kumamaru, Y. Otani, and Y. Yamamoto, Bull. Chem. Soc. Japan 40:429 (1967).
127. W.S. Zaugg and R. J. Knox, Anal. Chem. 38:1759 (1966).
128. U. Westerlund-Helmerson, Perkin-Elmer Corp. Atomic Absorption Newsletter 5:97 (1966).
129. D.A. Roe, P.S. Miller, and L. Lutwak, Anal. Biochem. 15:313 (1966).
130. T.C. Rains, personal communication, 1967.
131. J.A. Dean and W. J. Carnes, Analyst 87:743 (1962).
132. P.T. Gilbert, Proceedings of the Xth Colloquium Spectroscopicum Internationale, 171 (1963).
133. D.C. Manning, Perkin-Elmer Corp. Atomic Absorption Newsletter 5:127 (1966).

Spectrographic Analysis of Inhaled Air Pollutants in Lung Tissue

Sheila Elton, Joanne Szajnar, and Ralph Smith

School of Medicine
Wayne State University
Detroit, Michigan

The lungs of animals exposed to air pollutants in city air and animals exposed to clean filtered air only were analyzed and the results compared. A semiquantitative emission spectrographic method for the determination of 18 trace metallic elements and 5 major constituent elements of lung-tissue ash was used.

Analysis of the lungs was based on the method of Tipton et al. Dry ashing was used to concentrate the metals in the tissue prior to dc-arc excitation of sample and internal standard. Direct-current arc excitation in an argon—oxygen atmosphere by using an enclosed water-cooled Stallwood jet was employed.

No significant difference was found in the concentration levels of the elements studied for the two groups.

INTRODUCTION

This paper describes a general semiquantitative method developed for the determination of trace-metal concentrations in lung tissue and its application to the analysis of lung tissue of animals that had been exposed to automobile exhaust products and those exposed only to clean air. The exposure of these animals was accomplished in specially constructed chambers. The purpose of this program was to determine if there was any significant difference in the concentrations of the various trace metals found in the lung tissue of the exposed and control groups.

Emission spectroscopy was selected as a rapid, sensitive method which would give all the desired information with adequate precision considering the variations between individual samples. The dc-arc technique, which is frequently applied to trace-metal analysis of medical and biological materials, was utilized because of its inherent sensitivity and because of the variety of sample types it can accommodate.

For the analysis of tissue samples for a large number of elements by emission spectrography, the method of Dr. Isabel Tipton and her co-workers [1–10] was chosen as being most readily adaptable to the purposes of this study. In this procedure dry ashing is employed to concentrate the metallic elements present in microgram amounts in the tissue by a factor of about 100 and to reduce the samples to a common matrix prior to dc-arc excitation of the sample plus the internal standard. According to Mitchell [11], dry ashing is preferred because of the difficulties associated with the purification of reagents used for wet ashing. However, whereas Tipton performed the arcing in air, controlled-atmosphere dc-arc excitation with an enclosed Stallwood jet was selected for the work reported in this paper in order to reduce arc wander, diminish matrix effects with an attendant increase in reproducibility of results, increase spectral sensitivity, and suppress cyanogen bands.

In the present study, certain elements were determined by anode excitation [12], and others, the more volatile elements, were arced as the cathode. Three palladium lines and two indium lines were used as internal standards. However, even with the appropriate choice of homologous line pairs, it is known that sample composition may have an appreciable effect on both line intensities and the intensity ratio of analytical to internal standard line. Hence, it is important to make up standards in an artificial matrix that closely approximates the sample matrix in chemical, mineralogical, and physical composition.

The depressant effect of the alkali metals on many of the more volatile elements has been frequently noted. Arcing in a controlled atmosphere such as argon can effectively reduce these matrix effects.

Controlled-Atmosphere Excitation

Studies published as early as 1950 by Vallee [13–15] on the influence of the noble gases on emission spectra in the dc arc indicated the potential value of the controlled atmosphere in spectrochemical techniques. For the majority of the spectral lines investigated, the signal-to-noise ratio was found to be more favorable in argon and helium than in air. This increase in the signal-to-noise ratio is a basic factor in increasing the sensitivity of the emission spectrographic determination of trace amounts of metals encountered in biological materials. In addition, by the use of a gas-tight stand such as that constructed for Vallee's work, the band spectra of CN are suppressed [16].

In comparing the effect of argon and helium atmospheres, it has been found that, along with the extremely prolonged volatilization observed for samples arced in pure argon (about 11 min at 6 A) as opposed to those arced in pure helium, there is also a decided enhancement of the lines of many elements burned in the pure argon environment. This enhancement was especially true for the more volatile elements, e.g., zinc, lead, and silver.

The Stallwood Jet

The Spex water-cooled Stallwood jet was used for the work described in this paper. The water-cooled version along with the use of deep-cupped electrodes keeps the unburned portion of the sample cooler and thus prolongs the time of volatilization of trace elements by presenting the arc column with a slow stream of material. This enhances the sensitivities achievable and promotes greater reproducibility. Improved precision with the Stallwood jet has been noted by several spectrographers, including Joensuu [17] and Mitteldorf [18].

EXPERIMENTAL

Sample Preparation

The frozen lung, which had been stored in a polyethylene bag, was thawed, rinsed in distilled water, trimmed of fat and connective tissue, and then diced. The cutting of tissue was done with a hard-grade stainless steel knife on a Plexiglas surface. The samples were then put in acid-washed platinum dishes, placed in a cold muffle furnace, and gradually brought up to 460°C, at which temperature they were ashed for 16 hr or more until a white ash was obtained. To ensure a uniform mixture of the ash, a few Teflon beads were added, and the polyethylene bottle in which the sample was stored was shaken on the Pica Mill for 3 to 5 min. For very small amounts of ash, the sample was mixed by grinding with either an agate or Teflon pestle.

Apparatus

The spectrograph used in this investigation was a dual-grating Bausch and Lomb instrument. This instrument is equipped with one

grating blazed for 6000 Å with 600 lines/mm, reciprocal dispersion
of 8 Å/mm in the first order, and another grating blazed for 3000 Å
with 1200 lines/mm, reciprocal dispersion of 4 Å/mm in the first
order. The arc source used was an NSL Spec Power. Densitometer
readings were made with an NSL Spec Reader. An NSL developing
machine was used for processing the Eastman No. 3 plates on which
the spectra were photographed.

The Spex Industries water-cooled, enclosed Stallwood jet provided
controlled atmosphere conditions. The 8 liter/min flow of 25% oxygen
and 75% argon mixture was regulated with a Linde oxygen type of
flowmeter with toggle switch.

A light chopper of this laboratory's design was located immediately
in front of the slit turret to further reduce the light intensity of the
filtered portion of the spectral lines. The fast response of the SA No. 3
plates, especially to the more intense spectral lines, necessitated the
use of the chopper in order to expand the metallic-element concentration ranges that would give satisfactory densitometer readings.

Preparation of Standards

The major constituents of lung tissue are potassium, phosphorus,
sodium, and chlorine with smaller amounts of iron, calcium, and magnesium.   Only spectroscopically pure chemicals, products of the
Johnson-Matthey Company, were used in the preparation. Sodium
chloride, potassium dihydrogen phosphate, magnesium oxide, ferric
oxide, and calcium carbonate were ground together in an agate mortar
for 30 min.   This mixture was transferred to a platinum dish and
ignited at 500°C for 24 hr.   This ignition converts the potassium
dihydrogen orthophosphate to potassium metaphosphate. The element
composition values of this synthetic-tissue ash which served as the
matrix for the standards are given in Table I.

Standard Mixtures

Series of standards were prepared by making known additions of
the desired metallic elements to the artificial-tissue ash matrix. In
the A and B series of standards, only spectroscopically pure materials
were used.   These standard mixtures were all heated in a manner
almost identical with that of the tissue samples in order to compensate
for any volatilization or other losses that may have occurred in the
tissue.

TABLE I

Composition of Synthetic-Tissue Ash Matrix

Compound	Element	Concentration in ash, %
KPO_3	K	17.40
	P	13.90
NaCl	Na	16.20
	Cl	25.00
MgO	Mg	1.08
$CaCO_3$	Ca	1.08
Fe_2O_3	Fe	1.08

The *A* series of standards contained Al, Be, Bi, Cd, Co, Cr, Mn, Mo, Ni, Sn, Sr, Ti, V, and Zn in the 1 to 100 ppm range, Ag and Ba in the 0.1 to 10 ppm range, and Cu in the 10 to 1000 ppm range. A solution containing 1 mg of each metal (0.1 mg for Ba and 10 mg for Cu) in 200 ml was prepared by the addition of the appropriate amounts of barium and strontium carbonates, and beryllium, bismuth, cadmium, copper, manganese, molybdenum, tin, titanium, vanadium, and zinc oxides. The solvent was 0.5 M sulfuric acid, previously purified by dithizone extraction. Another solution containing 1 mg Pb in 100 ml was prepared by the addition of the appropriate amount of lead oxide to a 0.25-M NaOH solution which had been purified by passing the solution through an ion exchange column. A third solution of 0.1 mg Ag in 100 ml was made by the appropriate addition of silver oxide to ammonium hydroxide. A mixture containing 10 mg each of Al, Cr, and Ni in 1 g of the previously prepared matrix was made by the addition of the appropriate amounts of the metal oxides to the matrix. This mixture was ground in an agate mortar and ignited at 500°C for 24 hr in a platinum dish. The desired series of standards was obtained by successive additions of these solutions and the mixture to the synthetic-tissue ash matrix with grinding, oven drying at 110°C, and ignition at 500°C in platinum dishes in a muffle furnace. Ignition at this temperature not only compensates for volatilization but also provides further mixing since the material is molten at this temperature.

The *B* series of standards contained Ba, Cd, and Mn in the 20 to 600 ppm range. By following a procedure similar to that given for the *A* series, a solution containing 10 mg each of Ba, Cd, and Mn and 100 mg of Zn in 100 ml was prepared from the appropriate amounts of barium carbonate and cadmium, manganese, and zinc oxides. Likewise, a solution of 10 mg Pb in 100 ml of 0.25 *N* NaOH and a mixture

TABLE II

Composition of Standards

Series A matrix synthetic-tissue ash
Calculated composition, μg metal element/g artificial-tissue ash

	Ag	Al	Ba	Be	Bi	Cd	Co	Cr	Cu	Mn	Mo	Ni	Pb	Sn	Sr	Ti	V	Zn
A 1	0.1	1	0.1	1	1	1	1	1	10	1	1	1	1	1	1	1	1	1
2	0.2	2	0.2	2	2	2	2	2	20	2	2	2	2	2	2	2	2	2
3	0.4	4	0.4	4	4	4	4	4	40	4	4	4	4	4	4	4	4	4
4	1.0	10	1.0	10	10	10	10	10	100	10	10	10	10	10	10	10	10	10
5	2.0	20	2.0	20	20	20	20	20	200	20	20	20	20	20	20	20	20	20
6	4.0	40	4.0	40	40	40	40	40	400	40	40	40	40	40	40	40	40	40
7	10.0	400	10.0	100	100	100	100	100	1000	100	100	100	100	100	100	100	100	100

Series B matrix synthetic-tissue ash
Calculated composition, μg metal element/g artificial-tissue ash

	Al	Ba	Cd	Pb	Mn	Ni	Sn	Ti	Zn
B 1	200	20	20	...	20	200	200	200	200
2	400	40	40	...	40	400	400	400	400
3	1000	100	100	...	100	1000	1000	1000	1000
4	2000	200	200	200	200	2000	2000	2000	2000
5	4000	400	400	400	400	4000	4000	4000	4000
6	6000	600	600	1000	600	6000	6000	6000	6000

Series C matrix, variable amounts of KH_2PO_4, NaCl, MgO, Fe_2O_3, and $CaCO_3$
Calculated percent composition

	Ca	Fe	Mg	Na	K	P
C 1	0.2	0.2	0.2	5.0	24.8	19.6
2	0.6	0.6	0.6	8.0	21.9	17.4
3	1.0	1.0	1.0	15.0	16.2	11.9
4	1.5	1.5	1.5	10.0	19.8	15.6
5	2.0	2.0	2.0	20.0	10.9	8.6
6	0.2	5.0	0.2	0.0	26.4	20.9
7	2.0	10.0	2.0	19.5	8.0	6.3
8	1.5	5.0	1.5	6.0	32.0	12.7
9	1.5	5.0	1.5	12.0	15.2	11.2
10	3.0	5.0	0.6	10.0	16.9	13.4
11	4.0	5.0	0.6	10.0	16.2	12.9

containing 100 mg each of Al, Ni, Sn, and Ti in 1 g of synthetic matrix were prepared. A second series of standards was obtained by a procedure completely analogous to that given for the first series.

The C series of standards was prepared for the determination of K, P, Na, Ca, Mg, and Fe. The proper amounts of reagent-grade potassium dihydrogen phosphate, sodium chloride, calcium carbonate, magnesium oxide, and ferric oxide were mixed and ground for 30 min in an agate mortar. Since trace-level concentrations were not determined with this series, it was thought that reagent-grade chemicals were adequately pure.

The composition of the standards is given in Table II.

Preparation of Graphite–Internal–Standard Mixtures

To obtain a uniform mixture of palladium in graphite, it is necessary first to dissolve the palladium chloride, next to add enough graphite to make a slurry, and then to evaporate to dryness.

It was found that palladium chloride is only moderately soluble in acetone and decomposes in ethanol on heating. An aqueous slurry is undesirable since graphite then climbs the sides of the beaker, which makes a quantitative preparation difficult.

Therefore, the best method of preparation was found to consist of the initial dissolving of the palladium chloride by the drop–by–drop addition of a small amount of concentrated hydrochloric acid. This step was followed by the addition of graphite and enough reagent-grade acetone to make a slurry, which was then evaporated to dryness on a water bath with stirring. A concentrated mixture of 5000 μg was first made and then subsequently diluted to 450 μg with the addition of graphite by the concentrated acid–acetone method.

A measured volume of concentrated indium solution (1000 μg) prepared from the metal was pipetted into an acid-washed beaker, and enough graphite to give a 40-μg mixture was added. As before, acetone was added to give a slurry, which was then evaporated.

Alternately, an amount of indium chloride was dissolved in a few drops of concentrated hydrochloric acid and then graphite and enough acetone to give a slurry were added. A mixture containing 1000 μg of indium was first made and subsequently diluted to 40 μg with the addition of graphite, as before.

When the sample was made the cathode, a mixture of graphite and 40 μg of indium was used as the internal standard. To increase the evenness of burning and to function as a fluxing agent, aluminum sulfate was added to this mixture in a 1:2 ratio to graphite.

When the sample was made the anode, a mixture of graphite and 450 μg of palladium and 160 μg of indium was used as the internal standard. This mixture was prepared by appropriate dilutions of the two concentrated mixtures with graphite.

Spectrochemical Procedure

The appropriate amounts of sample ash and graphite—internal-standard mixture were weighed on a Roller—Smith torsion balance, placed in plastic Wig-L-Bug containers, and mixed for 1 min on a Wig-L-Bug amalgamator.

The lower electrode contained the sample and was a $^3/_{16}$-in. crater undercut electrode. The counter electrode was made from $^3/_{16}$-in. graphite rod. The sample electrode was filled with the aid of a plastic cup funnel, the powder being tamped down with a plastic rod.

All samples and standards were run in duplicate. Throughout the burning of each sample, the length of the analytical gap was maintained at 5 mm for anode excitation and 10 mm for cathode excitation, monitored by an image of the electrodes projected onto a screen. To strike the arc, the electrodes were touched together, then quickly drawn 5 mm (or 10 mm) apart, measured by the projected image. The equipment and operating conditions are summarized in Table III.

RESULTS AND DISCUSSION

The addition of oxygen to the gas mixture is made necessary by the practical consideration of length of burning time (11 min in pure argon as opposed to 100 sec in a 75:25 mixture or argon and oxygen for biological materials at 6 A). Even though the addition of oxygen to argon increases the spectral background owing to the incandescence in the arc of small solid particles of carbon freed by the electrode reaction of carbon and oxygen, the signal-to-noise ratio is still more favorable in the argon—oxygen mixture than it is in air for most spectral lines [13].

It was found that, in controlled-atmosphere excitation, ion lines are often much stronger than neutral atom lines. For example, in the argon—oxygen atmosphere, the Sr 4077.7 Å line was more sensitive and reliable than the Sr 4607 Å line; the opposite was true for the arc in air. Likewise, the ion line Ti 3685 Å was more sensitive than the neutral atom line Ti 3653 Å. The use of the argon—oxygen atmosphere

TABLE III

Spectrographic Equipment and Operating Conditions

Sample excitation	Anode	Cathode layer
Elements determined	Al, Ba, Be, Ca, Co, Cr, Cu, Fe, K, Mg, Mn, Ni, P, Sr, Ti, V	Ag, Bi, Cd, Mo, Pb, Sn, Zn
Spectrograph	NSL (Bausch & Lomb) dual grating	Same
Slit width	20 μ	Same
Line height	7	Same
Split filter	No. 4 (8% transmission-filtered portion)	No. 3 (32% transmission-filtered portion)
Light chopper	Reduces per cent transmission of filtered portion approximately 50%	Same
Current	6.5-A d. c. (electrodes open)	11 A d. c. (electrodes open)
Exposure time	100-sec (to completion) total time, 2-sec pretime	15-sec total time, 0.5-sec pretime
Igniter	Off	Off
Photographic record	Eastman SA·No. 3 plates	Same
Densitometer	NSL Spec Reader	Same
Electrodes	Lower, anode: 3/16-in. crater, National, AGKSP, L-4000; upper, cathode: 3/16-in. graphite rod, National, AGKSP, L-3806, made 1/8-in. at tip	Lower, cathode: 3/16-in. crater, National, AGKSP, L-4000; upper, anode: 3/16-in. graphite rod, National, AGKSP, L-3806, made 1/8-in. at tip
Internal standard	Palladium 450 μg, + indium, 160 μg	Indium, 40 μg
Sample size	6-mg sample + 12-mg graphite internal standard	6-mg sample + 9-mg graphite internal standard and $Al_2(SO_4)_3$ (2:1 ratio)
Arc gap	5 mm	10 mm
Portion of arc used	Central portion, focused on gratings	Layer 1 mm above cathode focused on gratings
Primary aperture	1.5 mm	1 mm
Dual grating	Upper grating: 2600–3500 Å; lower grating: 7300–9200 Å (3600–4600 Å in second order)	Same
Developing time	3.5 min at 69°F	Same
Atmosphere control apparatus	Water-cooled enclosed Stallwood jet (Spex Industries, Inc., No. 9027)	Same
Inert atmosphere	Argon—oxygen in a 75%–25% mixture (Matheson Co.)	Same
Gas flow rate	8 liters/min	Same

S. ELTON, J. SZAJNAR, AND R. SMITH

TABLE IV

Concentrations of the Elements in Lungs of Two Exposure Groups of Animals

Element	Clean air			Ambient air		
	Number animals	Range	Mean	Number animals	Range	Mean
		(μg/g ash, %)			(μg/g ash, %)	
Aluminum	12	40-952	170.0	10	20-925	249.0
Barium	12	<0.1-7.1	2.9[f]	10	0.4-22	5.6
Beryllium	12	<1	<1.0	10	<1	<1.0
Bismuth	12	<1	<1.0	8	<1	<1.0
Cadmium	12	<10	<10.0	8	<10	<10.0
Chromium	12	<1-2.8	1.7[d]	10	<1-4.5	1.9[a]
Cobalt	12	<2-2.2	2.0[b]	10	<2-2.5	2.0[b]
Copper	12	71-200	117.0	10	52-170	114.0
Lead	12	3.5-19.5	10.4	8	3.8-103	23.5
Manganese	12	6.1-33.6	16.0	10	2.6-20.7	13.4
Molybdenum	12	<1-7.5	4.8[a]	8	3.9-10.7	5.5
Nickel	12	2.5-16	6.8	10	<1-11.5	5.1[a]
Silver	12	<0.1-0.9	0.15[g]	8	<0.1-0.75	0.25[d]
Strontium	12	1.6-10.7	5.6	10	2.4-7.9	4.8
Tin	12	<1-7.5	2.5[h]	8	<1-4.5	0.9[d]
Titanium	12	<2-15.7	5.0[i]	10	<2-9.2	5.2[e]
Vanadium	12	<4	<4.0	10	<4	<4.0
Zinc	12	725-2550	1480.0	8	1125-2475	1750.0
Calcium	12	0.35-1.79	0.93	10	0.25-1.41	0.72
Iron	12	0.31-2.92	1.44	10	0.55-1.91	1.16
Magnesium	12	0.37-2.24	1.19	10	0.33-1.90	1.00
Phosphorus	12	6.2-14.9	10.8	10	8.9-12.3	10.6
Potassium	12	9.7-19.4	14.5	10	7.7-16.6	13.0
Per cent ash in wet tissue	12	0.90-1.15	1.04	10	0.93-1.32	1.11
Per cent ash in dry tissue	12	5.14-5.98	5.59	10	4.87-6.35	5.63

[a]Mean includes 1 value of < 1.0 taken as zero.
[b]Mean includes 9 values of < 2.0 taken as zero.
[c]Mean includes 5 values of < 0.1 taken as zero.
[d]Mean includes 5 values of < 1.0 taken as zero.
[e]Mean includes 6 values of < 2.0 taken as zero.
[f]Mean includes 1 value of < 0.1 taken as zero.
[g]Mean includes 7 values of < 0.1 taken as zero.
[h]Mean includes 6 values of < 1.0 taken as zero.
[i]Mean includes 5 values of < 2.0 taken as zero.

also enhanced the sensitivity of Ag, Be, Pb, Sn, and Zn. This enhancement was especially notable in the cases of the selected Zn neutral atom and the Be ion lines. Also notable was the great improvement in the determination of phosphorus by excitation in the controlled atmosphere over excitation in air.

Recovery studies were performed by analyzing a well-mixed sample of human lung tissue to which known amounts of 18 elements had been added. The accuracy found here compares well with that reported from the use of a photoelectrically recording spectrometer [19]. The accuracy was determined with an average of 9.5% error for the elements. The precision determined from 14 runs of one sample has been found to have an average coefficient of variation of 10 for the elements.

Given in Table IV are the analytical results for a limited number of animals from two different groups. The first group was exposed to clean, filtered air for periods of approximately 700 to 900 days, and the other group was exposed to ambient air for a similar period of time. The ranges of concentration found and the mean concentration for each element for each group are presented here.

It appears from the data, as computed in Table V, that there is no significant difference in the concentration levels in at least the cases of lead, manganese, iron, and zinc between the two groups at the 95% confidence level. A greater volume of data would be desirable for making more accurate calculations; however, we can say that the trend seems to indicate no significant differences in the trace or essential element concentrations in the lungs of the animals of the clean air and ambient air groups. From the broad ranges of concentrations expected and found, it is doubtful that more precise chemical assay would reveal any significant differences.

TABLE V

Representative Calculations for Differences between the Groups

*	Lead		Manganese		Iron		Zinc	
	CA	AA	CA	AA	CA	AA	CA	AA
s^2	41.60	1051	69.9	52.1	0.854	0.328	51,400	222,000
$s_{\bar{x}}^2$	3.47	117	5.84	5.21	0.071	0.0328	4,280	22,200
$S_{\bar{x}_1 - \bar{x}_2}$	10.54		3.325		0.3224		163.00	
t	1.24		0.78		0.875		1.70	
$t_{0.05}$ (tables)	2.101		2.086		2.086		2.101	

*Where $t = (\bar{x}_1 - \bar{x}_2)/S_{\bar{x}_1 - \bar{x}_2}$.

REFERENCES

1. I.H. Tipton, W.D. Foland, F.C. Bobb, and W.C. McKorkle, ORNL-CF, 53-8-4 (1953).
2. I.H. Tipton, R.L. Steiner, W.D. Foland, J. Mueller, and M. Stanley, ORNL-CF, 54-12-66 (1954).
3. I.H. Tipton, M.J. Cook, R.L. Steiner, W.D. Foland, D.K. Bowman, and K.K. McDaniel, ORNL-CF, 56-3-60 (1956).
4. I.H. Tipton, M.J. Cook, R.L. Steiner, J.M. Foland, K.K. McDaniel, and S.D. Fentress, ORNL-CF, 57-2-2 (1957).
5. I.H. Tipton, M.J. Cook, R.L. Steiner, J.M. Foland, K.K. McDaniel, and S.D. Fentress, ORNL-CF, 54-2-3 (1957).
6. I.H. Tipton, M.J. Cook, R.L. Steiner, J.M. Foland, K.K. McDaniel, and S.D. Fentress, ORNL-CF, 57-2-4 (1957).
7. I.H. Tipton, M.J. Cook, R.L. Steiner, J.M. Foland, K.K. McDaniel, and S.D. Fentress, ORNL-CF, 57-11-33 (1957).
8. I.H. Tipton, M.J. Cook, J.M. Foland, J. Rittner, M. Hardwick, and K.K. McDaniel, ORNL-CF, 58-10-15 (1958).
9. Marvin J. Seven, ed., Metal Binding in Medicine, J.B. Lippincott Co., Philadelphia, Pa. (1960), p. 27.
10. H.M. Perry, Jr., I.H. Tipton, H.H. Schroeder, and M.J. Cook, J. Lab. Clin. Med. 60:245 (1962).
11. J.H. Yoe and H.U. Koch, Jr., eds., Trace Analysis, John Wiley and Sons, Inc., New York (1957), p. 398.
12. J.H. Yoe and H.U. Koch, Jr., eds., Trace Analysis, John Wiley and Sons, Inc., New York (1957), p. 363.
13. B.L. Vallee, C.B. Reimer, and J.R. Loofbourow, J. Opt. Soc. Am. 40:751 (1950).
14. B.L. Vallee and R.W. Peattie, Anal. Chem, 24:434 (1952).
15. S.J. Adelstein and B.L. Vallee, Spectrochim. Acta 6:134 (1954).
16. L.T. Steadman, Phys. Rev. 63:322 (1943).
17. O.I. Joensuu and N.H. Suhr, Appl. Spectry. 16:101 (1963).
18. A. Arrak and A.J. Mitteldorf, Appl. Spectry. 13:85 (1959).
19. R.E. Nusbaum, E.M. Butt, B.A. Gilmour, and L. DiDio, Am. J. Clin. Pathol. 35:44 (1961).

Statistical Applications

What Can Be Detected?*

W. L. Nicholson

Mathematics Department
Battelle Memorial Institute
Pacific Northwest Laboratory
Richland, Washington

The characteristics of an analytical method are dynamic. The state of the apparatus and of the operator, the laboratory environment, and the level and type of contaminants present in the sample are examples of such characteristics. Detectability, which concerns functioning of an analytical method near its natural limit of sensitivity, is particularly dependent on dynamic characteristics. At the instant of analysis, past history only determines these characteristics in a frequency sense. Probability theory provides a tool for deciding what can be detected by using frequency-type information to define the characteristics of the analytical method.

Within the framework of classical statistics, detection limits are defined in terms of error rates. Economic, political, or, in general, some worldly considerations must determine what are acceptable error rates. Again, probability theory provides the tool for modeling the error-rate concept.

INTRODUCTION

The determination of the sensitivity of a quantitative chemical analysis method is a statistical problem. Characteristics of an analytical method are necessarily dynamic. At the moment of analysis, the state of the apparatus and of the operator, the laboratory environment, and the level of contaminates in chemical reagents, all are examples of such characteristics. Usually, the exact state of such characteristics can only be inferred from past experience. Probability theory provides a frequency-type description of dynamic characteristics.

The question of what can be detected concerns the functioning

*This paper was prepared on work performed under U.S. Atomic Energy Commission Contract AT(45-1)-1830. Dr. Nicholson is Senior Research Associate, Mathematics Department, Pacific Northwest Laboratory, operated by Battelle Memorial Institute for the U.S. Atomic Energy Commission, Richland, Washington.

of the analysis method near its natural limit of sensitivity. Here, experimental data can be thought of as being generated by several sources of comparable magnitude. All but one are irrelevant to the purpose of the analysis. The effect of dynamic characteristics of the apparatus, etc., is an example of such irrelevant sources. The sum total of these irrelevant sources is here designated "the background source." The single source of interest is designated the "sample source." Detection of a sample source of specified strength in the presence of a background source of specified strength fits naturally into the framework of classical statistical theory. This theory provides an objective approach for construction of a detection criterion which weighs experimental evidence and selects one of the alternatives of positive sample source and background-only source. The error rates associated with a detection criterion (the frequency with which positive sample sources are not detected and background-only sources are erroneously flagged as "positive") are simple probability calculations. Error rates can be used to investigate the relative worth of several detection criteria and to select an optimal method of analysis from the detectability standpoint. The threshold of sensitivity of an analytical method can also be described in terms of error rates by introducing the concept of the "minimum detectable sample source." This paper elaborates on these ideas with the use of appropriate probability and statistical theory. The author's statistical consulting experience has been primarily in radiochemistry, which limits the scope of examples. However, the general approach used in the analysis of the examples should apply equally to the evaluation of any quantitative method near its natural limit of sensitivity.

FREQUENCY DISTRIBUTIONS

A quantitative method is available for estimating a specific property of a sample. Only for the sake of explicitness, suppose that the method estimates the concentration of a particular substance in the sample. Let c be the true concentration in the sample. Analysis of the sample with the quantitative method results in a sample gross determination of c.

The beginning point in the evaluation of the sensitivity of a quantitative method is the modeling of gross determinations when, in fact, c is zero. The hypothetical population of all possible such gross determinations can be described with a frequency distribution function. The first line of Fig. 1 is such a function. The total area under the curve

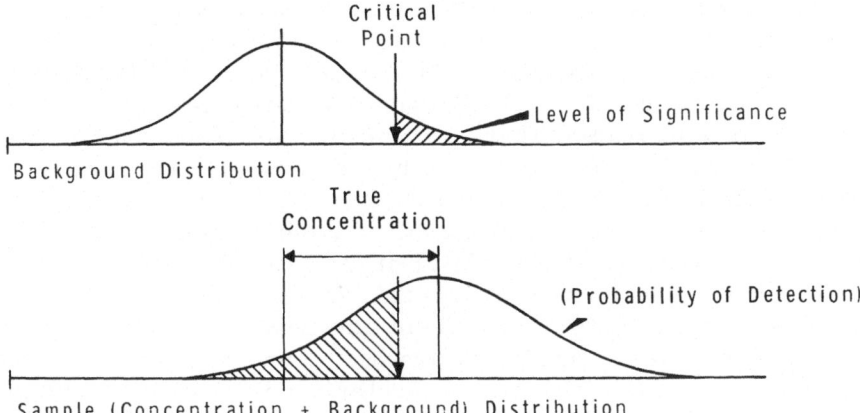

Fig. 1.   Determination of minimum detectable concentration with known background distribution.

is unity.   The area under the curve between any two specified points is the probability that an individual gross determination will fall between these two points.   This frequency curve is here designated the "background distribution."   In many cases, the background distribution has a positive bias; i.e., the average gross determination when the true concentration is zero is a positive quantity. Examples of factors which could result in such a positive background-distribution mean include reagent contamination during a chemical separation prior to spectrometric analysis and background radiation in radiochemical analysis.   The standard deviation of the distribution (numerically proportional to the width at half maximum) reflects variability in such factors as well as the dynamic character of the apparatus and the laboratory environment plus all remaining nondescript fluctuations. These fluctuations are usually lumped together as random error.   The two examples above of positive biasing features are variable, so that their contribution to the mean of the background distribution would actually be their long-run average value as opposed to their specific value in any given gross determination.   With proper calibration apparatus variation should average to zero.   By definition, the random error effects perturb individual gross determinations from their true value but in the long run average to zero.   Examples of contributors to the random error component of variability include measurement error due to imprecise optical readings, power supply fluctuations, effects of minor ambient-temperature variability, and imprecise setting of apparatus controls.

When the true concentration C is positive, the gross determination frequency distribution shifts to the right. The second line of Fig. 1 illustrates such a distribution. In general, the shift of the mean is equal to the value of C. The standard deviation of the distribution almost always increases. There are basically two mechanisms responsible for this increase; often, both are present. The first is the inherent stochastic character of the substance being measured. For example, if C is the concentration of a discontinuous and heterogeneously distributed phase, individual samples have true concentrations which are distributed around the average value of C. Radioactivity behaves similarly. In any finite time period, the actual number of disintegrations is distributed around the underlying true average disintegration rate. The second mechanism is associated with the basic measuring process of the apparatus. Usually random error increases with the magnitude of the measurement. For example, any logarithmic measurement scale induces a constant percent random error when read optically.

DETECTION CRITERION

The state of knowledge about the background distribution determines the form of the statistical procedure for deciding whether C is positive. Such a procedure is here called a "detection criterion." (Detection criterion is simultaneous with "hypothesis test" in the statistical literature.) Suppose that the exact mathematical form of the background distribution is known. The *modus operandi* of a detection criterion is to assume that a sample gross determination is an observation from this background distribution unless there is strong evidence to the contrary. Since C positive always shifts the distribution to the right, the only evidence favoring a positive C is having the gross determination fall in the extreme right tail of the background distribution. "Extreme" is defined in terms of tail area. This tail area is the probability that a zero-concentration sample will be judged positive based on a single gross determination. The probability of such an erroneous judgment is commonly called the "level of significance." (In statistical literature the level of significance is usually expressed in percent.) With the background distribution completely specified, the left-hand end of a tail area is known once the level of significance is given. This end point is called the "critical point." It defines the detection criterion. The situation is illustrated in the first line of Fig. 1. The vertical arrow locates the critical point. The area of the shaded portion to the right of the critical point is the level of significance. The specific form of the detection criterion is:

Judge that $C = 0$ if gross determination $\leq$ critical point

Judge that $C > 0$ if gross determination $>$ critical point

It is important to realize that, because of the random nature of background, there is always the possibility that a zero-concentration sample will be judged positive. Consideration of the consequences of such an erroneous judgment may help set the level of significance.

Frequently, background information is strictly empirical. A history of gross determinations from zero-concentration samples provides a frequency histogram which is an estimate of the background distribution. The above detection-criterion approach can be approximated with an estimated critical point by using this frequency histogram. An alternative which is often more attractive is to use a background corrected *net* determination and an estimate of the standard deviation of this net determination.

Suppose B is an estimate of the mean of the background distribution. Then,

Net determination = gross determination $- B$

Suppose S is an estimate of the standard deviation of the net determination. The detection-criterion statistic is

$$\frac{\text{Net determination}}{S}$$

Distributions of this statistic replace those of the gross determination in the argument. The first line of Fig. 2 is an illustration of the distribution of the detection-criterion statistic for zero-concentration gross determinations. The mean of the distribution is zero, and the standard deviation is approximately unity. The second line of Fig. 2 is the distribution of the same statistic for positive-concentration gross determinations. The mean is positive, and the standard deviation, in general, is larger than for the zero-concentration case. The argument now continues as for the case of a known background distribution of the detection criterion. The exact location of the resulting critical point will in general not be known because the standard deviation of the distribution still involves the unknown background-distribution parameter. Usually a good approximation to the distribu-

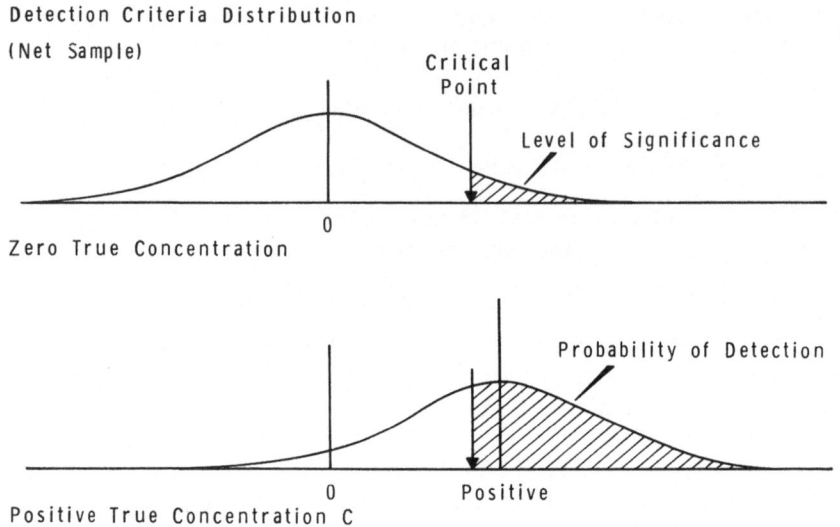

Fig. 2.   Determination of minimum detectable concentration with estimated background
distribution.

tion is the normal.   Critical points for specified tail areas are read
from appropriate tables.

The advantages of the net–determination approach are twofold.
The dependence on unknown background–distribution parameters is
minimized.   More important is the fact that the net–determination
approach works with a minimum of background information as opposed
to a rather complete history which is needed to construct a histogram.
Often the mean of the background distribution fluctuates from day to
day so that it is impossible to compile a lengthy history.   With the net–
determination approach one background determination will suffice.

MINIMUM DETECTABLE CONCENTRATION

With either the background known or the net–determination approach, the sensitivity of the method is investigated by calculation
of the probability of detection as a function of the true concentration
c.   The second lines of Figs. 1 and 2 illustrate this calculation.   The
probability of detection of the positive concentration c based on a
single determination is the area under the curve to the right of the
critical point (in statistical literature this area is usually called
"power" when c is positive).   The term "detection coefficient" can
be used for the numerical value of this probability.   A precise statement

of sensitivity can now be given in terms of level of significance, detection coefficient, and the concept of "minimum detectable concentration." Suppose a detection criterion is defined in terms of a specific level of significance; then the minimum detectable concentration is the concentration C for which the detection criterion has a probability of detection equal to the specified detection coefficient. These ideas can be summarized as:

Quantities
   Level of significance = α
   Detection coefficient = β
   Minimum detectable concentration = C_0

Defining relationships
   Prob {judge $C > 0$ when in fact $C = 0$} = α
   Prob {judge $C > 0$ when in fact $C = C_0$} = β

A common practice is to determine minimum detectable concentration for a detection coefficient of 0.5. This is analogous to the practice in the pharmaceutical industry of defining drug effectiveness in terms of the ED–50 (median effective dose), i.e., the drug dose which is effective in 50% of the cases. In our situation, the minimum detectable concentration is that true concentration which is detected in 50% of the individual determinations.

It is important to distinguish between detection of a given true concentration and estimation of the same concentration. A detection criterion is only concerned with the question of whether or not the true concentration is positive. In general, this criterion gives no information concerning the actual value of the true concentration. Thus the question of estimability is completely outside the framework of determining sensitivity and calculation of minimum detectable concentration.

DETECTION OF RADIOACTIVITY, AN EXAMPLE

A simple counting problem in radiochemistry illustrates the general ideas discussed above. A sample is counted for a fixed time period in a single-channel instrument to determine the sample radioactivity. The data consist of a gross count x recorded over time t. A background estimate for the same instrument consists of a count y recorded over time s. Let C be the true average counting rate for the radioactivity source within the sample and B the true average background counting rate for the instrument. Assume that both C and B are time invariant. Thus, the radioactive decay rate does not diminish, and background intensity is constant during the course of the experiment. The net counting-rate estimate of C is

$$\hat{C} = x/t - y/s$$

If $C = 0$, the best estimate of the variance of $\hat{C}$ is

$$x + y/st$$

On applying the general formulation involving net determinations, the form of the detection criterion is

$$\hat{C}/\sqrt{(x + y)/st}$$

When $C = 0$, the distribution of this criterion is approximately normal with zero mean and unit variance independent of the value of B. Critical points can be read directly from a table of the normal distribution. Thus, e.g., with the level of significance equal to 5%, the detection criterion is

$$\text{Judge that } C = 0 \text{ if } \hat{C} \leqq 1.645 \sqrt{x + y/st}$$

$$\text{Judge that } C > 0 \text{ if } \hat{C} > 1.645 \sqrt{x + y/st}$$

This criterion is optimal in the sense that, for a given level of significance and a given detection coefficient, it has the smallest minimum detectable concentration. Again, it is stressed that the criterion is not connected with estimation of the true counting rate. When $\hat{C}$ is used to estimate C, the appropriate standard-deviation estimate is $\sqrt{x/t^2 + y/s^2}$.

Figure 3 shows the lower left-hand corner of the sample space for this counting problem. The horizontal scale is the total background count y. The vertical scale is the sample gross count x. As the counts are necessarily integers, the only possible experimental outcomes are the integer lattice points, as indicated in Fig. 3. The dotted line is the expected relationship between x and y when $t/s = 1/3$ and $C = 0$. The detection criterion amounts to a division of the sample space into two regions, as indicated by the solid curve in Fig. 3. The lower region contains all the experimental outcomes for which the decision criterion judges $C = 0$. The upper region contains all those experimental outcomes for which the judgment is $C > 0$. Thus, e.g., if $x = 4$ and $y = 2$ when the background count is taken over a period three times as long as that of the sample count, the judgment is $C > 0$. The probability of detection is now the probability that the experimental point falls in the upper region. This probability can be calculated as soon as the average

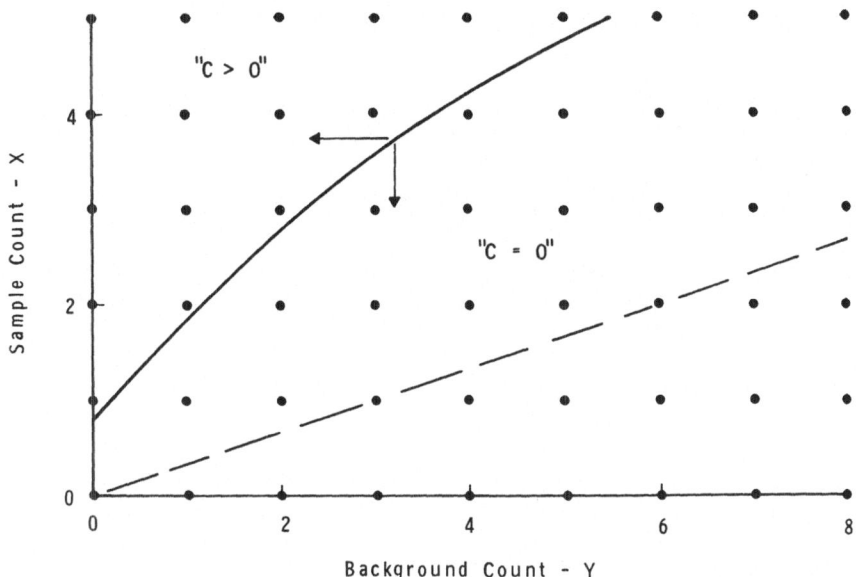

Fig. 3. Sample space dichotomy based on detection criterion; $\alpha = 0.05$, $t/s = 1/3$

sample count and the average background count are given. Figure 4 is a graph of the probability of detection as a function of the average sample count for several levels of average background counts. With true sample and background counting rates C and B, the average net sample count is tC and the average background count contribution to the gross sample count is tB. The standard deviation of this background contribution is $\sqrt{tB}$. The horizontal scale measures this average net sample count in units of the standard deviation of the average background contribution to the gross sample count. The curves in Fig. 4 represent the relation between probability of detection and the true sample count for background contributions of $tB = 1$, 2, 5, and infinity. The curve for infinite background results from a limiting mathematical treatment which, in practice, can be used to determine the probability of detection for any background contribution greater than 50. The dotted lines in Fig. 4 show that, with a detection coefficient of 0.5, the minimum detectable sample count is approximately twice the square root of the average background contribution to the sample count. This statement holds over the entire range of backgrounds from 1 to infinity. For detection coefficients close to 1, the minimum detectable sample count ranges from 3.5 to 7 times the square root of the average background, depending upon the numerical value of this background.

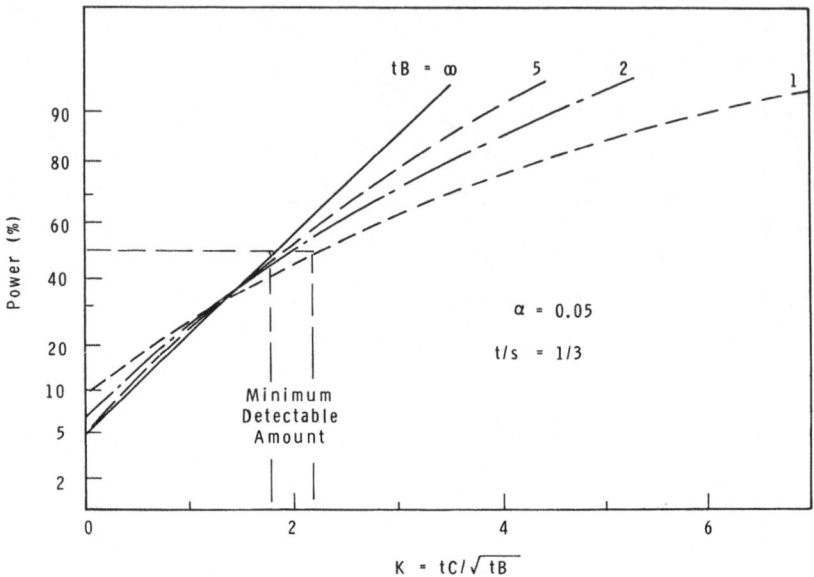

Fig. 4. Power curves for net counting-rate detection criterion.

With a family of power curves (such as those in Fig. 4) which describe the probabilistic nature of a detection criterion, experimental conditions can be set to meet a required minimum detectable sample counting rate. For example, suppose the above detection criterion is to be used when $C = 1$ and $B = 2$ cpm. Suppose further a detection coefficient of 0.9 is required. A simple calculation shows that the 5 curve in Fig. 4 gives a lower detection coefficient than 0.9. The necessary curve lies between the 5 curve and the infinity curve. Interpolation gives a sample counting time T of approximately 30 min. Thus, 30 sample counts can be detected with probability 0.9 in the presence of 60 background counts with the aid of a background counting-rate estimate based on 180 counts.

MINIMUM DETECTABLE CONCENTRATION FOR MULTISOURCE PROBLEMS

A problem of particular interest to spectroscopists is the calculation of minimum detectable concentration when the quantitative analysis resolves a mixture into its several components. Now, the probability of detection of any given component is a function of what else is present in the mixture; or, looking at the problem another way, the sum total of all irrelevant components in the mixture can be thought of as the background. The behavior of a detection criterion for the component of interest must be determined over a representative set of all possible backgrounds—mixtures of the other components. Once this multidimensional characterization is completed, minimum detectable concentration can be defined in terms of the mixture of the other components, which makes detection most difficult.

Minimum–detectable–concentration calculations are in general quite messy for mixture problems. Without the digital computer, they would not be feasible. An example is treated here which, though transparent, illustrates the general approach in a real problem. The example is a simplified version of the minimum–detectable–concentration calculation in gamma-ray spectroscopy. Consider a situation where a mixture of gamma-ray spectra is resolved into three components by using the method of simultaneous equations. (Simultaneous equations are mathematically elementary but practically inferior to other approaches.) Figure 5 shows the response matrix for the simultaneous equation solution. Each line of the matrix distributes the total counts from a single radionuclide into three energy bins. Thus, 70% of the counts for radionuclide B that are in the energy range covered by the analysis fall in bin 2. The response matrix is typical of the simultaneous-equation approach, where bins are selected to include a major portion of the photopeak of a given radionuclide and at the same time minimize overlap with photopeak regions of the other spectra.

The simultaneous-equation approach admits estimates of the true total count for each of the radionuclides and estimates of the standard

Radionuclide	Energy Bin		
	1	2	3
A	0.9	0.1	0
B	0.2	0.7	0.1
C	0.2	0.2	0.6

Fig. 5. Gamma-ray spectroscopy response matrix.

deviation of each of these total count estimates. Suppose that we are interested in minimum detectable concentration for radionuclide B. Let N_A, N_B, N_C be the expected total counts for radionuclides A, B, and C. Let $\hat{N}_B$ be the estimate of the total count for radionuclide B, and let S_B be the estimate of the standard deviation of N_B. Use as a detection criterion

$$\text{Judge } N_B = 0 \text{ if } \hat{N}_B \leqq 2S_B$$

$$\text{Judge } N_B > 0 \text{ if } \hat{N}_B > 2S_B$$

When N_A, N_B, N_C are large, the level of significance associated with this detection criterion is 2.5%. The probability of detection of a given amount of radionuclide B can be calculated as a function of N_A and N_C. Define a function of three variables

$$F(N_A, N_B, N_C) = \text{Prob } \{\hat{N}_B > 2S_B\}$$

which is the probability of detecting N_B, given N_A and N_C. To investigated minimum detectable concentration, a plot can be made of the relationship

$$F(N_A, N_B = K, N_C) = 0.5$$

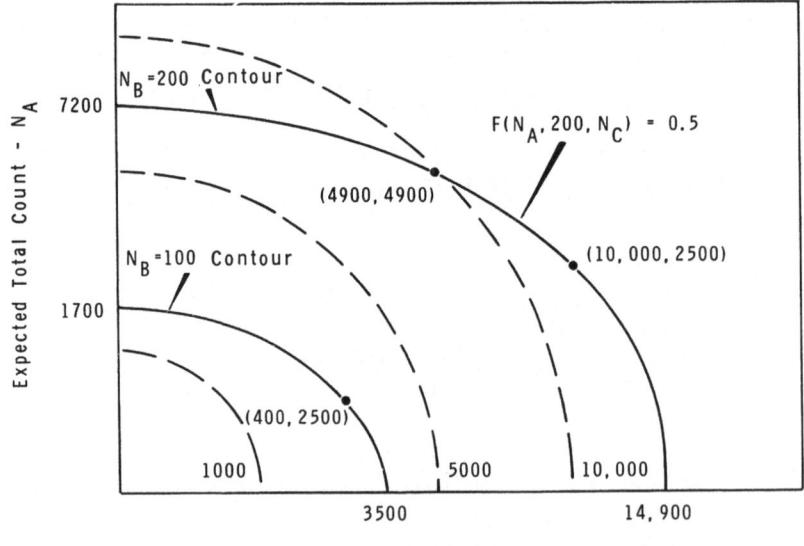

Fig. 6. Minimum detectable amount of radionuclide B; $\alpha = 0.025$, $\beta = 0.5$.

Such a plot shows the relative amounts of radionuclides A and C which give a minimum detectable concentration of radionuclide B equal to K with detection coefficient of 0.5. Figure 6 is a graph of this relationship for the two values of $K = 100$ and $K = 200$. The two scales are linear in the square root of the total count; thus, a straight line $N_A + N_C =$ constant appears as a circle. Three such circles are contained in Fig. 6 as dotted curves. The two solid curves give the relationship between N_C and N_A, for which the minimum detectable concentration of radionuclide B is 100 and 200, respectively. If these curves were circles and hence symmetrically placed with respect to the dotted curves, minimum detectable concentration would depend only on the sum of the other two radionuclides in the mixture. The form of the curves shows that it is more difficult to detect radionuclide B in the presence of radionuclide A than in the presence of radionuclide C. In fact, as the ratio $N_A/N_C =$ decreases, the sum $N_A + N_C$ increases, for which the minimum detectable concentration is a constant. The three marked points on the $N_B = 100$ contour illustrate this fact. The ratios are ∞, $4/25$, and 0, while the sums are 1700, 2900, and 3500, respectively. With use of the "mixture-which-makes-detection-most-difficult" approach, minimum detectable concentration would be defined in terms of an $(N_A, 0)$ mixture as the background.

REFERENCES

The article is an application of the Neyman and Pearson theory of testing statistical hypotheses [1], which is considered in several elementary textbooks on statistics [2–4]. An excellent mathematical treatment of the theory is given by Lehmann [5]. The example on the detection of radioactivity is taken from the author's AEC Research and Development report [6], an abridged version of which has appeared in the open literature [7]. The example on minimum detectable concentration was constructed specifically for this article.

1. J. Neyman and E.S. Pearson, "On the use and interpretation of certain test criteria for purposes of statistical inference," Biometrika, 20A:175–240; 20A:263–294 (1928).
2. W.J. Dixon and F.J. Massey, Introduction to Statistical Analysis, McGraw-Hill, New York (1957).
3. W. Allen Wallis and Harry V. Roberts, Statistics: A New Approach, Macmillan, New York (1956).
4. John E. Freund, Modern Elementary Statistics, 2nd Ed., Prentice-Hall, Englewood Cliffs, N.J. (1960).
5. E.L. Lehmann, Testing Statistical Hypotheses, Wiley, New York (1959).
6. W.L. Nicholson, "Fixed time estimation of counting rates with background corrections," HW-76297 (1963).
7. W.L. Nicholson, "Statistics of net-counting-rate estimation with dominant background corrections," Nucleonics, August 1966.

Statistics Help Evaluate Analytical Methods

C. L. Grant

Engineering Experiment Station
University of New Hampshire
Durham, New Hampshire

The literature of analytical chemistry is filled with methods which were well conceived experimentally but were poorly evaluated. Many of the difficulties experienced in interlaboratory test programs are caused by such methods. Proper description of the capabilities of an analytical method requires the use of objective rather than intuitive data analysis. Some objective evaluation procedures will be described with emphasis on the concepts and their relationship to the best means of data collection.

Once the systematic errors in a procedure have been eliminated, some simple statistical techniques can provide a quantitative description of the remaining random errors. The opportunity to assign the proportions of this total variability to different stages of the procedure is particularly useful. With this information, reduction of the total random error can be accomplished in an efficient manner when necessary.

INTRODUCTION

The material presented in this paper is neither new in concept nor original with me. The sole purpose of this presentation is to encourage wider and more effective use of statistical techniques in analytical chemistry. Calder [1] pointed out

Each observation in analytical chemistry, no less than any other branch of scientific investigation, is inaccurate in some degree, and while the accurate value for the concentration of some particular constituent in the analysis material cannot be determined, it is reasonable to assume that the accurate value exists, and it is important to estimate the limits between which this value lies. It is, therefore, desirable—nay, imperative—that the chemist should be familiar with the elements of statistical method, not only from the point of view of consistency in the general presentation of analytical results but in order to derive reliable estimates from the observational data.

The analytical chemist does not want to convert analytical problems into statistical problems for a statistician to solve. Instead, the chemist should develop some facility with the basic concepts of statistics so he can do a better job of experimental design and analysis in his everyday work.

Statistical analysis does not overcome the handicap of faulty data collection. In the application of tests of significance, freedom from systematic bias is assumed. If we consider the term "experimental design" includes all aspects of the data collection process, then we can say that proper experimental design is necessary if we hope to benefit from statistical analysis of the data. It is indisputable that the statistical evidence resulting from the analysis of data obtained from a carefully designed experiment provides a much sounder basis for inference than could possibly result from data collected in a haphazard manner. Furthermore, it is quite possible that an imperfect or inefficient analysis of a well-designed experiment will yield conclusions which are substantially correct but the converse is almost never true.

ESTIMATING THE PRECISION OF A METHOD

In the American Society for Testing Materials' (ASTM) "Suggested Practices for Use of Statistical Methods in Spectrochemical Analysis" [2] precision is defined as: "The extent of agreement of a series of measurements with their average as frequently measured by the standard deviation. It is essential to express the conditions under which the data have been obtained." How often do we hear analytical methods described and precision data quoted with no description of how the data were obtained ? At most meetings it is heard rather often. Even when conditions are described, the values obtained are often referred to as the precision of the method when they really represent only the variation in one or two steps of the method.

Just as an experimental mean is an estimate of a true mean, a standard deviation calculated from a limited amount of data is an estimate of the true standard deviation. For the moment, let us assume that we have a set of replicate measurements without worrying about how the data were obtained. Let us further assume that the variations represent random error only and that the data are expressed in a form such that the size of the random error does not vary systematically with the magnitude of the results. The dispersion of the distribution of y values (assuming the data approach a normal distribution) is given by the true standard deviation σ_y where

$$\sigma_y = \sqrt{\frac{\Sigma (y - \mu)^2}{n}} \tag{1}$$

Here, μ is the true value for the quantity being measured, and n is the number of replicate measurements. The difficulty is that we do not know the true mean μ when we have a sample of values and so we must use the experimental mean $\bar{y}$ in place of μ. It can be shown that this substitution provides an estimate of the true standard deviation which is biased on the low side. To counteract this bias, we replace n with $n-1$. Thus, we have the standard deviation calculated from a sample of values and denoted by S_y, given by

$$S_y = \sqrt{\frac{\Sigma \, (y - \bar{y})^2}{n - 1}} \qquad (2)$$

where $\bar{y}$ is the sample mean and $n-1$ is known as the degrees of freedom (n in formula (1) is also called the degrees of freedom since, in both cases, it is the number of independent deviations).

Formula (2) is very cumbersome to use and presents a special problem in significant figures if $\bar{y}$ happens to be a repeating fraction. From a simple algebraic manipulation, either of the following two equivalent forms result:

$$S_y = \sqrt{\frac{\Sigma \, y^2 - (\Sigma \, y)^2/n}{n - 1}} = \sqrt{\frac{n \, \Sigma \, y^2 - (\Sigma \, y)^2}{n(n - 1)}} \qquad (3)$$

Here, $\Sigma \, y^2$ is the sum of the squares of each replicate observation, $\Sigma \, y$ is the sum of the observations, and n is the total number of replicates.

The computations can often be simplified if the original data are coded by adding or subtracting a constant amount from each value. This transformation increases or decreases the mean, but the deviations of individual values from the mean are unaltered and, therefore, the standard deviation is unchanged. When very large numbers or negative values are involved (as often happens with logarithmic transformations), coding is particularly helpful.

Many investigators determine the precision of an analytical method by obtaining between 10 and 20 replicate measurements on one bulk standard or sample. From a computation point of view, calculation of the standard deviation is very simple, in this case, with the use of formula (3). If each replicate is an independent entity and is carried through the entire analytical process, the precision estimate includes sample heterogeneity for that particular sample or standard plus the random errors in the preparation and determination steps. However, if he uses 10 aliquots of a solution which has been through all preparation steps prior to determination, the precision estimate includes only the error associated with the actual determination step. The values obtained in these two ways can be expected to differ appreciably.

A precision estimate that is often more useful can be obtained from duplicate or triplicate measurements on a series of samples spanning the whole concentration range for which the method is recommended. By using samples rather than standards, the precision estimate will include the normal variability due to typical sample heterogeneity for the particular material involved. Of course, the replicates must be entirely separate entities for each sample and should be carried through the complete procedure. When duplicates are used, S_y can be calculated from

$$S_y = \sqrt{\frac{\Sigma \, d^2}{2n'}} \tag{4}$$

where $\Sigma \, d^2$ is the sum of the squares of the differences between the duplicate values of each set and n' is the number of sets of duplicates. Since this precision estimate applies to the whole useful concentration range, the data must be in such a form that there is no direct correlation between the difference d and the concentration. If the differences tend to increase in proportion to the magnitude of the value, a transformation to logs will usually elliminate this correlation. This will be illustrated later.

For triplicate or more replicates on each of several samples, a generalized standard deviation can still be obtained by using formula (3) and pooling the results as described in most statistics texts.

Often, the standard deviation is expressed on a relative percentage basis as the coefficient of variation, c, where

$$c = \frac{S_y(100)}{\bar{y}} \tag{5}$$

This has the advantage of expressing the precision in terms of a percentage of the value being estimated, but we must remember that both S_y and $\bar{y}$ are estimates and subject to error.

Let us now consider what these quantities represent by considering Fig. 1. The top diagram is the idealized normal distribution of individuals for some population. The important point to note is that approximately one-third of the area under the curve lies outside the limits of $\mu - 1\sigma$ to $\mu + 1\sigma$. This means that about one-third of the estimates can be expected to deviate from the true mean by more than one standard deviation. Going a step further, approximately 5% will deviate by more than 1.96σ. However, only about 0.3% can be expected to deviate by more than 3σ. Clearly, we should expect a certain portion of our estimates to deviate from the mean by an amount much greater than one standard deviation. Thus, the common idea that we should discard values which deviate by more than 1σ is entirely incorrect.

Similarly, the policy of shifting calibration curves from day to day on the basis of two to four data points is *often incorrect* because the variations may be due entirely to random error. On the other hand, if the variations can be shown to fall outside reasonable limits of random error (at whatever probability level the experimenter desires), the adjustment is justified or the cause of the systematic shift should be located.

When samples of several individuals are taken and the distribution of sample means is plotted, we expect the dispersion of the distribution to be reduced as n, the number of individuals in each sample, is increased. From the central-limit theorem, we find that the standard deviation of sample means, $\sigma_{\bar{y}}$, is related to the standard deviation of individuals by

$$\sigma_{\bar{y}} = \frac{\sigma_y}{\sqrt{n}} \tag{6}$$

This is illustrated in the remainder of Fig. 1 for cases where $n = 4, 9,$ and 16. Clearly, increasing the number of replicates in order to improve our confidence in the mean is helpful when n is small but becomes a rather inefficient process as n becomes large.

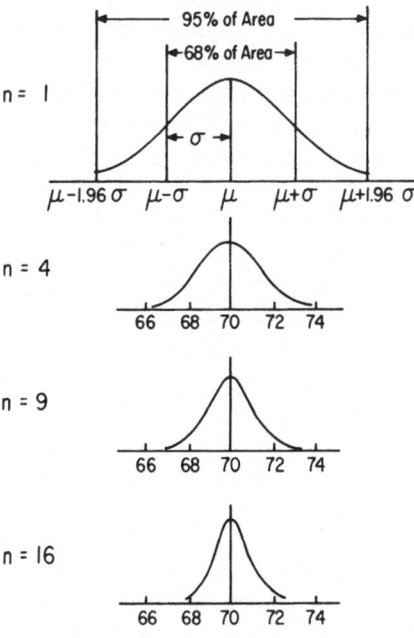

Fig. 1. The normal distribution.

In the previous two paragraphs, we have been discussing the areas under the normal curve (and, hence, probabilities) as though we had the true standard deviation σ_y. As explained earlier, we normally estimate this quantity as S_y from limited data. As a consequence of the uncertainty in this estimate, we really should use the t distribution rather than the normal distribution. Since the variability in our estimates of the standard deviation should decrease as n increases, there must be a different t distribution for each sample size. As the sample size becomes very large, the t distribution approaches the normal distribution. However, for small samples, the deviation is appreciable. For example, if $n = 5$, we must use $\pm 2.78\, S_y$ for 95% probability (compared with $\pm 1.96\, \sigma_y$). Tables of the t distribution are given in most statistics texts.

The routine use of an analytical method generally involves two or more replicate determinations on each sample to be analyzed. In order to place boundaries on the range of acceptable values associated with each experimental mean, it is helpful to use the concept of confidence intervals. In the calculation of these confidence intervals, it is important to use a value for S_y which includes all of the variation in operation under routine conditions. For example, if two discrete samples are selected from each portion of material to be analyzed and one determination is recorded for each sample, then S_y should be calculated from measurements obtained in the same way. This does not mean that S_y must be only from duplicates, but it does mean that each replicate measurement should be one determination on a discrete sample and that several samples should be used in the evaluation because sample heterogeneity will undoubtedly vary. For a specific example of 95% confidence limits, the general expression is

$$\mathrm{Prob}\left[\bar{y} - \frac{tS_y}{\sqrt{n}} \leq \mu \leq \bar{y} + \frac{tS_y}{\sqrt{n}}\right] = 0.95 \tag{7}$$

This expression is read as follows: All postulated values for the true mean μ which lie between $\bar{y} - (tS_y/\sqrt{n})$ and $\bar{y} + (tS_y/\sqrt{n})$ are acceptable at the 95% confidence level. The value of t used in this expression is governed by the number of degrees of freedom $(n-1)$ used to obtain S_y and the probability level. For 95% confidence limits, there is still a 5% chance that a sample could be obtained with the mean $\bar{y}$ and come from a population whose true mean is outside of the limits shown. If we cannot afford this large a risk, we must use correspondingly larger values of t for the confidence level desired. Conversely, lower confidence levels can also be used. If the boundaries are too wide for the confidence level desired, we must either improve the method (sampling and analysis) in order to reduce S_y or we must increase n.

In an earlier section, it was mentioned that data can sometimes be transformed if the standard deviation varies systematically with the quantity being measured. If we introduce the term "variance" to represent the square of the standard deviation (S_y^2), then we are saying that the variances should be homogeneous. This is illustrated in Table I, where the variances of intensity ratios and log intensity ratios for six replicate measurements on each of five yttrium standard solutions are recorded. These data were from an emission spectrochemical method for the determination of yttrium. Inspection of the variances of the intensity ratios suggests that they vary directly with the concentration of yttrium and, hence, with the magnitude of the intensity ratios. The fact that the variances are not homogeneous was confirmed by applying Bartlett's test [3, 4]. Inspection of the variances for the same data in logarithmic form suggests a small residual trend toward systematic variation. However, Bartlett's test on these variances leads to the conclusion that the hypothesis of homogeneity cannot be rejected at the 95% probability level. Thus, log intensity ratios are preferred for statistical analyses rather than intensity ratios.

Linnig and Mandel [5] have presented a very interesting discussion of the determination of the precision of an analytical method involving a calibration curve. They emphasize that, when there is scatter in the calibration data, the precision of analysis for an unknown will be considerably poorer than indicated from several repeat determinations on the same sample. In such circumstances, they believe that the precision estimate should include "(1) replication error, (2) scatter about a particular calibration line, and (3) variability among calibration lines." The interested reader should refer to this article for the details of calculation methods.

TABLE I

Variances for Intensity Ratios and Log Intensity Ratios*

Concentration	S^2 (Intensity ratios)	S^2 (Log intensity ratios)
5.0	0.35800	0.00147
2.5	0.08660	0.00102
1.0	0.00870	0.00049
0.5	0.00319	0.00060
0.1	0.00035	0.00081

*By Bartlett's test, the S^2 values for log intensity ratios constitute a homogeneous group but the S^2 values for intensity ratios do not at the 95% level of probability.

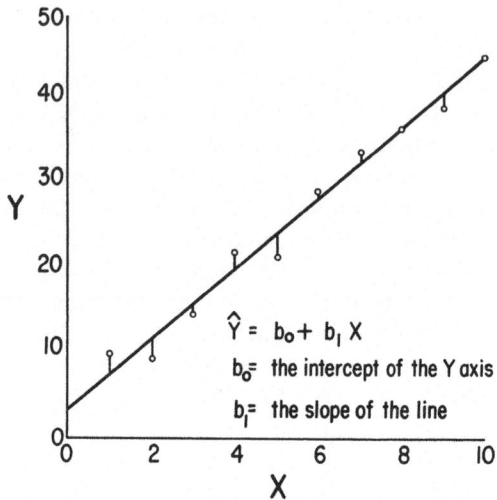

Fig. 2. A linear function illustrating the principle of least squares.

Another question which often arises in connection with functional relationships between variables is the proper location of a straight line. When scatter is minimal, the "eyeball" method is usually adequate but often a more objective procedure is needed. The least-squares method of fitting a curve is illustrated in Fig. 2. The "best-fitting" line is the one which minimizes the sum of the squares of the vertical distances between the points and the line. The line can be described by the equation

$$\hat{y} = b_0 + b_1 x \qquad (8)$$

where $\hat{y}$ is the y value predicted by the equation for a given value of x, b_0 is the intercept, and b_1 is the slope. It can be shown from calculus that a minimum value for $\Sigma(y - \hat{y})^2$ is obtained when

$$b_1 = \frac{n \Sigma xy - \Sigma x \Sigma y}{n \Sigma x^2 - (\Sigma x)^2} \qquad (9)$$

and

$$b_0 = \bar{y} - b_1 \bar{x} \qquad (10)$$

Equation (8) is referred to as the regression of y on x, where x is the independent variable and y is the dependent variable, i.e., the value of y depends on the value of x. Often, the choice of which variable is dependent is determined by which values we wish to predict for a specified value of the other variable. In either case, it should be realized that the method assumes no error in the independent variable.

Furthermore, if we calculate the regression of x on y, we should then minimize $\Sigma(x - \hat{x})^2$ and the curve would be located in a somewhat different position than for the regression of y on x.

We often wish to calculate confidence intervals around the slope, the intercept, and the average predicted value $\hat{y}$. The equations for these calculations can be found in many texts, but a small ASTM manual [6] is especially recommended.

Sometimes we are uncertain whether a straight line adequately describes a set of data. Should we draw a curve? If replicate measurements are made, we can isolate the portion of the total variability which is due to "lack of fit" of the linear model and compare it with the replication variance. In this manner, we can decide if the lack of fit is significant on a statistical basis. If so, the data can often be transformed to a function which is adequately described by a straight line.

LOCATING SOURCES OF VARIABILITY

In order to improve efficiently the precision of analysis, it is necessary to determine the magnitude of the several sources of variability which contribute to the total variance of a system. In Table II, a hypothetical case is presented where the variances have the desirable property of being additive, i.e., the total variance is equal to the sum of the variances due to each cause system. The total variance is considered to consist of three parts, sampling, preparation, and determination. If we assume the values shown in case I for the three sources of variation, then the standard deviation of the total

TABLE II

Sources of Variability

S_s = std. dev. of the sampling step
S_p = std. dev. of the preparation step
S_d = std. dev. of the determination step
S_t = std. dev. of the total analysis

$$S_t = \sqrt{S_s^2 + S_p^2 + S_d^2}$$

	S_s	S_p	S_d	S_t
Case I	6.0	4.0	3.0	7.8
Case II	6.0	2.0	1.0	6.4
Case III	4.0	4.0	3.0	6.4
Case IV	4.0	2.0	1.0	4.6

analysis is 7.8. If we now work very hard to reduce the variability, due to preparation and determination, we might find the situation in case II. Despite our efforts, the standard deviation of the total analysis has only been reduced to 6.4. On the other hand, if we tackle the major problem, which is sampling, and reduce its variability without altering the preparation or determination, an equivalent improvement in the total variation results (case III). At this point, it is reasonable to work on the preparation and determination steps, as shown in case IV.

Analytical chemists often argue that they have no control over the sampling operation and, therefore, the above sequence is not practical. Still, the same analysts will devote long hours to the improvement of preparation and determination steps when the actual benefit in the total analytical program is nil. This is not to imply that sampling is always the major source of variation; only that it sometimes is. Consequently, I feel that every analyst has the responsibility to determine the variability contributed by all steps of an analysis, including sampling. Only then can he do an intelligent job of reducing the total variability of analyses.

Time limitations prevent a comprehensive discussion of how to design experiments to isolate the magnitude of the variance associated with several cause systems. However, detailed examples of the necessary steps are given in the literature [1, 7–10].

ACCURACY

Accuracy is defined as "A quantitative measure of the variability associated with the relating of an analytical result with what is assumed to be the true value" [2]. One of the difficulties in using this definition is that often the true value is not known, although the increasing availability of certified standards is doing much to alleviate this problem. Still, in many analyses, particularly for trace elements, accuracy is estimated by analyzing samples before and after the addition of a known quantity of analyte because no standards are available. Unfortunately, this procedure is improperly employed in many cases.

Consider the analysis of a foliage sample for some trace element. One method might involve the dry ashing of the material to destroy all organic material, dissolution of the resultant ash, and extraction of the trace element from an aliquot of the aqueous solution by a complexing agent prior to the determination step. Analytical errors can occur in all of these steps. Consequently, if the recovery experiments are conducted by adding an aqueous solution of the analyte in inorganic form to a second aliquot of the ash solution, the ashing and dissolution steps have been eliminated as possible sources of error. Experiments using radioactive tracers have shown that this can lead to serious misconceptions about the accuracy. A better procedure would be to add a metallo-organic form of the element to a second portion of original foliage sample and carry out the total procedure.

Another aspect of the accuracy problem inherent in the suggestion above is the assumption that the sample is homogeneous with respect to the analyte. If this is not the case, analyses based on subsamples do not accurately represent the original material. For some types of samples, heterogeneity may be of interest and more information may result from selected subsamples than from the homogenized material. Clearly the approach must match the needs of the problem.

When analytical results obtained by a given method are found to deviate by a meaningful amount from the assumed true value, some form of bias must be present. This may take the form of systematic bias or variable bias. One example of systematic bias would be a small "blank" which would tend to cause all the values obtained by a method to be high by a constant amount. If a plot is made for several samples of the results by a given method versus the assumed true values, the curve should pass through the origin and have a slope of 1.0. By fitting the least-squares linear model, we can readily test the hypothesis that the intercept passes through the origin. If it is found to deviate from 0 at some accepted probability level, we can conclude that a systematic bias exists. Such an objective test is usually much better than the subjective judgment of the experimenter.

Variable bias can take several forms and be caused in several ways. The amount of bias may be related to the concentration of analyte present. In such a case, a plot of the experimental results versus the true values will yield a curve which may pass through the origin but with a slope which differs from 1.0. Another form of variable bias occurs when the results are too high at one end of the concentration range and too low at the other end. This can result from an incorrect calibration curve or from several other causes. Still another type of variable bias occurs when a second component in the sample affects the analysis (often called an "interelement effect"). If the interfering concentration of the component does not vary systematically with the analyte, a completely variable bias may be noted. In the worst possible case, combinations of two or more of these effects exist.

From the foregoing, it should be clear that statistical analysis of data can help guide the experimenter to the cause of his problem. Obviously, the analyst must still think clearly and understand the foundations of his procedure to benefit from any analysis. The techniques suggested here can only assist the analyst in reaching reliable conclusions in the most efficient manner.

REFERENCES

1. A.B. Calder, Anal. Chem. 36:9, 25A (1964).
2. Methods for Emission Spectrochemical Analysis, ASTM, Philadelphia, Pa. (1964), p. 218.
3. W.J. Dixon and F.J. Massey, Jr., Introduction to Statistical Analysis, McGraw-Hill Book Company, New York (1957).
4. G.W. Snedecor, Statistical Methods, The Iowa State College Press, Ames, Iowa (1956).
5. F.J. Linnig and John Mandel, Anal. Chem. 36:13, 25A (1964).
6. "Fitting Straight Lines," ASTM Spec. Tech. Pub. No. 313, Philadelphia, Pa. (1962).
7. W.J. Youden, Ind. Eng. Chem. 43:2059 (1951).
8. H.B. Vincent and R.A. Sawyer, J. Opt. Soc. Amer. 32:686 (1942).
9. H.T. Shirley, E. Elliott, and J.J. Meeds, J. Iron Steel Inst. 157:391 (1947).
10. H.T. Shirley, A. Oldfield, and H. Kitchen, J. Iron Steel Inst. 166:329 (1950).

Optimization of Analytical Methods Using Designed Experiments

R. K. Skogerboe

Chemistry Department
Cornell University
Ithaca, New York

To develop any analytical method, answers to the following questions are required:

1. What variables influence the measurement?
2. What are the best conditions for making the measurement?
3. What degree of regulation is required for each of the respective variables?

Because most analyses are affected by several variables, some of which may be interdependent, obtaining definitive answers via classical methods of experimentation is both time-consuming and complex, if not impossible. Experimental designs have been developed, however, which are capable of efficiently providing objective answers. The basic concepts of the design and analysis of such optimization experiments will be discussed with graphic illustrations and a minimum of mathematics. Applications of these methods to various spectrochemical optimization problems will be presented.

In 1933, Lundell published a perceptive and somewhat philosophical paper entitled "The Chemical Analysis of Things as They Are." In this evaluation of the status of analytical chemistry, Dr. Lundell states, "There is an increasing tendency to consider chemical analysis as dealing with only one or two variables instead of the dozen or more that are often involved. Of course, there is good reason for this loss of the analytical viewpoint since a system containing ten or twenty diverse variables can hardly be handled on a strictly scientific basis" [1]. Even though this statement was made over three decades ago, the situation still exists, perhaps even to a higher degree, because of the increasing complexity of instrumentation now popularly used for analysis. The inability to handle diverse systems scientifically was

TABLE I

Questions Relevant to an Optimization Problem

1. What measurement parameter (function) best indicates optimization?
2. What variables influence the selected response parameter? (Can any of these be ignored?)
3. Which variables are interdependent?
4. What is the best combination of variable levels? (Are there others?)
5. What degree of regulation is required for the respective variables?

generally true 30 years ago but, in view of the rapidly developing statistical field known as experimental design, this situation no longer exists although it is popularly accepted. Experimental techniques have now been formulated which allow the objective evaluation of highly complex systems with a very high level of efficiency. The bases for the application of these techniques to chemical analysis problems will be considered here.

Every analysis is culminated by measuring some instrumental response which is functionally related to the amount and the identity of analyte present in the sample. Ideally, the conditions for measuring this response should be such that the response is highly specific and stable. This concept, then, offers a practical definition of the term optimization, i.e., the selection of experimental conditions which maximize the efficiency of the analysis. The questions which must be answered to accomplish optimization are listed in Table I, and each of these will be considered in turn [2].

CHOICE OF MEASUREMENT FUNCTION

Choosing the best response function obviously depends upon the problem in hand. In trace analysis problems, sensitivity is most often the object of methods development projects, so it may be well to examine this aspect. In the analytical sense, sensitivity refers to the ability to discern a small change in the concentration (or amount), and there are two factors which affect it: first, the relationship between the measured signal and the concentration and, second, the reproducibility with which the measurement can be made. Mathematically, sensitivity can be defined by

$$\gamma = \frac{dI/dc}{s_I} \tag{1}$$

where dI/dc is the slope of the analytical curve and s_I is the standard deviation of the signal measurement [3, 4]. This definition is completely consistent with the verbal definition and demonstrates how the slope and the measurement precision affect the ability to determine a small change in concentration (see also Fig. 1).

Implicit in this definition is the fact that the best analysis will be accomplished when the signal per unit of concentration is maximized and the standard deviation of measurement is minimized [3]. In the majority of instances, the relationship between signal intensity and some controlling variable passes through a maximum, as shown in Fig. 2. Obviously, the choice of the level of x maximizing the intensity is desirable in two respects. First, the slope is maximized, and, second, because there is usually an inherent limit to the degree with which x can be regulated (e.g., ±5%), the best measurement precision is often obtained on the plateau, as demonstrated in the figure. In essence, sensitivity can frequently be maximized by maximizing the signal intensity. There may, however, be situations where this is not true. For example, there is considerable disagreement in the field of emission spectrography on whether the emission intensity or the signal-to-background ratio should be optimized. Without discussing the relative merits of these two possibilities, the most obvious solution can be obtained by a brief experimental investigation of the system at hand.

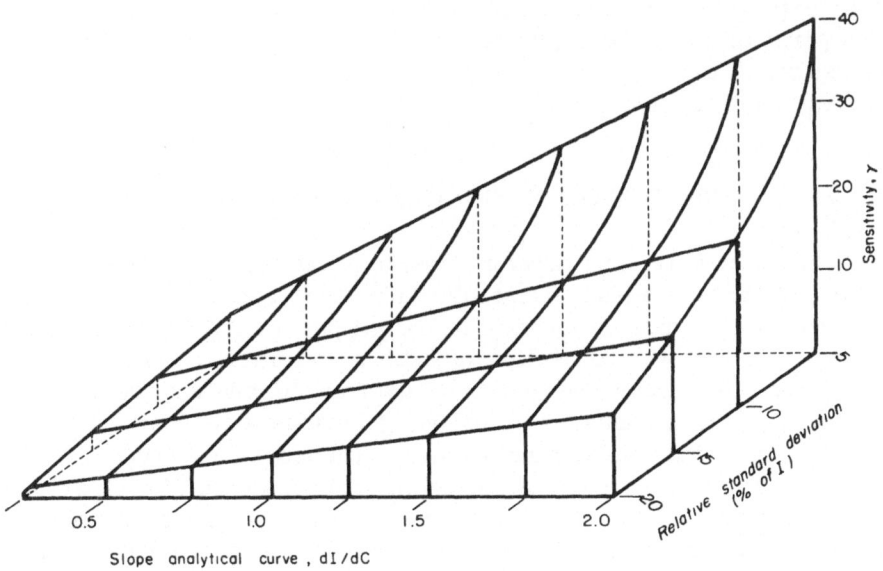

Fig. 1. Sensitivity as a function of curve slope and measurement precision.

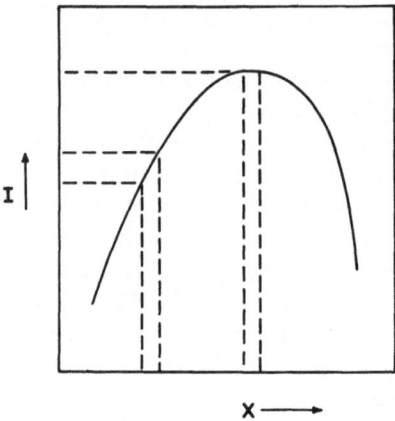

Fig. 2. Dependence of the intensity variation on the optimization of the variable level.

The fact that the detection limit is ultimately determined by the ratio of the analysis signal to the statistical fluctuation in the background or blank signal should also be kept in mind [4, 5, 6].

There are many other types of measurement functions that might be considered for optimization. The production-line control analyst may be forced to consider obtaining maximum sensitivity for a minimum expenditure of time or cost or both. The choice of the maximization parameter may be highly diverse depending upon the system to be investigated, the purpose of investigation, and other technical demands placed upon the analyst.

SELECTION OF SIGNIFICANT VARIABLES

This is a screening problem that, in practice, proceeds in somewhat indistinct stages. First, the knowledge and experience of the experimentalist must be invoked to list all potentially important variables. The actual preparation of such a list offers at least two advantages: It forces the persons involved to analyze the problem thoroughly—the advantages here being obvious—and examination of the list may indicate distinct natural grouping of certain variables which were not previously recognized. The existence of such groups can frequently be useful in the simplification of experimental problems, regardless of the type of experimentation used. A major problem associated with such a list is that of deciding which variables are potentially significant.   Because knowledge and experience may be in short supply, a good rule of thumb would be: If there is doubt, check it out—experimentally.

Given the list of variables, other decisions are required before conducting the experimental screening. The best form for expressing each variable should be determined, i.e., if a variable is known to produce an effect of a square root or a logarithmic nature, it would be best to express the variable in that form. This is not an absolute requirement, however.   The experimentalist should also make an "educated guess" concerning the range to be studied for each variable. At this point, the actual screening experiment can be conducted.

Perhaps the most widely used experimental designs for this purpose are the 2^n factorials [2, 7, 8]. A factorial design consists of all combinations of several sets of factors (or variables, or treatments). A 2^n factorial thus includes all combinations of n variables, each of which is examined at two levels; for example, for a 2^4 factorial, there are 16 combinations of the four variables, each at two levels. Designs for one, two, and three variables are illustrated in Fig. 3. To simplify both the formulation and the analysis of the designs, it is usual to code the high and low levels of each variable as +1 and -1 respectively.   The coding equations given in the figure are set up so that substitution of the actual values of the variable levels used produces ± 1. Because it is possible for the variable levels chosen to straddle a maximum and indicate a very small effect, a center point is often added to the design as a partial safeguard. Rather than replicate all combinations, it is fairly common practice to repeat only the measurement at the center point to obtain an experimental error estimate that is useful in the analysis of the significance of the results.

At first, it might appear that formulating the designs in more than three dimensions would be a problem. Notice, however, that the coding technique produces an easily identifiable and expandable pattern, where the signs of the coded variables alternate in integral powers of 2, beginning at zero for the first variable.

To illustrate the advantages of factorials over the classical (single-factor) method of experimentation, which studies one variable at a time, consider the data given in Fig. 4 and Table I. The response values shown in the figure are the flame intensities (in millivolts) of a boron atomic line obtained from a study of the effects of the oxygen-to-fuel flow ratio and the total gas-flow rate for the flame. If single-factor experimentation were used, a minimum of three of these runs would be required to study the two variables. It should be obvious that any variable level chosen for initiating the experiment would produce

2^n FACTORIAL DESIGNS

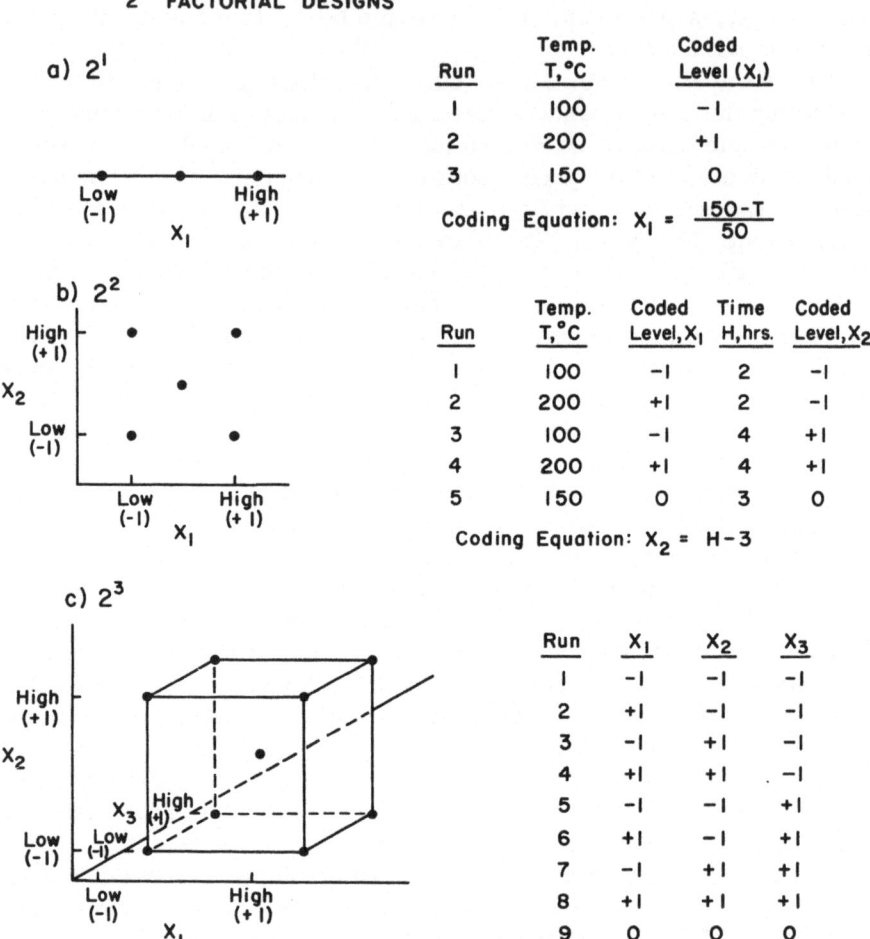

a) 2^1

Run	Temp. T, °C	Coded Level (X_1)
1	100	-1
2	200	$+1$
3	150	0

Coding Equation: $X_1 = \dfrac{150-T}{50}$

b) 2^2

Run	Temp. T, °C	Coded Level, X_1	Time H, hrs.	Coded Level, X_2
1	100	-1	2	-1
2	200	$+1$	2	-1
3	100	-1	4	$+1$
4	200	$+1$	4	$+1$
5	150	0	3	0

Coding Equation: $X_2 = H - 3$

c) 2^3

Run	X_1	X_2	X_3
1	-1	-1	-1
2	$+1$	-1	-1
3	-1	$+1$	-1
4	$+1$	$+1$	-1
5	-1	-1	$+1$
6	$+1$	-1	$+1$
7	-1	$+1$	$+1$
8	$+1$	$+1$	$+1$
9	0	0	0

Fig. 3. Factorial designs for one-, two-, and three-variable systems.

the same general result (an intensity increase) for experimentation in the y direction, but the result obtained from experimentation in the x direction certainly depends on the initial level of the oxygen-to-fuel ratio. By adding a fourth point to complete a 2^2 factorial, it is possible to average the effects of changing the levels of both variables. Thus, increasing the total flow rate from 6 to 6.5 liters/min decreases the intensity by an average of $(-1.2 \quad 0.7)/2 = -0.25$ mV, and increasing the ratio from 1.0 to 1.2 produces an average intensity increase of

(1.1 + 3.0)/2 = 2.05 mV. The addition of a fourth point allows covering the extremes of the system and greatly increases the knowledge gained. Because each effect has been estimated from the average, the error has also been reduced. By averaging across the diagonals (ignoring the center point), it is also possible to obtain a measure of the degree of interaction (interdependence) between the two variables. The value of this measure here would be (3.6 + 5.4)/2 −(4.7 + 2.4)/2 = 0.95 mV. The existence of interdependence means effectively that it is impossible to make any general statement about the effect of one of these variables on the response without specifying the level of the other. It also indicates that optimization of this system, using the single-factor method of experimentation, would result in a slower convergence on the optimum compared with the rate of convergence in the absence of interaction.   In fact, for systems exhibiting a high degree of interaction between variables, the single-factor method may never converge on the optimum [7–9].

To estimate curvature, other than that caused by interdependence, the center-point response value is compared with the average of the responses at the peripheral points. The curvature estimate is then 4.0−(3.6 + 4.7 + 5.4 + 2.4)/4 ≈ 0.0 and is usually referred to as a lack-of-fit measure. Roughly translated, significant measures of curvature or interaction indicate that the response surface in this region cannot be adequately described by a linear equation.

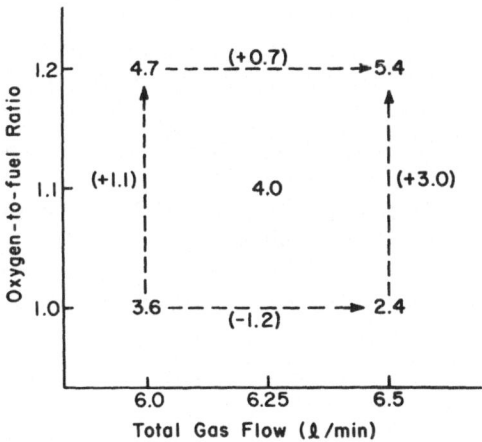

Fig. 4. Flame emission intensity of atomic boron.

TABLE II

The Effects of Oxygen-to-Fuel Flow Rate Ratio and the Total Gas
Flow Rate on the Flame Excitation of Atomic Boron

		Design matrix				Response matrix			
Run	x_1	$\dfrac{\text{Oxygen}}{\text{Fuel}}$	x_2	Total flow, liters/min	x_1x_2	y, mV	x_1y	x_2y	x_1x_2y
1	-1	1.0	-1	6.0	+1	3.6	-3.6	3.6	+3.6
2	+1	1.2	-1	6.0	-1	4.7	+4.7	-4.7	-4.7
3	-1	1.0	+1	6.5	-1	2.4	-2.4	+2.4	-2.4
4	+1	1.2	+1	6.5	+1	5.4	+5.4	+5.4	+5.4
5	0	1.1	0	6.25	0	4.0	0.0	0.0	0.0
				Summation		20.1	+4.1	-0.5	+1.9
				Effects			+2.05	-0.25	+0.95

$$i\text{th effect} = \frac{2\sum_i x_i y}{\sum_i x_i^2}$$

It must be emphasized that all the peripheral measurements have
been repeatedly used in the interpretive analysis. This characteristic
of experimental designs is known as "maximum efficiency" for obvious
reasons. The results are certainly only estimates of the true effects
simply because measurement errors exist. When the experimental
error is not known, an estimate can be obtained by replication and this
can be used to calculate the probability that the estimated effects are
real rather than the result of random variation [2, 7–9]. For flame
spectrophotometry, the measurement error is typically less than 5%,
and, since all the effects computed above (except the curvature effect)
deviate from the average by amounts greater than this, they may be
regarded as significant.

Having examined this design graphically, consider the general
method of analysis outlined in Table II for the same data. The interaction column x_1x_2 is generated by multiplying the coded values of x_1
and x_2 in each row. The columns x_1y, x_2y, and x_1x_2y are obtained by
multiplying the respective coded values and responses in the same
row. The effects for each of the variables are then calculated by

$$i\text{th effect} = \frac{2\sum_i x_i y}{\sum_i x_i^2} \qquad (2)$$

The solutions obtained in this manner agree with those computed in the
analysis of the figure. This analysis is based on techniques common to

matrix algebra, where the coding technique has produced a unit matrix that simplifies solution.

This simple example demonstrates the use of factorial experiments in the screening process; i.e., by applying this approach, the variables which do produce an appreciable effect and do not interact with other variables can be eliminated from subsequent investigations. Screening based on this type of experimentation effectively and efficiently provides answers to both questions two and three presented in Table I, and optimization can now be considered.

BEST VARIABLE LEVELS

An experimental design technique known as the "method of steepest ascent" is popularly used to provide the solution to this problem [9]. Figure 5 demonstrates the general principle of this method for a one-variable system. The factorial designs discussed above are used to fit

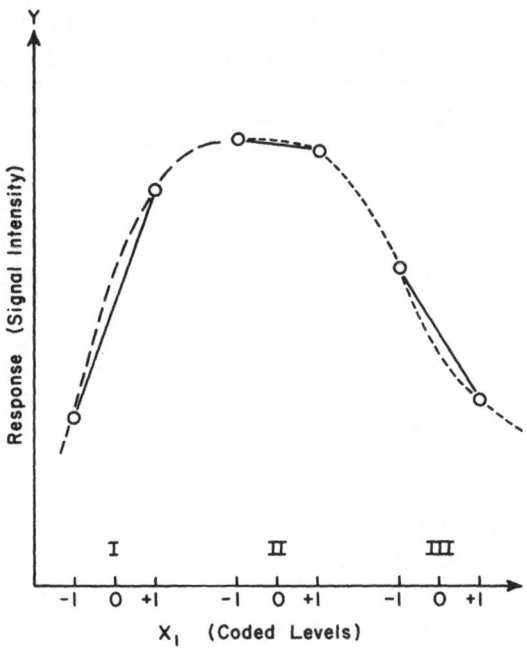

Fig. 5. Optimization by the method of steepest ascent.

a linear equation of the form

$$\hat{y} = b_0 + b_1 x_1 + \cdots + b_n x_n \tag{3}$$

to the response surface in the region of experimentation for an n-variable system. In the equation, $\hat{y}$ represents the estimated response, b_0 is the average value of y obtained from the experiment, and the b_i are linear coefficients (effects) determined for the respective variables. The signs of the coefficients indicate the direction in which the variable levels must be changed to approach the maximum. The magnitudes of the coefficients are indicative of the slope of the response curve in that region, but the method does not indicate how much to change the variable levels for successive experiments. This is left to the judgment of the experimentalist. Consecutive experiments are simply run in the directions indicated until the signs of the coefficients change (as between experiments I and III) or until the magnitudes of the coefficients approach zero (experiment II). When either of these conditions occur, the general position of the maximum has been bracketed and subsequent experimentation is run to specify its position and nature more precisely. Before discussing this aspect, however, the characteristic efficiency of the design methods should be reiterated.

The factorial experiment used for the initial screening of variables can frequently be used for the first phase of the steepest ascent. In the flame spectrophotometric example given, the variable effects computed indicated the general directions in which the variable levels must be moved to approach the optimum, even though some curvature was detected. There is an element of danger in ignoring the existence of curvature within an experimental region because this may indicate that the minimum was contained within the region. However, even if it is, subsequent steepest-ascent experiments run in the direction indicated will "point" back to the original region and verify the position of the maximum.

In most instances, knowing the general position of the maximum is not adequate, and it is desirable to specify its exact location and to describe its nature (or behavior) in the optimal region. The purpose of this phase is threefold. It completes the answer to question four, it provides information required to answer question five, and it might also be useful in elucidating any mechanistic or theoretic relationship between the response and the variables.

Central composite designs are one type easily used to estimate the coefficients (constants) of a higher-order equation required to describe a surface in the region of a maximum. Usually a curvilinear response surface can be adequately described by a second-order equation of the form

$$y = b_0 + b_1 x_1 + b_2 x_2 + \cdots b_n x_n + b_{11} x_1^2 + \cdots + b_{rm} x_n^2 + b_{12} x_1 x_2 + \cdots + b_{1n} x_1 x_n$$
$$+ \cdots + b_{23} x_2 x_3 + \cdots + b_{2n} x_2 x_n + b_{n-1,n} x_{n-1} x_n \tag{4}$$

where n is the number of variables.

Central composite designs are simply factorials to which sufficient experimental points (known as star points) have been added to permit estimation of the coefficients of the second-order terms in the equation. Figure 6 illustrates such a design for a two-variable system. Note that the star points are actually those of a second 2^2 factorial rotated 90° about the central point. Hence, the star points lie on a circle passing through the peripheral points at all combinations of zero and the radius of the circle by using the coded variables. This is known as the "criterion of rotatability," and the advantage derived from this property is that the standard error is the same for all points in the experimental region. This means that replication of only one point is required to estimate the standard errors associated with the others. The method of analysis of data obtained by using these designs follows the same pattern used for the factorial data in Table II. The formulation and analysis of these designs and those discussed previously are set out in "cookbook" detail in texts [7, 8].

Because these utilize the factorial, another advantage from the standpoint of efficiency is frequently realized. Suppose, e.g., that one of the steepest-ascent experiments indicates that the optimum combination of variable levels is within or near the region of experimenta-

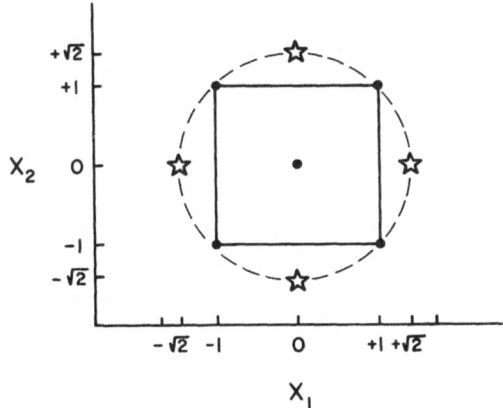

Fig. 6. Geometric configuration of a central composite design in two variables.

TABLE III

Response Surface Methods Applied to the Flame Excitation of Atomic Boron

| | Steepest-ascent experiments | | | | Central composite experiment* | |
| | First experiment | | Second experiment | | Third experiment | |
Variables	Levels, low–high	Calculated effects	Levels, low–high	Calculated effects	Levels, low–high	Levels at maximum
Oxygen/fuel ratio, x_1	1.0–1.2	1.0	1.3–1.4	−0.30	1.2–1.3 (x_1)	1.35
Total gas–flow rate, liter/min, x_2	6.0–6.5	0.3	7.0–7.2	0.03	7.2†	7.2†
Flame region viewed, mm above burner, x_3	17.0–20.0	0.1	22.0–24.0	−0.20	19.0–22.0 (x_2)	19.0
Sample feed rate, ml/min, x_4	1.0–1.5	−1.4	1.0–1.3	−0.02	1.0–1.1 (x_3)	1.1
Slit width, μ, x_5	20.0–30.0	1.1	40.0–50.0	−0.20	30.0–40.0(x_4)	35.0
Slit height, mm, x_6	15.0–20.0	−0.05	12.0–14.0	0.02	20‡	20.0‡

*Equation calculated from central composite data:

$$y = 3.88 - 0.18\,x_1 - 0.84\,x_2 + 0.29\,x_3 + 0.05\,x_4 + 0.03\,x_1^2 - 0.23\,x_2^2 - 0.50\,x_3^2 - 0.11\,x_4^2 - 0.03\,x_1 x_2 - 0.06\,x_1 x_3 - 0.05\,x_1 x_4 + 0.07\,x_2 x_3 - 0.10\,x_2 x_4 + 0.57\,x_3 x_4.$$

†Fixed at upper limit.

‡Fixed at upper limit for lack of effect.

tion (experiment II, Fig. 5). Star points can be added to the factorial design, which will produce this indication, and the surface can then be fitted without rerunning the original factorial points.

When the quadratic surface has been fitted, the position of the maximum (or minimum) is determined by differentiating the fitted equation for each variable in turn, setting the differentials equal to zero, and solving for the value of each coded variable at the maximum.

Before presenting an actual application of these techniques, it is desirable to consider experimentation with many-variable systems. Obviously, for more than four variables, the number of experimental points required becomes unwieldy; for example, a 2^6 factorial requires 64 measurements, even without replication. By introducing a design principle known as "confounding," however, a fraction of the total design—a fractional factorial—can be run and analyzed. Time does not permit the discussion of confounding except to indicate that some experimental capability is sacrificed when fractional factorials are used. Certain interaction effects become "intermixed" in such a way that the ability to obtain individual estimates for these terms is lost. When some of the variables under investigation are known to be independent, the fractional designs are easily arranged so the intermixing occurs for these variables and this negates the problem. Even when the intermixing is unavoidable the fractional designs still offer improved precision, higher efficiency, and greater experimental capability than the classical method.

To demonstrate the application of the method of steepest-ascent and central composite designs, consider the data in Table III. In the flame spectrophotometric literature, there is sufficient evidence to document the importance of variables such as the choice of the fuel and oxidant gases, the fuel and oxidant flow rates, the total gas-flow rate, the choice of solvent, the sample feed rate, the droplet-size distribution within the flame, the flame region viewed by the monochromator, and the populations of potential flame reaction participants. It has been shown [4] that, for a particular choice of flame type and solvent, these variables which affect the "observed excitation energy" of the flame can be regulated directly or indirectly by controlling the oxygen-to-fuel ratio, the total gas-flow rate, the sample feed rate (by using an external pumping system), and the region of the flame viewed. In addition, the width and height of the monochromator slit can affect the observed emission intensity; so these were included in the optimization experiments which were run with an oxacetylene flame and an organic solvent.

TABLE IV

Analysis of Variance for the Fitted Equation of Table III

Source of variation	Sum of squares	Degrees of freedom	Mean squares	Computed F	Tabular F, 95%.	Conclusions
1st-order terms, linear	20.10	4	5.03	50.3	6.2	Significant at 95% confidence level
2nd-order terms, quadratic	14.16	10	1.42	14.2	3.2	Significant at 95% confidence level
Lack of fit	0.73	10	0.07	0.7	3.2	The equation fits the surface
Error	0.61	6	0.10			
Total	34.12					

Because the steepest-ascent factorial for 6 variables would require 64 runs, $\frac{1}{4} \times 2^6$ fractional factorials requiring only 16 runs were used to estimate the first-order effects. In the first experiment, the linear effects due to varying slit height and the flame region were found to be insignificant, but the data also suggested the existence of appreciable curvature within the experimental region. This implied that the levels chosen for one or more of the variables straddled the maximum, so the levels were moved in the directions indicated to verify this possibility. The second set of data confirms that the slit-height effect is negligible and that the total flow-rate effect has been reduced to insignificance. Both variables were subsequently fixed at their maximum levels. The signs of the coefficients for the remaining variables were all changed, which indicated a bracketing of their optimal levels near those selected for the central composite experiment. The equation calculated from the last experiment (presented at the foot of the table) produced the optimal levels indicated in the last column when solved. These results verify the straddling of the maximum for both the flame region and the sample feed rate, as suggested by the first experiment.

Table IV presents the analysis of the variance data for the final calculated equation. The sum of squares for each source of variation is calculated with simple methods presented by Davies [7] and Cochran and Cox [8]. Mean squares are computed by dividing the sum of the squares by the number of degrees of freedom in each instance. The computed F statistics are obtained by dividing each mean square by the mean square for error. Comparison of this value of F with a standard table value for equivalent degrees of freedom and the confidence level required permits an objective formulation of the conclusions. For example, when $F_c > F_t$ the contributing effects are significant at that confidence level. When $F_c < F_t$ the effects are not significant. The latter situation for the lack-of-fit term in Table IV actually indicates that the equation does fit the surface. In essence, the analysis indicates that a fair degree of curvature or interaction (or both) exists about the maximum and consequently implies the necessity of closely regulating the variables involved. For a more comprehensive treatment of the analysis of the fitted surface, see Davies [7] or Cochran and Cox [8].

There are a number of other types of experimental designs for solving optimization problems, but time and space do not permit consideration of these except to indicate that the majority of them utilize factorial experimentation. The examples discussed above indicate only a few of the possibilities for further application in analytical chemistry. Efficiency and objectivity are two prime advantages which accrue from this type of application, and extremely diverse systems can be readily optimized, e.g., optimization problems involving as many as 15 variables have been solved in the author's laboratory. None of the techniques covered are new but rather have been used extensively by agricultural scientists, biologists, and chemical engineers, among others, for a number of years. At least two textbooks [7, 8] are notable in that they present optimization concepts in an intelligible fashion and include simple instructions for the formulation and analysis of these concepts.

REFERENCES

1. G. E. F. Lundell, Ind. and Eng. Chem., Anal. Edition 5:221 (1933).
2. C. L. Grant, Canadian Spectroscopy 8:47 (1963).
3. J. Mandel and R. D. Stiehler, J. Res. Natl. Bur. Std. A53:155 (1954).
4. R. K. Skogerboe, Ann T. Heybey, and G. H. Morrison, Anal. Chem. 38:1821 (1966).
5. V. A. Fassel and D. W. Golightly, Anal. Chem. 39:466 (1967).
6. H. Kaiser, Z. Anal. Chem. 216:80 (1966).
7. O. L. Davies, Design and Analysis of Industrial Experiments, Hafner Publishing Co., Inc., New York (1960).
8. W. G. Cochran and G. M. Cox, Experimental Designs, John Wiley and Sons, Inc., New York (1960).
9. G. E. P. Box and K. B. Wilson, J. Roy. Stat. Soc. B 13:1 (1951).

Nuclear Applications

Some Recent Developments in Organic Scintillators

Donald L. Horrocks

Chemistry Division
Argonne National Laboratory
Argonne, Illinois

A general review of the basic concepts of the scintillation process in organic systems will be presented. Special emphasis will be given to liquid scintillators. Discussion will include energy transfer, lifetime, quenching, and excimer properties of organic scintillator systems.

Some practical applications of organic scintillators will be discussed. Low-level counting, pulse-shape discrimination, and gas counting with organic scintillators will be reviewed.

Finally, some recent developments in the synthesis of new organic scintillators will be presented. Special attention will be given to work describing the prevention of self-quenching and the development of a new class of organic scintillators, those which are in a glassy state at room temperature.

In a swiftly developing study such as organic scintillators it is desirable to have periodical reviews of some of the recent developments. However, since its conception in 1947 by Prof. H. P. Kallmann [1], the field of study of organic scintillators has become so vast that it is almost impossible for anyone to present a discussion on all of the recent developments. With this in mind, the following presentation will be limited to a discussion of only a small part of the recent developments in organic scintillators. An attempt will be made to discuss the following:

1. Studies of some organic compounds that form glasslike masses that scintillate and are stable at room temperature
2. Some correlations between the molecular structure of an organic scintillator and self-quenching

TABLE I
Some Data on Organic Scintillator Which Form a Glasslike State

Compound	Abbreviation	Melting point, °C	Stability	315 mμ light excitation		Electron excitation	
				Scintillation yield*	Glasslike / dil soln	RPH†	Glasslike / dil soln
	EPI-N5i	50	> 9 months	1.06	0.79	125	1.20
	TMQP	74	Infinitely	1.00	0.80	92	0.95
	PI-3P	125	6 months	0.51	0.71	74	0.76
	PI-NB	144	5 months	0.73	1.05	67	0.75
	PI-N5a	86	1 day	0.55	0.87	56	0.62

*Normalized to 1.00 for the glasslike sample of TMQP.
†Relative pulse height [15]. Normalized to 100 for deaerated toluene solution of PPO (3 g/liter).

3. Some recently developed fast scintillators
4. Some newly developed applications of liquid-scintillation counters

GLASSLIKE ORGANIC SCINTILLATORS

In 1959, it was reported [2] that the compound $1^2, 2^3, 3^2, 4^3$-tetramethyl-p-quaterphenyl (which is a very good scintillation solute) when melted and recooled to room temperature would form a glasslike state that was stable for a very long time (at least several years). In a recent study [3], several more low-molecular-weight organic scintillator solutes were found to form a stable glasslike state at room temperature. The compounds studied were derivatives of the basic molecule 2-phenyl indole. The glasslike states were stable for various times, only 1 day for one of the compounds to more than 9 months for another compound, if they were maintained under evacuated conditions.

Several interesting observations were made from the results of the study of these organic glasslike samples. They include:

1. Some of the compounds in their glasslike state had a relative response to excitation by electrons that was about 25% higher than the maximum obtainable response of a gas-free toluene solution of PPO (3 g/liter).
2. Some of the compounds had a higher light yield for excitation with electrons when in the glasslike state than the maximum obtainable as a solute in a toluene solution.
3. The fluorescence spectra of the glasslike state and of a dilute solution of some of these compounds were quite different, while those of others were almost identical.

A few of the compounds reported in this study are listed in Table I. The compound N-isoamyl-3,2'-ethyleno-2-phenyl indole, EPI-N5i, is one of three similar compounds, all ethyleno-bridged derivatives of 2-phenyl indole, which had a higher relative response to electron excitation while in their glasslike state than as a solute in a dilute toluene solution. Also the glasslike states of these three compounds had a higher relative response to electron excitation than the standard PPO–toluene liquid scintillator (3 g/liter, gas-free).

Figure 1 shows some fluorescence spectra for excitation with 315 mμ of light. The spectra are for the compound in a glasslike state and in a dilute toluene solution (mole fraction of about 0.001). In some

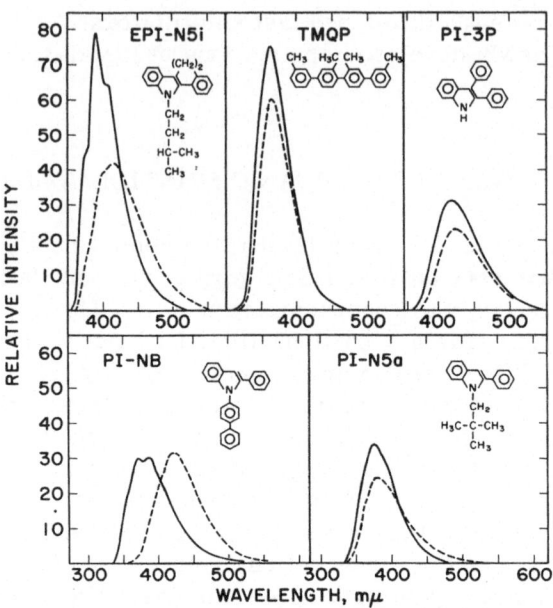

Fig. 1. Fluorescence spectra of several compounds in their glasslike state (---) and in a dilute toluene solution for excitation by 315 mμ of light.

cases, the general shape of the spectra of the glasslike state and the dilute solution are almost identical and the peak of the spectra differ by only 5 to 10 mμ. In other cases, there are very pronounced differences between the fluorescence spectra of the glasslike state and of the dilute solution of a given compound. For the compound N-biphenylyl-2-phenyl indole, PI-NB, the fluorescence of the glasslike state peaks at about a 45-mμ longer wavelength than the dilute solution. This shift corresponds to a difference of 0.4 eV between the levels which are emitting the observed fluorescence. The magnitude of this shift could be due to the formation of excimers [4] which fluoresce. There were also those compounds for which the shift between the peaks of the fluorescence spectra of the glasslike state and the dilute solution was too small to be considered due to the formation of excimers. These shifts were only about 20 to 40 mμ, or 0.2 to 0.3 eV. These could be the result of weak interactions between solute molecules in the glass-like state such as those associated with a solvent shift [5] or exciplex formation [6].

An efficient glasslike organic scintillator would have several advantages over organic crystals or solutions (liquids or plastics). A few are:

1. Elimination of solvent–solute energy-transfer processes of solutions
2. Ease of preparation of pure scintillators in odd shapes and large volumes
3. Reduction of diffusion-controlled quenching processes

It should also be pointed out that if a defect, such as a fissure or a bubble, should appear in the scintillator, it could be removed by simply remelting the glasslike state and recooling to room temperature.

There are some disadvantages associated with these glasslike states (which may not be present with another series of organic compounds):

1. The stability of the glasslike state is not always predictable,
2. The stability was found to be dependent upon the surroundings; contact with air usually reduced the lifetime of the glasslike state.
3. Of the glasslike states investigated, none were rigid in the sense of being able to support themselves. They will flow at room temperature, some very slowly.

SELF-QUENCHING—STRUCTURE CORRELATIONS

The fluorescence of toluene solutions of the compound 2-phenyl indole is strongly quenched with increasing 2-phenyl indole concentra-

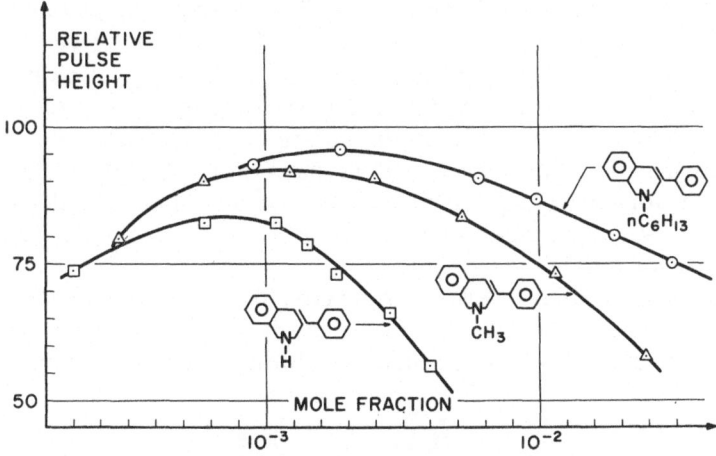

Fig. 2. Relative scintillation efficiency for electron excitation as a function of concentration for N-substituted derivatives of 2-phenyl indole.

tion at high concentrations. The process is called self-quenching. Figure 2 shows the relative response of toluene solutions of 2-phenyl indole to excitation by electrons as a function of the concentration. The increase in relative response at low solute concentrations is due to an increase in the transfer efficiency of absorbed energy from the primary absorber (the solvent) to the fluorescing species (the solute). At some concentration, a maximum relative response is obtained, and further increases in solute concentration produce decreases in the relative response. The decrease (self-quenching) involves the inter-action and exchange of energy between the excited solute molecule and an unexcited solute molecule. This leads to a decrease in probability of fluorescence of the excited solute, which diminishes the relative amount of light per unit of energy absorbed. Some of the processes which occur are illustrated by the following equations:

$$A + \text{energy} \longrightarrow A^* \qquad\qquad \text{absorption}$$

$$A^* \longrightarrow A + h\nu \qquad\qquad \text{fluorescence}$$

$$A^* + A \longrightarrow 2A + en \qquad\qquad \text{self-quenching}$$

$$A^* + A \longrightarrow A_2^* + en \qquad\qquad \text{excimer formation}$$

$$A_2^* \longrightarrow 2A + h\nu' \qquad\qquad \text{excimer fluorescence}$$

Self-quenching and excimer formation compete with the fluorescence of the excited solute molecule, which decreases the amount of fluores-cence. However, the amount of light measured could be the sum of monomer and excimer fluorescence depending on the wavelength of the two fluorescences and the response of the measuring device (usually a multiplier phototube) to the two wavelength spectra. For this reason, it is usually advisable to have additional information on the spectrum of the light as a function of the solute concentration. Another piece of information which would aid in the determination of the extent of excimer formation in a sample is the decay time. The decay time of excimer is much longer than the decay time of monomer (about 7.9 to 1.6 nsec for PPO).

Recently a study [7] was undertaken to find out how the basic mole-cule of 2-phenyl indole could be altered by substitution in the appro-priate positions so as to reduce or eliminate the self-quenching. The basic molecule is

Alkyl (normal and branched) and aryl groups were added at the 1 position (on the N), and the effects on self-quenching were measured. In another set of experiments, alkyl (methyl) and aryl (phenyl) groups were substituted in the 3 position for a study of the effect which led to a restriction of the rotation of the 2-phenyl group. Finally, a series of compounds in which the 3 to 2' positions were bridged through methylene, ethylene, and propylene groups. Tables II, III, and IV summarize some of the data of this investigation.

From these data it was concluded that self-quenching occurs only if the solute satisfies the following criteria:

1. The basic chromophor is coplanar or can assume a coplanar configuration.
2. Two molecules can overlap completely and in a special orientation (mirror image).

It was found that certain completely unprotected organic solutes, such as 1,1'-dinaphthyl, had no self-quenching [8] ability. Upon examination of the molecular model, it was often found that these compounds were

TABLE II

Self-Quenching and Structure Correlations

R_1	R_2	Self-quenching
$-H$	$-H$	Very strong
$-CH_3$	$-H$	Strong
$-n-C_6H_{13}$	$-H$	Yes
$-CH_2-C(CH_3)_3$	$-H$	Yes
$-$⟨○⟩	$-H$	Yes
$-$⟨○⟩$-$⟨○⟩	$-H$	Little
$-CH_3$	$-CH_3$	None
$-H$	$-$⟨○⟩	Little
$-CH_3$	$-$⟨○⟩	None

TABLE III

Self-Quenching and Structure Correlations

n	R_1	Self-quenching
1	$-H$	Strong
1	$-CH_3$	Yes
1	$-CH_2-C(CH_3)_3$	Very little
2	$-H$	Strong
2	$-CH_3$	Little
2	$-n\text{-}C_3H_7$	Little
2	$-n\text{-}C_4H_9$	Little
2	$-CH_2-CH_2-CH(CH_3)_3$	Little
2	$-\bigcirc$	Little
3	$-H$	Strong

nonplanar. Other molecules with certain substituents in particular positions, such as the compound $1^1, 4^4$-di(2-butyloctyloxy)-p-quaterphenyl (QP-G12), showed no self-quenching, owing to the shielding effect of the substituted groups [8]. It was with these findings in mind that certain substitutions were made on the 2-phenyl indole molecule in an attempt to eliminate or at least reduce the self-quenching process.

The substitution of alkyl groups of differing length in the 1 position (on the N) produced a decrease in self-quenching with an increase in the length of the alkyl chain (at least up to C_6). Figure 2 shows the effect of the length of the alkyl group. Some other results indicate that increasing the length of the alkyl group to greater than C_6 does not decrease the self-quenching much more. In one case, a branched alkyl group, neopentyl, was substituted in the 1 position, but this did not eliminate the self-quenching.

Substitution of a group in the 3 position was shown to eliminate self-quenching in the molecule. It is believed that the methyl or phenyl group in the 3 position sterically hindered the 2-phenyl group so that it could not obtain a coplanar configuration with the rest of the chromophor.

A final series of compounds investigated in that study was one in which the 3 and 2' positions were linked by a bridge. The methylene-group bridge fixed the 2-phenyl and indole groups in a coplanar arrangement. There was no freedom of rotation about the bond between the 2-phenyl and indole groups. There was very strong self-quenching for the methylene-bridged compound. There were evidences that increasing the number of carbon atoms in the bridge decreased the self-quenching, at least up to C_3. It was concluded that the increasing length of the bridge caused a distortion of the molecule by forcing the 2-phenyl group into a nonplanar configuration with respect to the indole part of the molecule.

FAST SCINTILLATORS

One practical consequence of a fast scintillator is that the faster the scintillator decay time is, the less is the effect of oxygen quenching. For a scintillator with a decay time of 1 nsec or less, it is theorized

TABLE IV

Self-Quenching and Structure Correlations

	Self-quenching
(structure: four methyl-substituted phenyl rings linked)	None
RO—⬡—⬡—⬡—⬡—OR $R = -CH_2-\overset{\overset{H}{|}}{C}-n\text{-}C_6H_{13}$, with $n\text{-}C_4H_9$	None
$R = -CH_2-\overset{\overset{H}{|}}{C}-n\text{-}C_8H_{17}$, with $n\text{-}C_6H_{13}$	None
(fused bicyclic ring structure)	None
⬡—(oxazole ring)—⬡ (PPO)	Yes

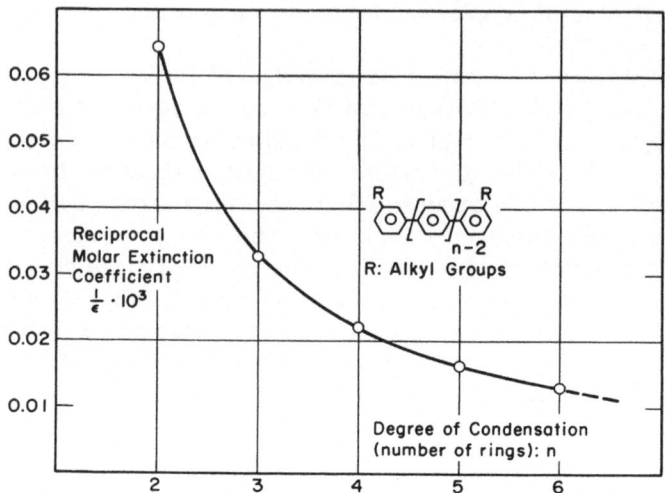

Fig. 3. Variations of reciprocal of molar excitation coefficient of the p-oligophenylene series on the degree of condensation (from Wirth et al. [10]).

that oxygen quenching would be negligible in air-saturated liquid scintillator solutions. Also, as the decay time is shortened, the probability of other competing processes (quenching, excimer formation, and intersystem crossing) will be decreased.

According to Forster [9], the decay time of fluorescence decreases as the molar extinction coefficient increases. In a recent study by Wirth [10] of the p-oligophenylene series, some very fast scintillators were prepared. In general, it was observed that as the degree of condensation (the number of phenyl groups) increased, the decay time decreased and the molar extinction coefficient increased.

The natural decay time is inversely proportional to the molar extinction coefficient of the molecule according to the equation developed by Forster [9] and proved by Berlman [11]. Figure 3, which was presented by Wirth [10], shows a plot of the reciprocal of the molar extinction coefficients for a series with undisturbed conjugation system against the number of phenyl groups. It can be seen from this plot that the decrease in the natural decay time becomes smaller with increasing degree of condensation. Berlman [5] measured a decay time of about 0.8 nsec for the nonsubstituted p-quaterphenyl. From extrapolation according to Fig. 3, the sexiphenyl would have a decay time of about 0.6 nsec.

Lynch [12] reported on the effects of solute concentration on the decay time. By using the probability-sampling technique [13], variations in light intensity with time, following excitation by gamma rays, were measured. The decay curves were analyzed and information on three time constants was obtained; τ_1 related to energy transfer from solvent to solute, τ_2 related to the rapidly decaying component from the singlet state, and τ_3 related to the slower component which is probably due to excimers. According to the measurements, at very low concentrations, τ_1 was large and increased the value of τ_2. In this study, the decay times of the excited singlet state and excimer of PPO were measured as about 1.6 and 7.9 nsec, respectively. Also, it was observed that, at the lower concentrations of PPO, the value of τ_1 was inversely proportional to the mole fraction of PPO. Figure 4 shows the typical data obtained by Lynch [12] for higher concentrations of PPO.

SOME RECENT APPLICATIONS

Solubility of Gases in Organic Solvents

Organic scintillator solutions can be used for the measurement of the solubility of certain gases in the solvents of the scintillator solu-

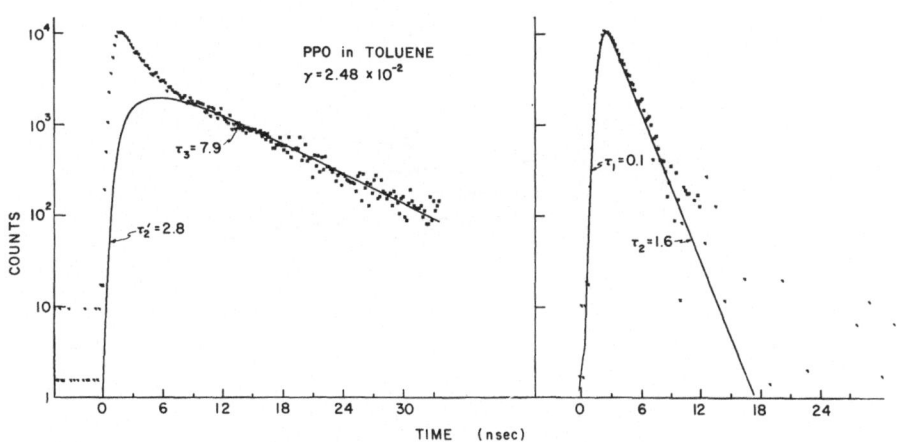

Fig. 4. Measured intensity of light from PPO—toluene solution as a function of time. The curve at left shows data points with background subtracted and theoretical slow curve superimposed. The curve at right shows data points, following subtraction of slow component, and theoretical curve for τ_2 and τ_1 (from F. J. Lynch [12]).

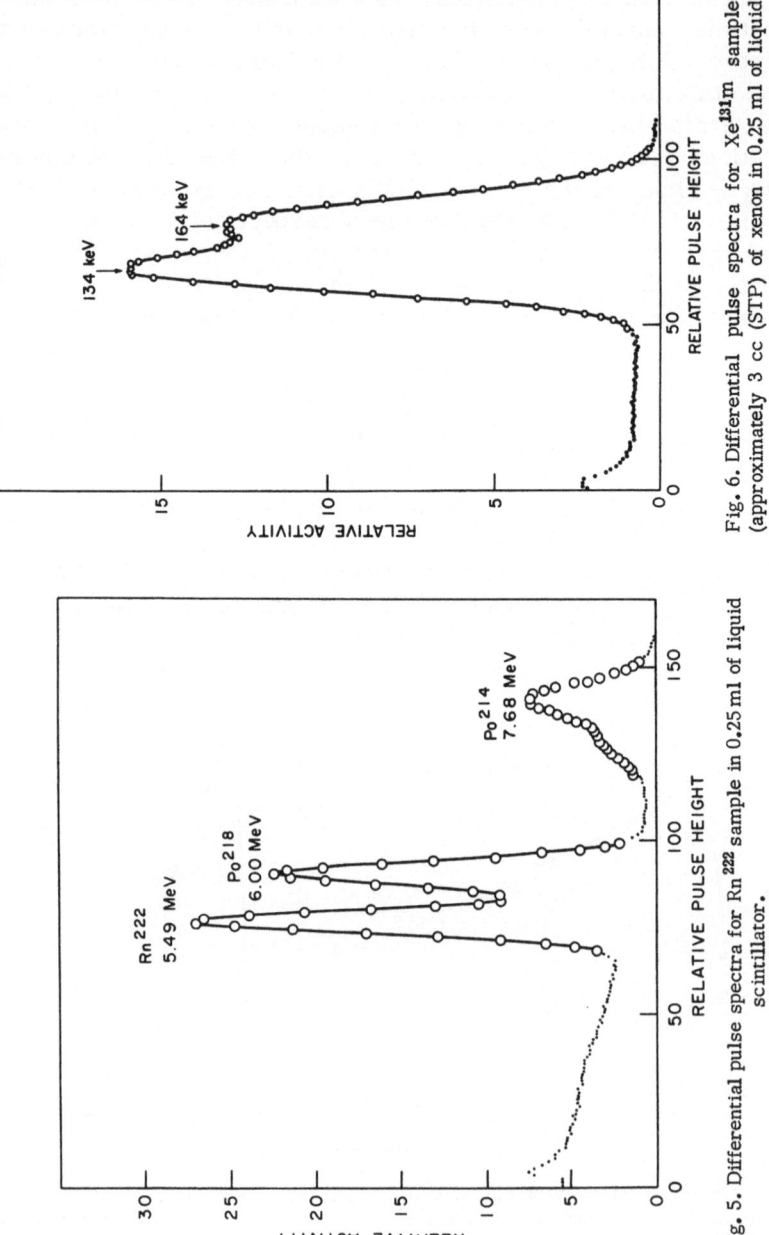

Fig. 6. Differential pulse spectra for Xe[131m] sample (approximately 3 cc (STP) of xenon in 0.25 ml of liquid scintillator.

Fig. 5. Differential pulse spectra for Rn[222] sample in 0.25 ml of liquid scintillator.

TABLE V

Solubilities of Noble Gases in Toluene-Base Liquid Scintillator

Gas	K, distribution ratio $-15°C$	$27°C$
Kr	0.9	
Xe	5.0	3
Rn	32.0	

tions. By using appropriate radioactive tracers, the concentration of a gas in a solvent (with added scintillator solutes) can be determined by measuring the amount of fluorescence produced. Those solvent-and-gas systems which have appreciable solubilities can also be used as the basis of a method for the determination of the amount of a radioactive nuclide in a given gas sample.

The solubilities of three of the noble gases, radon, xenon, and krypton, in toluene were measured by this technique recently [14]. The amount of the appropriate radioactive tracer that dissolved in a toluene-base liquid scintillator was measured as a function of the space above a constant volume of the liquid scintillator solution. The solubility of radon (which exists only as radioactive nuclides) was measured as Rn^{222}. The solubilities of xenon and krypton were measured by using the naturally abundant gases tagged with the radioactive nuclides Xe^{131m} and Kr^{85}, respectively. The activity of the given nuclide dissolved in the solution was measured with varying amounts of space above the solution. A plot of the volume of the space, V_g, against the reciprocal of the measured activity, A_g, gave a straight line. The x-axis intercept of such a plot was used to calculate the distribution ratio K by the equation

$$K = \frac{-\text{ intercept}}{V_s} = \frac{a_s}{a_g}$$

where a_s and a_g are the specific activity in the scintillator solution and the space above the solution, respectively, and V_s is the volume of the scintillator solution. Table V lists the solubilities that were measured with this technique. The values obtained agree very well with other published determinations.

With the measured solubilities of these gases in toluene, liquid scintillator solutions can be used to measure the amount of a radioactive nuclide of Rn, Xe, or Kr. Figures 5, 6, and 7 show the pulse-height spectra obtained with samples containing Rn^{222}, Xe^{131m}, and Kr^{85}. By minimizing the space above the scintillator solution, conditions can be obtained so that 99% or better of the gas will be dissolved

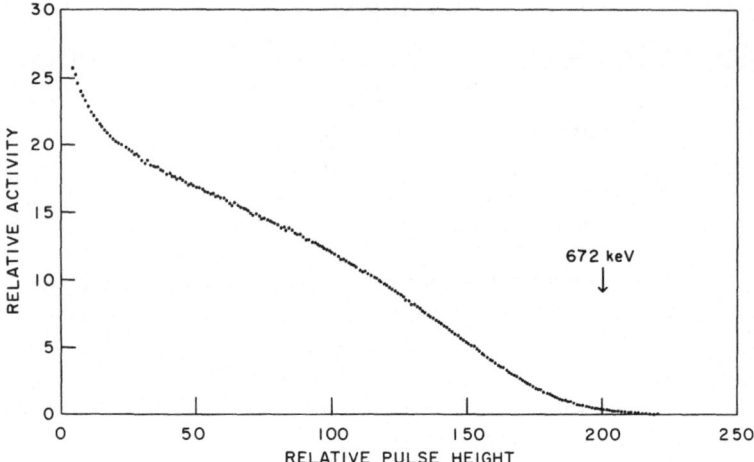

Fig. 7. Differential pulse spectrum for Kr^{85} sample in 10 ml of liquid scintillator.

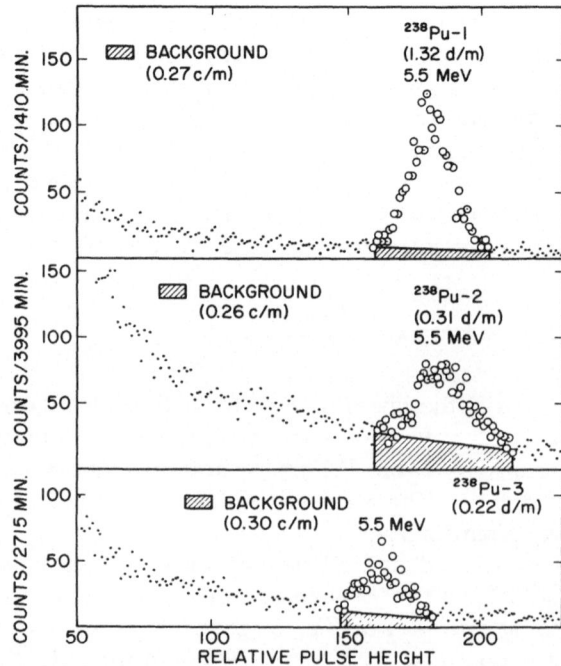

Fig. 8. Differential pulse spectra for samples of Pu^{238} in 0.25 ml of liquid scintillator. Backgrounds indicated by shaded areas. Pu-1,1.32 d/m; Pu-2,0.31 d/m; Pu-3,0.22 d/m.

in the solution.   Under these conditions, the counts measured are the disintegration rate to within better than 1%.

Low-Level Alpha Counting

The recent introduction to the market of low-noise, high-quantum-efficiency multiplier phototubes that can be operated at room temperatures made possible the design of a simple single-channel liquid scintillation counter for the determination of as little as 0.1 disintegration/min of an alpha active nuclide [15]. Such a multiplier phototube is the RCA-8575 tube, which has a bialkali photocathode of about 28% quantum efficiency and low noise when operated at room temperature. In a recent work [15], samples of alpha-active nuclides were dissolved in liquid scintillator solutions [by the preparation of the di-(2-ethyl-hexyl)phosphoric acid complexes], and the differential pulse spectra of the amount of fluorescence was measured with a multichannel analyzer. Figures 8 and 9 show some typical pulse spectra obtained in this manner.

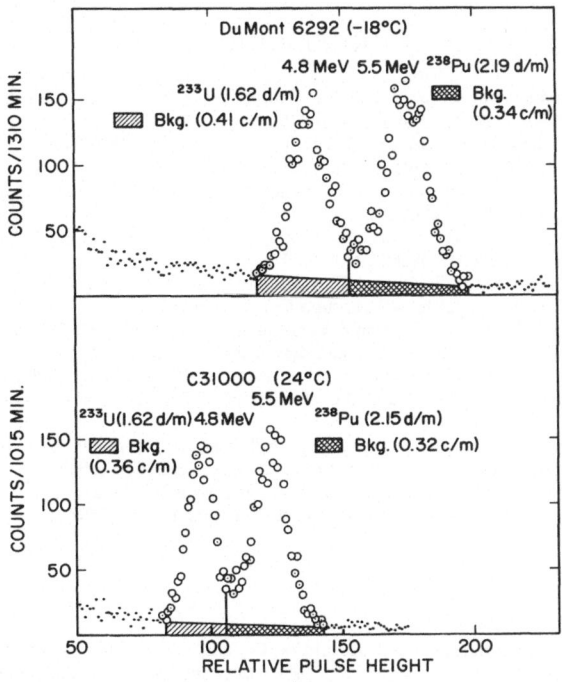

Fig. 9. Differential pulse spectra for sample containing both Pu^{238} and U^{233}.

After stripping the background, the number of pulses under a given peak is the disintegration rate. More than one nuclide can be measured with one sample provided the energies of the alpha particles are different enough to be resolved by the counting system. About a 6% energy resolution was obtained with the experimental arrangement employed in that work.

Liquid scintillator systems have several properties which make them useful for alpha disintegration rate determinations:

1. The counting efficiency is 100% for alpha particles, and
2. background counting rates are low.
3. No sample self-absorption or backscattering occurs, as is associated with deposited samples.
4. Alpha particles of different energy can be measured in the same sample with an approximately 6% energy resolution.
5. Samples are easily recovered for other purposes.
6. Inexpensive chemicals and equipment are used.

REFERENCES

1. H.P. Kallmann, Natur und Technik, July, 1947; Z. Naturforsch. 2a:439, 642 (1947).
2. W. Kern and H.O. Wirth, Kunststoffe—Plastics 6:12 (1959).
3. D.L. Horrocks and H.O. Wirth, J. Chem. Phys. 47:3241 (1967).
4. B. Stevens and T. Dickinson, J. Chem. Soc. 5492 (1963).
5. I.B. Berlman, Handbook of Fluorescence Spectra of Aromatic Molecules, Academic Press, Inc., New York (1965).
6. M.S. Walker, T.W. Bedner, and R. Lumry, J. Chem. Phys. 45:3455 (1966).
7. D.L. Horrocks and H.O. Wirth, to be published.
8.* D.L. Horrocks and H.O. Wirth, "Scintillation Measurements at Very High Solute Concentrations; Self-Quenching—Structure Correlations."
9. Th. Forster, Fluoreszenz Organic Verbindungen, Göttingen (1959), p. 158.
10.* H.O. Wirth, F.U. Herrmann, G. Herrmann, and W. Kern, "On the Correlations Between Constitution and Scintillation Properties in the p-Oligophenylene Series."
11.* I.B. Berlman, "Factors Relevant to a Short Fluorescence Lifetime of an Aromatic Molecule: Empirical Data."
12.* F.J. Lynch, "New Liquid Scintillators with Faster Response and Higher Efficiency."
13. L.M. Bollinger and G.E. Thomas, Rev. Sci. Instr. 32:1044 (1961).
14. D.L. Horrocks and M.H. Studier, Anal. Chem. 36:2077 (1964).
15. D.L. Horrocks, Intern. J. Appl. Radiation Isotopes 17:441 (1966).

---

*Presented at the International Symposium on Organic Scintillators, Argonne National Laboratory, Argonne, Illinois, June 20—22, 1966; proceedings to be published as a supplement to the January, 1968, issue of Molecular Crystals, Gordon and Breach Science Publishers, Inc., New York.

Accelerator Systems for Activation Analysis—A Comparative Survey

J. R. Vogt

Battelle Memorial Institute
Columbus Laboratories
Columbus, Ohio

Accelerator systems for neutron activation analysis have become increasingly common in recent years. Some of these are simple systems designed specifically for activation analysis, while others are primarily sophisticated instruments for nuclear physics research, which may be adapted for the activation method. Most of these accelerators have used the radio frequency or Penning ion sources for deuteron production. Recently, however, increased attention has been focused on the Duo-Plasmatron ion source, which is capable of producing much larger beam currents. The beam from the Duo-Plasmatron ion source can be magnetically and electrostatically analyzed to produce an intense monatomic beam at the target.

This paper compares the different types of accelerator systems and discusses vacuum systems, targets and target cooling, neutron flux monitoring, and sample transfer systems.

INTRODUCTION

Although neutron activation analysis is most widely known for the high sensitivity which is possible when using a high-flux nuclear reactor, the proliferation of the activation method as a routine analytical tool is largely dependent on the development of high-intensity, moderately priced accelerator systems as a source of neutrons. This is a result of the high cost of nuclear reactors and the fact that a large fraction of routine analytical work requires rapid analysis of large numbers of samples for major and minor components. Thus, while subnanogram analysis with a reactor may be the most unique feature of activation analysis, the possibility for rapid, accurate, and often nondestructive analyses of major and minor constituents at a relatively low cost has led to an increasing demand for accelerator systems.

161

Van de Graaff

The first major applications of accelerators as neutron sources for activation analysis were based on the Van de Graaff [1–3]. Much of the early developmental work using these accelerators was done by Atchison and Beamer at the Dow Chemical Company [3]. The distinguishing feature of a Van de Graaff accelerator is the continuous transfer of electric charge from a low-voltage dc power supply to a hemispherical high-voltage terminal by means of a rapidly moving insulated belt. The potential of the terminal rises as charge is sprayed on the belt by a corona discharge, until breakdown of the gaseous or solid insulation occurs or until the charging current equals the sum of the load current taken by the accelerator tube, the current through the column resistors, and the corona current from the high-voltage terminal. The high-voltage terminal is insulated from the pressure vessel surrounding the generator by a suitable compressed gas. Gases with large molecules are best for this purpose since they are the best dielectrics. Most of the organic gases which could be used in this respect have the disadvantage of being subject to decomposition in an electrical discharge. The resultant materials corrode most materials used in accelerators. In addition, organic gases are subject to carbonization and leave traces on the insulator surfaces, which increase the probability of sparking. Use of nitrogen avoids the fire risk of high-pressure air with only a slight loss in dielectric strength. If nitrogen is used alone, the pressure is usually between 16 and 27 atm. Nitrogen with some CO_2 added is the most common insulator used for these accelerators. Electronegative gases such as CCl_4 and freon are much better dielectrics than air or nitrogen; however, the vapor pressures of CCl_4 and freon are too low for the gases to be used alone at the pressures required, because they will liquefy. Freon mixed with nitrogen at 10 atm is one of the cheapest mixtures. The best gas is sulfur hexafluoride [4], but it is too expensive for general use with Van de Graaff accelerators.

The typical single-stage Van de Graaff accelerator permits a relatively high voltage, 0.4 to 5.5 MeV, but has a relatively low beam current, 70 to 400 μA. For these reasons, the production of neutrons for activation analysis with these accelerators is usually based on nuclear reactions which exhibit a high neutron yield for bombarding energies in the range of 1 to 5 MeV. The more common nuclear reactions used to produce neutrons for activation analysis applications are given in Table I. Of these reactions, the one most often used with Van de Graaff accelerators is the $Be^9(d,n)B^{10}$ reaction because this provides

TABLE I

Reactions Used for Neutron Production

Reaction	Threshold energy, Mev	Q value, Mev	Neutron energy at threshold, Mev	Typical neutron energies	
				Bombarding energy	Neutron energy, Mev
D (d,n) He^3	. . .	+3.27	2.45	200 keV	2.94
T (d,n) He^4	. . .	+17.6	14.04	200 keV	14.81
T (p,n) He^3	1.02	-0.76	0.064	2 MeV	1.2
Be^9 (d,n) B^{10}	. . .	+4.3	1.0-4.5	2 MeV	1.0-6.0

the highest neutron yield per microampere of beam current at bombarding energies above 1 MeV. For most applications, the neutrons produced are moderated with water or paraffin to provide neutrons approaching thermal energies. In this way, an approximately thermal neutron flux of 5×10^6 n cm$^{-2}\mu$A^{-1} can be obtained at 2 MeV. Because accelerators of this type are quite expensive and because the maximum beam current obtainable is relatively low, they are not widely used for activation analysis.

Cockcroft—Walton

The name Cockcroft—Walton is often applied to all neutron generators which are not of the Van de Graaff type, although, technically, it applies only to those accelerators using voltage-multiplier circuits similar to those used by Cockcroft and Walton [5]. High-voltage supplies using voltage-multiplier or transformer-rectifier circuits are particularly well suited for accelerators producing energies up to approximately 1 MeV. The lack of moving parts and the excellent electrical insulation which can be achieved by immersing the entire power supply in oil make properly designed high-voltage supplies of this type inherently more reliable at voltages of 1 MV, or less than those of the Van de Graaff type. More important, however, is the fact that they permit beam currents in the milliampere range.

Because of the relatively high beam currents which can be obtained, accelerators of this type generally utilize those nuclear reactions which provide high neutron yield at a relatively low voltage and a high beam current. Thus the D(d,n)He^3 and T(d,n)He^4 reactions are used to produce neutrons of 3 and 14 MeV, respectively. At 200 kV and a beam current of 2 mA, the D(d,n)He^3 reaction will have a yield of about 4×10^9 n sec^{-1} and the T(d,n)He^4 reaction will have a yield of about 2×10^{11} n sec^{-1}.

Because these accelerators generally have power supplies producing only 100 to 250 kV, it is not absolutely necessary to insulate the high-voltage terminal or the accelerating column. However, since accelerators of this type are often operated by technicians or students, there is a trend toward insulating these machines anyway for safety purposes. This may be done by immersing the high-voltage terminal and accelerating column in oil or by surrounding them by a pressure dome filled with gas. The pressure required to insulate potentials of 100 to 250 kV is rather low, so low-vapor-pressure gases such as freon can be used. Also, because the amount of gas necessary is low compared with that required to insulate a 2-MeV Van de Graaff, it is economically practical to use sulfur hexafluoride. The use of gas as an insulating medium has two obvious advantages over oil. First, it is much less messy when it becomes necessary to do maintenance work on the terminal or accelerator column, and, second, it does not form the carbon deposits during electrical discharges which tend to increase the probability of sparking. Sulfur hexafluoride has excellent dielectric and heat-transfer properties and is nontoxic and noncorrosive. Its electrical breakdown products are toxic and corrosive, but these are completely and easily removed by keeping activated alumina in the pressure dome [6].

There are two methods which may be used with this type of accelerator to supply the necessary power for the ion source and extractor in the terminal. The first is to incorporate all the necessary power supplies in the tank housing the main high-voltage supply. The second is to include an isolation transformer in the main high-voltage tank which supplies 115-V alternating current to separate power supplies in the accelerator terminal. The first method results in a compact terminal and, if suitably designed, can be less expensive. The second method produces a bulky terminal but tends to result in a more reliable and more easily maintained system.

ION SOURCES

Radio Frequency

Most radio-frequency (rf) ion sources in current use are based on the design of Moak, Reese, and Good [7]. The primary advantages of this type of ion source are a high percentage of deuterons, high stability, and low gas and power consumption. A comparison of the rf,

TABLE II

Characteristics of D^+ Ion Sources

Type	Typical beam current, mA	Beam D^+ component, %	Gas consumption, cm^3/hr	Typical time to failure, hr	Cooling
Radio frequency	1−3	70−90	2−10	200	Air
Penning or PIG	1−5	10−30	15−25	10,000	Freon[113]
Duo-Plasmatron	1−10	40−60	10−20	1,000	Freon[113]

Penning or PIG, and Duo-Plasmatron ion sources is given in Table II. A modification of the original design of Moak et al. is shown in Fig. 1 [8]. The principal differences between this and the design of Moak et al. are the use of an air-cooled aluminum extraction electrode and the substitution of a synthetic-sapphire sleeve around the extraction canal for the silica sleeve used in the original design. The use of the aluminum extraction electrode prevents the usual breakdown of the ion bottle in this area due to electron bombardment. This breakdown can be retarded by the construction of a glass shield within the ion bottle, but use of the air-cooled aluminum extraction electrode simplifies construction and decreases the cost of the ion bottle. The substitution of synthetic sapphire for the sleeve surrounding the extraction canal virtually eliminates sleeve deterioration problems. With an extraction-canal bore of 0.073 in., beam currents up to 2.2 mA have been obtained.

To produce a deuteron plasma, deuterium is introduced into the ion-source bottle through a hole in the aluminum base of the source. The flow of deuterium is regulated by a palladium leak, which also serves to purify the gas—a fact which is of importance because rf ion sources are quite sensitive to gas impurities. It is preferable to have both the palladium leak and the bottle of deuterium located in the accelerator terminal. This eliminates high-voltage isolation problems and permits a soldered copper line between the deuterium bottle and ion source. The copper line has two advantages over a system using some sort of plastic tubing. First, it eliminates poisoning of the leak by molecules from the plastic, and, second, it eliminates leakage of deuterium through the plastic. Experience indicates that the use of a solid-copper deuterium line will decrease gas consumption by an order of magnitude compared with a plastic line—from 25 cm^3/hr to 2 to 3 cm^3/hr [8]. The rf power requirements for ionization of the deuterium are reduced by producing an electromagnetic field around the base of the ion-source bottle with a solenoid. This increases the ionization

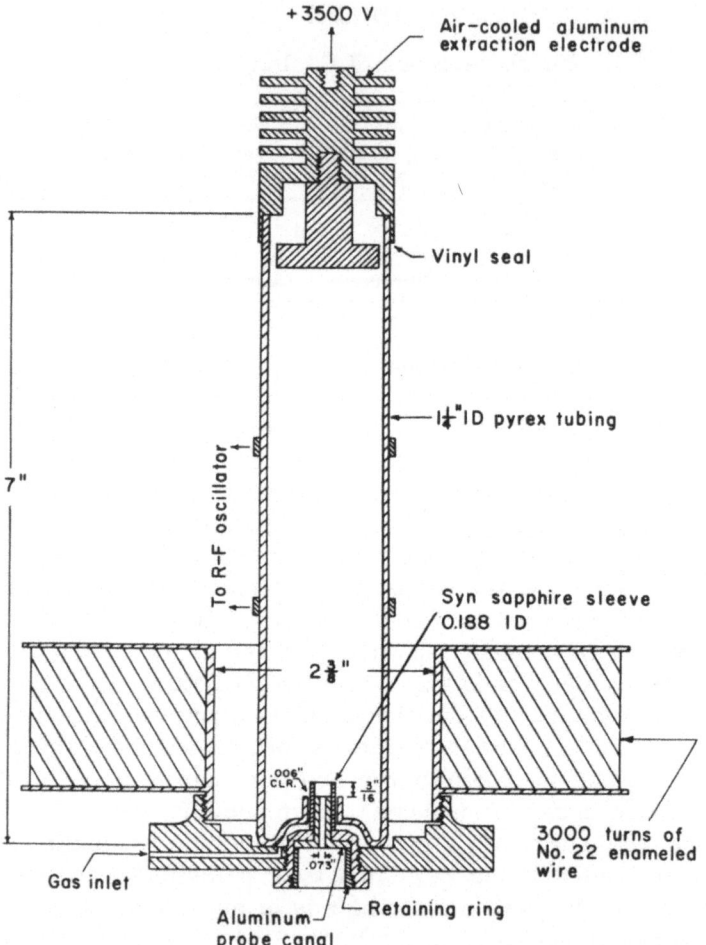

Fig. 1. Sectional view of a 98-Mc rf ion source.

probability owing to restriction of the electron paths in the source. After ionization, the deuterons are accelerated toward the probe canal by the extraction system. This system consists of an aluminum anode having a potential of +3500 V and a hollow, cylindrical, aluminum cathode which forms a canal through which the ions pass into the accelerating tube. The synthetic-sapphire sleeve is mounted over the canal. This sleeve shields the plasma from the metallic cathode and serves as a virtual anode when the extraction voltage is applied. Thus, a lens is formed over the canal opening, which focuses a large number of ions so that their path is straight enough to miss the walls of the canal. The electromagnetic field around the base of the source limits space-charge spreading in the canal and thus increases the output current.

Although the rf ion source produces the highest percentage of D^+ ions of any of the ion sources, it is also a rather sophisticated source. It, therefore, requires a more experienced operator to get top performance than do some of the other ion sources.

Penning or PIG

The Penning or Phillips Ionization Gauge (PIG) ion source has received considerable application as an ion source for neutron generators in recent years because of its simplicity, long life, and the high beam currents which can be obtained. Its effective use is complicated, however, by the low percentage of D^+ ions which can be obtained, usually 15 to 20%. Because the majority of the beam consists of D_2^+

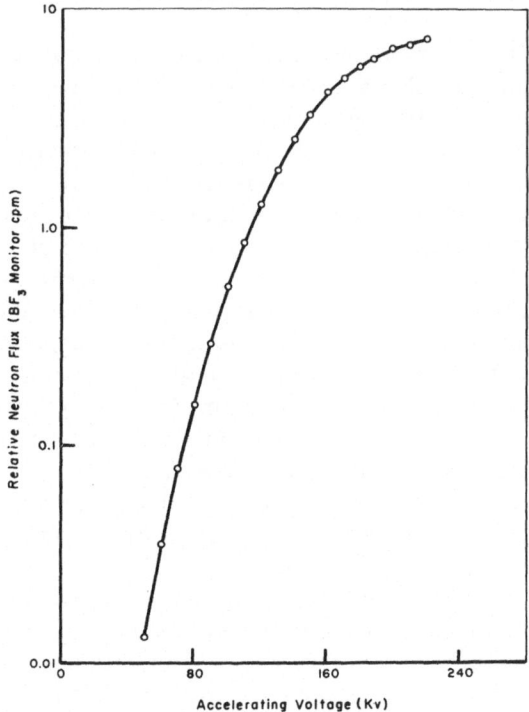

Fig. 2. Relative neutron yield versus accelerating voltage for an rf ion source.

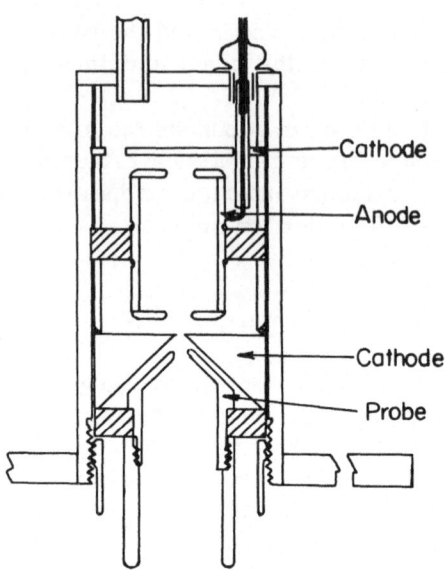

Fig. 3. Sectional view of a Penning ion source.

molecules, the mean-incident deuteron energy at the target is half the accelerating voltage. The variation of neutron output with accelerating voltage was determined by using a rf ion source for the $T(d,n)He^4$ reaction and is shown in Fig. 2. This indicates that a high D_2^+ percentage results in a large decrease in the possible neutron output, compared with the rf ion source, for a given accelerating voltage. At an accelerating voltage of 200 kV, the contribution of D_2^+ to the total neutron flux will be 69%, if a beam of 15% D^+ and 85% D_2^+ is assumed. In order to maintain a high neutron flux over a reasonable period of time, it is necessary to prevent carbon or other deposits from forming on the target so as to obtain the highest possible yield from the less penetrating D_2^+ portion of the beam. In other words, a very clean vacuum system is much more important for a Penning ion source than it is for a rf ion source at a given accelerating voltage. This generally means using an ion pump rather than an oil-diffusion pump. If this is done and the beam is spread uniformly over the target, the neutron yield and target life are quite good.

A typical Penning ion source is shown in Fig. 3. A solenoid coil—not shown in the figure—is placed around the source to provide a magnetic field of 500 to 1000 gauss. The cathodes are electrically connected, and the anode is positive with respect to both. Ionization of the deuterium gas is achieved by the successive acceleration and reflection of electrons which are present in the gas or released from the cathodes, from one end of the source to the other. In this way, a number of ion pairs are made by each electron, which results in a high ionization efficiency. The radial motion of the electrons and positive ions is restricted by the magnetic field. A plasma is formed which extends through the anode and enters the accelerating tube through a hole in one of the cathodes. The beam is then focused and accelerated by the probe [9, 10]. The best D^+/D_2^+ ratio is generally obtained with as large an anode current as possible and by using the probe voltage to control the beam current. A permanent magnet can be used instead of a solenoid, but this decreases the amount of tuning that is possible. To provide a stable beam current, the source should be run at a constant and moderate temperature and is therefore usually cooled by liquid freon. Either a palladium or mechanical leak may be used to control the flow of deuterium because the source will ionize anything and is not bothered by a few gaseous contaminants as is the rf ion source. However, a palladium leak will contribute toward a cleaner vacuum system and tend to minimize deposits on the target. In addition, the palladium leak is somewhat more convenient to use.

Duo-Plasmatron

The Duo-Plasmatron ion source, developed by Von Ardeene [11] and modified by Moak et al. [12], provides a phenomenal ion current with high gas and power efficiency. Von Ardeene's original source was designed to deliver proton currents from 0.1 to 1 A. Since beam currents of this magnitude cannot be used in accelerator systems because of space-charge limitations, Moak et al. developed a version of the source which will produce beam currents of 0.5 to 30 mA. A sectional view of this source is given in Fig. 4. With a source of this type, D^+ yields of 40 to 60% or higher are easily obtained. Were it not for its present high cost, the Duo-Plasmatron ion source would probably be the only ion source used in Cockcroft—Walton type accelerators designed for activation analysis. By use of a magnetic—electrostatic beam analysis system, a 10-mA Duo-Plasmatron source would deliver a beam consisting of essentially 100% D^+ ions at a current in excess of 5 mA to the target. The Duo-Plasmatron ion source is a three-electrode system with an intermediate electrode between the cathode and the anode. The intermediate electrode is made of a magnetic material and provides both electrostatic and magnetic focusing so that the plasma is formed only in the region where it is useful. The ion

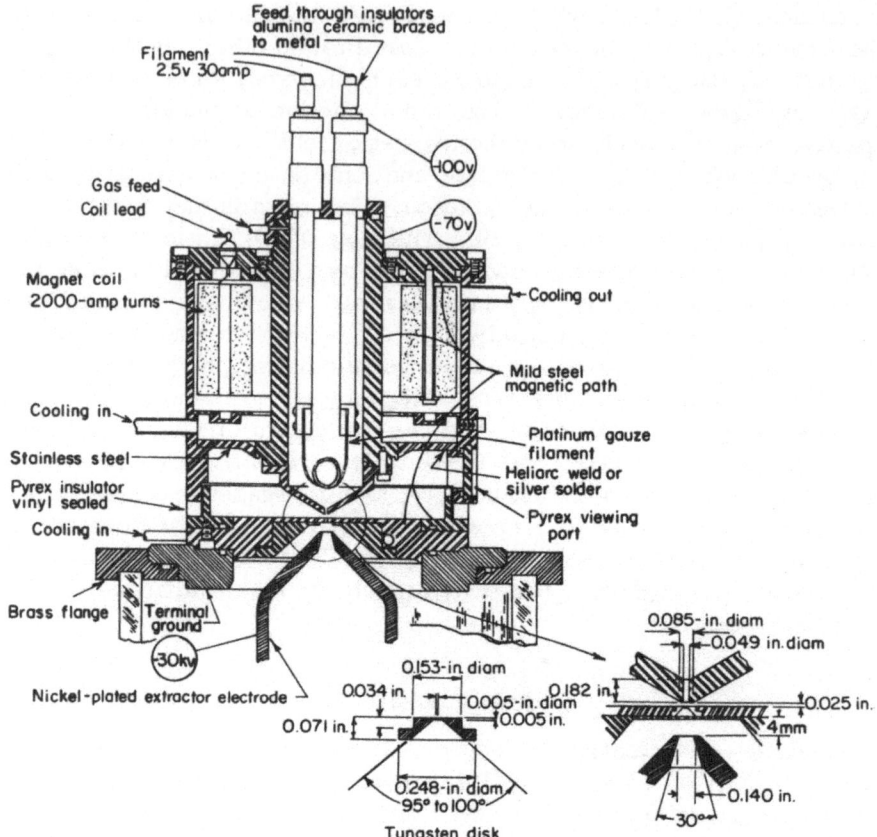

Fig. 4. Sectional view of a Duo–Plasmatron ion source.

density at the extractor hole in the anode is about 6×10^{14} ions/cm^3, which is much higher than the ion densities of about 10^{11} ions/cm^3 found in rf ion sources [12]. The plasma is not forced through the anode aperture but flows out because of the pressure differential. With the use of a BaO-coated platinum gauze filament, cathode life is greater than 1400 h. Ion sources of this type are generally cooled during operation with freon[113].

VACUUM SYSTEMS

Although mercury–diffusion pumps are generally used with large Van de Graaff accelerators, they are usually too expensive for most of the relatively low-cost Cockcroft–Walton accelerators used for activation analysis. For Cockcroft–Walton systems, an oil-diffusion pump or an ion pump is generally used. The oil-diffusion pump is very reliable, and its operating cost is negligible. In addition, tritium released from the targets can be steadily disposed of by venting the pump through a fume hood or a stack. The oil-diffusion pump has the disadvantage of causing carbon deposits on the target and within the accelerator from electrical breakdown of oil molecules. It is therefore best suited for systems using rf or Duo-Plasmatron ion sources where carbon build-up on the target is not as important. Deposition of oil molecules within the rf ion source during periods in which the accelerator is not in use interferes with the operation of the rf source, but the oil is readily "burned off" by 20 to 40 min of operation of the source.

Use of an ion pump eliminates the carbon problem and also the need for icing a trap. However, the pump elements must be replaced periodically, and the problems associated with tritium build-up in the ion pump can be considerable. An accelerator in routine daily use can easily accumulate 50 curies of tritium in the ion pump in a few months' time. This can be quite a problem when it becomes time to change the pump elements. Also, there is an ever-present possibility of a sudden release or regurgitation of tritium from the pump.

TARGETS AND TARGET COOLING

The usual targets for activation analysis with Cockcroft–Walton accelerators consist of tritium or deuterium occluded in a 1-mg/cm^2 layer of titanium on a 0.010-in.-thick copper backing. For tritium targets, the usual surface activity is 4 curies. Experiments with tritium targets using erbium instead of titanium have been conducted, but, while they result in somewhat better performance, they are expensive and are not in general use. Although the maximum cross section for the T(d,n)He^4 reaction occurs near 110 keV, the theoretical neutron yield with a thick target continues to increase very rapidly up to approximately 300 keV owing to deeper penetration of the deuteron beam in the titanium tritide.

Because the deuteron beam generates a considerable amount of heat in the target, it is necessary to cool the target to avoid an extremely rapid loss of tritium. Figure 5 shows a sectional view of a typical target-cooling assembly [13]. This cooling assembly is designed so that the water strikes the back of the target at a 12° angle. Tests

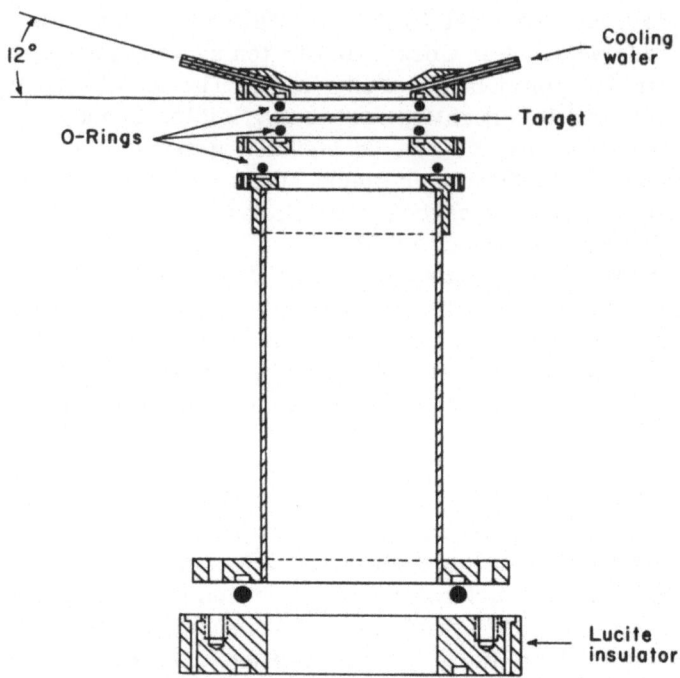

Fig. 5. Sectional view of a target-cooling assembly for high beam currents.

indicate that this minimizes the formation of air pockets and gives more effective cooling. A water flow rate of 600 to 700 cm^3/min effectively dissipates as much as 400 W of beam power at 200 kV and a beam current of 2 mA.

NEUTRON-FLUX MONITORING

Because activation analysis procedures generally compare the activity of a sample with that of a standard, absolute measurements of the neutron flux are not absolutely necessary. When some idea of the 14-MeV neutron flux is desired, the usual procedure is to measure the activity induced in copper by the $Cu^{63}(n,2n)Cu^{62}$ reaction. Thermal neutron-flux measurements are usually made using the $In^{115}(n,\gamma)In^{116m}$ reaction. What is more important from the standpoint of activation analysis is the comparison of the relative neutron flux on the sample and the standard. A number of different methods have been used to compare flux differences on samples and standards. These include the use of BF_3 counters [8], plastic scintillators [14], and the monitoring of N^{16} activity in target-cooling water [15]. Another approach, the reference-sample technique, has been developed, which attempts to eliminate the need for directly monitoring the neutron flux by irradiating the sample and standard simultaneously in a dual transfer system. This procedure generally provides the best precision [16].

When the sample and standard are not irradiated simultaneously and an appreciable decay occurs during the transfer and counting periods, it becomes necessary to normalize the number of counts obtained from both standards and samples for variations in neutron flux, irradiation time, delay time, and dead time. During a typical run, standards are periodically irradiated and counted, typically about one standard for every ten samples. This procedure requires that the neutron flux during a particular irradiation remain relatively constant, so that the integrated flux measured by the flux monitor will provide an accurate indication of the effective flux on samples and standards.

The moderated BF_3 counter is the most commonly used flux monitor. Its high sensitivity permits good counting statistics for short irradiation times. In addition to the high counting rates obtainable, the BF_3 counter has the advantages of stability, ruggedness, and insensitivity to γ rays. However, it does not provide an indication of differences in effective neutron flux due to flux attenuation in the sample, a source of error which becomes quite important for large samples or for samples containing elements with large neutron cross sections. To account for self-shielding effects, Anders [14] developed a flux monitor consisting of a small, plastic scintillator bead mounted on a photomultiplier tube. This detector is mounted in the geometrical shadow of the sample and monitors the neutron flux passing through the sample. By using this monitor, samples differing by a factor of 4 in their macroscopic cross sections have been analyzed for oxygen with a relative standard deviation of 1.5%. The use of target-cooling water to monitor the neutron flux was developed by Steele and Meinke [15]. The cooling water flows through a tube wrapped around a Geiger counter. The activity of N^{16}, produced by the $O^{16}(n,p)N^{16}$ reaction, is measured as an indication of the relative neutron flux. The primary advantage of this method is that it is inexpensive.

SAMPLE TRANSFER SYSTEMS

A number of systems designed to control the irradiation and transfer of samples, both single [8, 14, 15], and dual [16–18], have been described in the literature. These systems generally consist of plastic or metal tubing, in which the sample is transferred between the accelerator and counting rooms by compressed gas or a vacuum, along with the controls necessary to regulate and measure the irradiation time, delay time, and counting time. In dual transfer systems, where the sample and standard are irradiated and counted simultaneously, precise measurement of irradiation, delay, and counting times is generally unnecessary. However, such measurements can be quite important in single-sample systems. In particular, the actual irradiation time should be accurately controlled and measured in order to correct for differences between samples and standards.

Several methods for turning the neutrons on and off by controlling the deuteron beam can be used. These are plasma pulsing, pre- or postacceleration beam deflection, and beam "chopping." Plasma pulsing works best with the Penning ion source. Very rapid rise times can be obtained when the anode voltage is turned on. The rf ion source, on the other hand, will take a few moments to stabilize after the ex-

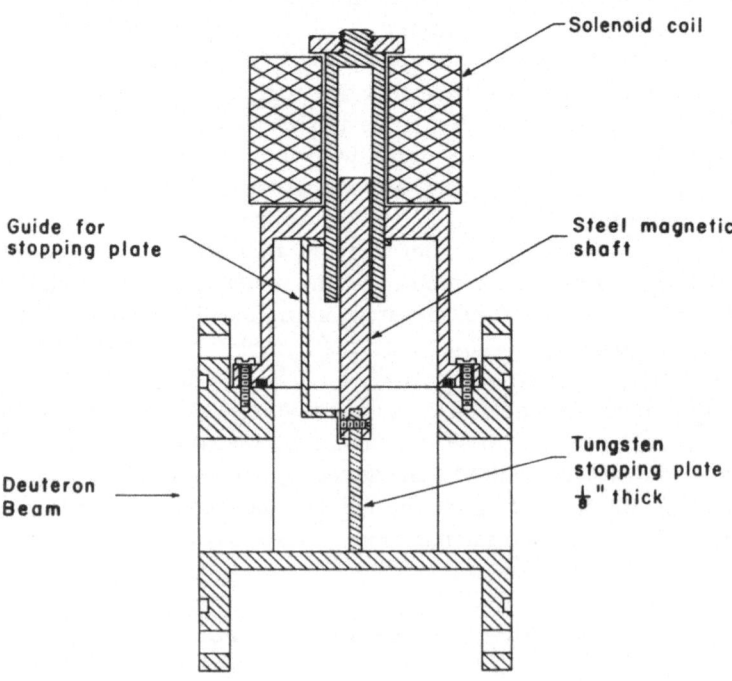

Fig. 6. Sectional view of a beam-stop assembly for millisecond control of a deuteron beam.

traction voltage is turned on. Preacceleration beam deflection is preferred when it is desirable to eliminate the neutron background, from the $D(d,n)He^3$ reaction, between irradiations. However, this procedure substantially lowers the usable beam current. Postacceleration deflection is widely used for activation analysis applications and permits accurate control of large beam currents. When using postacceleration deflection, the build-up of deuterium on the inside of the accelerator drift tube will result in a substantial neutron background from the $D(d,n)He^3$ reaction. In many cases, however, this is not objectionable. The least expensive way to turn the neutrons on and off in a time interval on the order of milliseconds is by postacceleration chopping of the deuteron beam with the use of a solenoid-operated beam stop. A cross-sectional view of such a beam stop is shown in Fig. 6 [8]. This beam stop is designed so that the solenoid coil is outside the vacuum system, while the movable beam-stopping plate and magnetic shaft are entirely within the vacuum system. In this way, the problem of vacuum sealing a movable shaft is eliminated. Heat generated by the deuteron beam causes the plate to become red hot so that it is necessary to water-cool the beam-stop assembly to avoid overheating its O rings. The high temperature of the beam-stop plate minimizes the build-up of deuterium on the plate, which would give rise to neutrons from the $D(d,n)He^3$ reaction.

SEALED-TUBE NEUTRON GENERATORS

The advent of high-intensity, sealed-tube neutron generators is probably the most significant step in recent years toward making activation analysis a truly routine analytical technique. These small accelerators are designed to be operated by people with little or no training in radioactive-contamination or high-vacuum techniques. Basically, they are Cockcroft—Walton type units with either a Penning or rf ion source. The ion source, accelerating tube, and target are all sealed in a compact tube. No vacuum pumping is required, and the tube is readily replaced as a unit. Because the tritium remains sealed in the tube, the usual contamination problems associated with target changing and vacuum pumps are eliminated. The small size of the unit greatly simplifies shielding problems. Accelerators of this type are now available with neutron yields of 10^{10} n/sec and tube operating lifetimes of 60 hr. The relatively low cost and the simplicity of operation of these units make them ideally suited for routine industrial applications.

REFERENCES

1. S. E. Turner, Anal. Chem. 28:1457 (1956).
2. V. P. Guinn and C. D. Wagner, Anal. Chem. 32:317 (1960).
3. G. J. Atchison and W. H. Beamer, Anal. Chem. 28:237 (1956).
4. W. Buechner, Rev. Sci. Instr. 18:764 (1947).
5. J. D. Cockcroft and E. T. S. Walton, Proc. Roy. Soc. (London) A136:619 (1932).
6. P. G. Ashbaugh, D. W. McAdam, and M. F. James, Report No. AECL-2183, Chalk River Nuclear Laboratories, Chalk River, Ontario (1965).
7. C. D. Moak, H. Reese, Jr., and W. M. Good, Nucleonics 9 (3):18 (1951).
8. J. R. Vogt, W. D. Ehmann, and M. T. McEllistrem, Int. J. Appl. Radiation Isotopes 16:573 (1965).
9. J. D. Gow and J. S. Foster, Jr., Rev. Sci. Instr. 24:606 (1953).
10. W. L. Bronner, K. W. Ehlers, W. W. Eukel, H. S. Gordon, R. C. Marker, F. Voelker, and R. W. Fink, Nucleonics 17 (1):94 (1959).
11. M. von Ardeene, Tabellen der Elektroneuphysik Ionenphysik und Ubermikroskopie, Deutscher Verlag der Wissenschaften, Berlin (1956).
12. C. D. Moak, H. E. Banta, J. N. Thurston, J. W. Johnson, and R. F. King, Rev. Sci. Instr. 30:694 (1959).
13. J. R. Vogt, U.S. Atomic Energy Commission Report No. ORO-2670-10 (1966).
14. O. U. Anders and D. W. Briden, Anal. Chem. 36:287 (1964).
15. E. L. Steele and W. W. Meinke, Anal. Chem. 34:185 (1962).
16. F. A. Iddings, Anal. Chim. Acta 31:206 (1964).
17. D. F. Rhodes and W. E. Mott, Anal. Chem. 34:1507 (1962).
18. A. Volborth, Fortschr. Mineral. 43:10 (1966).

Spectroscopy of Biologically
Significant Molecules

Spectroscopic Studies of Molecular Interaction in DNA Constituents

R.C. Lord and G.J. Thomas, Jr.

Spectroscopy Laboratory
Massachusetts Institute of Technology
Cambridge, Massachusetts

Stability of the DNA helix depends in part on the free energy of association of the DNA-base pairs. Similarly, the interaction between DNA and
certain biologically active molecules may be determined in large part by
the binding of the latter to the DNA bases. It is possible to study the base-
pair interactions in nonaqueous solvents, such as chloroform, by means
of infrared spectroscopy, and the results of a variety of such studies are
summarized. The selectivity of base pairing, effects of chemical substitution on base-pairing equilibria, the thermodynamics of base pairing, and
ultraviolet hypochromism in chloroform solution are discussed on the basis
of the infrared studies. For aqueous solutions, Raman spectra appear to
be a more powerful means of investigating associative equilibria, and a
number of interesting systems have been examined with this technique.
A survey of the Raman spectra of aqueous solutions of individual bases,
nucleosides, and nucleotides is given. The spectra of certain mixtures
of nucleosides are discussed, and results are reported from the spectroscopic study of nucleoside—metal ion complexes and of interaction
between organic phosphates and metal ions. The advantages of excitation
of Raman spectra by lasers are illustrated in preliminary fashion by
comparison of the spectrum of polyriboadenylic acid excited with mercury
blue light with that excited by a helium—neon laser.

INTRODUCTION

The specificity of interaction between the purine and pyrimidine
bases of DNA and RNA is believed to furnish the molecular basis for
the control of cellular chemistry by the nucleic acids. The interaction
is chiefly due to hydrogen bonding between donor NH groups and acceptor oxygen and nitrogen atoms in the bases. Specificity of interaction arises from the critical dependence of the strength of the

H bonding on the geometrical arrangement of the donor and acceptor sites and on their electronic structure. When the arrangement is such as to yield the most effective bonding, optimum interaction occurs, a situation sometimes called "electronic complementarity."

The pairing interaction between the bases is a relatively delicate matter because the energy difference between two solvated unpaired bases and the solvated base pair in aqueous solution is only a matter of a few kilocalories. Investigation of base pairing is thus not easy, and different techniques, including spectroscopic ones, are needed to study the process. The virtue of vibrational spectroscopy for this purpose is its ability to reveal directly the effects of hydrogen bonding on specific chemical groups—the donor and acceptor groups—in the bases and to permit quantitative measurement of molecular concentration.

There are two solvent media of special significance in studies of base pairing. The most important, of course, is water because the chemistry of living material takes place in an aqueous medium. However, it is often desirable to remove the large effects of hydrogen bonding between water and its solutes in order to examine the bonding between the latter. To do this a solvent is needed which lacks both donor and acceptor sites for hydrogen-bond formation. The nucleic-acid bases are generally not soluble to a useful extent in such solvents, but a compromise can be made by selection of a solvent of moderate polarity, such as chloroform. Chloroform has both donor and acceptor sites, but it is well known that only very weak hydrogen bonds are formed at these sites.

Since the hydrogen bonds between the various purine and pyrimidine derivatives are generally much stronger than those between chloroform and the bases, the base-pairing equilibria in this solvent are a reasonably satisfactory measure of the strength of hydrogen bonding between the bases. In particular, the strength of the bonding between bases in the helical form of DNA, where the bases are located within the helix and thus "protected" to a large extent from hydrogen bonding by water molecules, is probably about the same as that measured in chloroform solution.

This paper reviews recent work carried out in the authors' laboratory by means of infrared spectroscopy on base pairing in chloroform solution and reports on some investigations of association in water solution by Raman spectroscopy.

INFRARED STUDIES OF BASE PAIRING IN CHLOROFORM SOLUTION

Selectivity of Base Pairing

In the past few years, a number of infrared studies have been made of base pairing in chloroform solution [1–6].[*] These all showed very clearly that, in mixed solutions of equal concentrations of adenine, uracil, cytosine, and guanine derivatives, there is relatively little tendency for self-association and, in addition, that adenine associates preferentially with uracil and guanine with cytosine. In quantitative terms, the equilibrium constant for the adenine–uracil association is about thirty times larger than that for self-association of adenine and fifteen times larger than that of uracil [5]. Similar results hold for the guanine–cytosine system [2, 3], although quantitative evaluation of the equilibrium constants for this system and for the remaining base pairs A–C, A–G, C–U, and G–U has not yet been completed.

*In Ref. 4, the solvent is carbon tetrachloride.

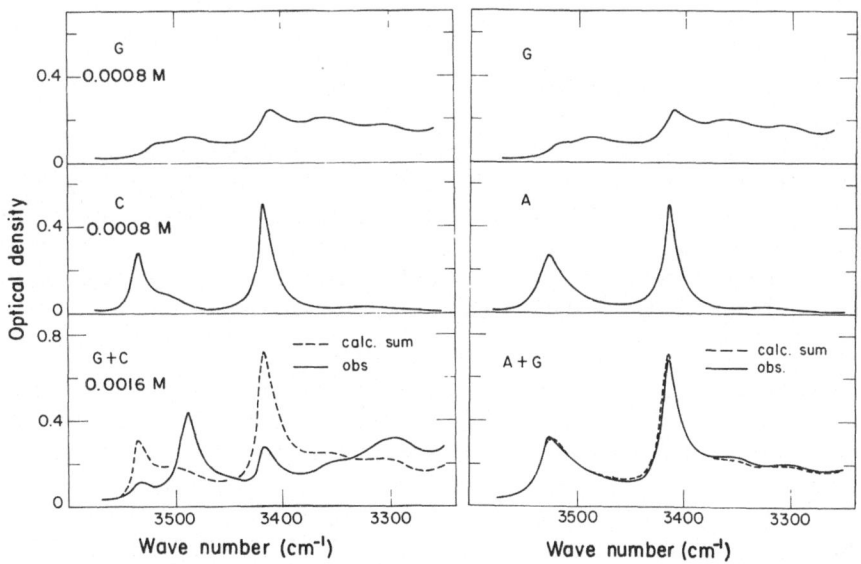

Fig. 1. Infrared spectra of the NH stretching region for derivatives of guanine and cytosine (left) and of guanine and adenine (right). The top two spectra in each case are of single solutes and the bottom of equimolal mixtures, as indicated. (Reproduced from Kyogoku, Lord, and Rich [3], where details are given.)

The specificity of the association can be illustrated in striking fashion by comparison of the spectra of the C–G system with those of the G–A system (Fig. 1). On the left of the figure, taken from Ref. 3, are shown- the spectra of pure derivatives of guanine and cytosine in chloroform and the spectrum of an equimolar mixture thereof, all at the same concentrations and optical paths. The spectra are presented with optical density as the ordinate, and, since the concentration of each component in the bottom spectrum is the same as that in the respective spectra above, the bottom spectrum should be the sum of the two others unless there is interaction between the two solutes. It will be seen that the observed spectrum of the equimolar G + C mixture (solid line) departs drastically from the curve expected on the basis of no interaction (dashed line).

On the right of Fig. 1, the corresponding spectra of the A–G system are shown under the same conditions. Here, however, the spectrum of the equimolal mixture is barely distinguishable from the sum of the component spectra, from which it is concluded that there is no appreciable interaction between A and G at the concentration used (0.0008 M). This same statement is valid even when the concentrations of A and G are ten times larger than those of Fig. 1 [3].

Factors Influencing Equilibria of Base Pairs

The foregoing demonstrations of selective base pairing are semi-quantitative except for the A–U system. The equilibrium between A and U has been studied [5] in chloroform solution over the temperature range of 4 to 58°C. From the equilibrium constants, the standard-state free-energy change $\Delta G°$ yields $\Delta H°$ and $\Delta S°$ in the usual way. The change of state to which these quantities apply is presumed to be

$$A(u.a.) + U(u.a.) = AU(u.a.)$$

where (u.a.) means unit activity in chloroform solution, assumed to be identical with 1 M. Thus the thermodynamic quantities are those for chloroform-solvated species. In particular, the value of $\Delta H°$ can be taken as a measure of the ΔH of hydrogen bonding between A and U only to the extent that ΔH of solvation of the two species A and U is the same as that for the single species AU. This may be correct to a fraction of a kilocalorie, but no proof is available. When the solvent is water, it surely is not correct and the interpretation of the results of similar equilibrium studies in water in terms of the hydrogen bonding of the base pairs is considerably more complex.

For the system 9-ethyladenine and 1-cyclohexyluracil in chloroform, the temperature dependence of the equilibrium constant gives
$\Delta H°$ of -4.3 ± 0.3, -4.0 ± 0.8, and -6.2 ± 0.6 kcal/mole of U–U, A–A,
and A—U formed, respectively, while $\Delta S°$ is -11 cal/mole-deg within
experimental error for all three dimers [5]. Thus the difference in
$\Delta H°$ between A–U and the other two dimers is not large, about 2 kcal.,
but the effect of this difference on K is quite striking, as mentioned
above.

In view of the close similarity of the cyclic H-bonded structures
that presumably prevail in A–A, U–U, and A–U, as suggested, e.g.,
by the crystal structure of A–U [7], the difference in $\Delta H°$ shows the
sensitivity of this quantity to small differences in geometry and charge
distribution.   To study the latter factor, a number of derivatives of
9-ethyladenine and 1-cyclohexyluracil were examined [6]. A summary
of the results of this study is given in Table I and, in structural fashion,
in Fig. 1, both adapted from Ref. 6.

It can be seen from Table I and Fig. 2 that the association constants
provide a sensitive and operationally meaningful probe of the contribution of charge distribution to the H bonding. For example, the change
from uracil to thymine by replacement of hydrogen with methyl in the
5 position increases K from 100 to 130 (the K-values are believed
reliable within 10 to 20%). This small difference is consistent with the
small electron-donating effect of the methyl group, which should make
the 4-carbonyl oxygen a slightly better acceptor for H bonding. Discussion of such effects on the K values has been given by Kyogoku et al. [6].

The relative constancy of the values of $\Delta S°$ of dimerization for AA,
UU, and AU suggests an approximate way of relating K to $\Delta H°$ for the
adenine and uracil derivatives of Table I. Since these derivatives differ

TABLE I

Association Constants Between Derivatives of Adenine
and Uracil,[*] K in liters/mole, t = 25°C, solvent = $CDCl_3$

	U	5,6-Dihydro-U	Thymine	5-Bromo-U
6-Amino-P (A)	100	30	130	240
2-Amino-P	45	10	55	75
2,6-Diamino-P	170	100	210	550
6-Methylamino -P	50	15	70	100
6-Dimethylamino-P	1.5			
8-Bromo-A	140			

*U = 1-cyclohexyluracil; P = 9-ethylpurine (data from Kyogoku et al. [6]).

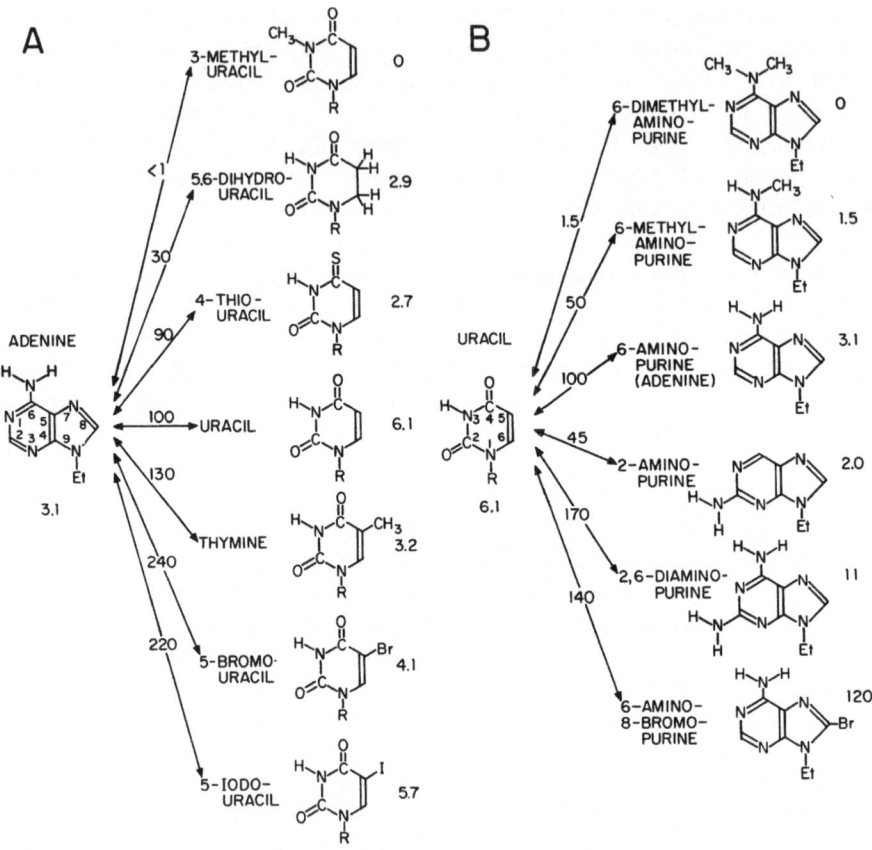

Fig. 2. Association constants for dimers of 9-ethyladenine, 1-cyclohexyluracil, and their derivatives: A, association of 9-ethyladenine with seven uracil derivatives; B, association of 1-cyclohexyluracil with six purine derivatives. Numbers on the arrows are A–U constants, those to the right of the structural formulas are self-association constants. Units are in liters per mole, $t = 25°C$, solvent = $CDCl_3$. (Reproduced from Kyogoku, Lord, and Rich [16].)

little in structure from 9-ethyladenine or 1-cyclohexyluracil, $\Delta S°$ for their dimerization reactions may be set equal to −11.4 cal/mole-deg, the average of $\Delta S°$ for AA, UU, and AU. Then it follows that, for 25°C,

$$\Delta H° = -1.36 \log_{10} K - 3.40$$

This equation gives K of the order of unity for a dimer with a single N—H···O or N—H···N bond ($\Delta H°$ approximates −3.0 to −3.5 kcal), about 10^2 for two such bonds ($\Delta H°$ approximates −5.5 to −6.5 kcal), and from 10^3 to 10^4 for three bonds ($\Delta H°$ approximates −8 to −9 kcal). Since most of the self-association constants are not far from unity, one concludes that, if self-association proceeds with the formation of two bonds, these are considerably weaker than those formed between A and U derivatives, presumably for steric or electronic reasons.

Since quantitative information is thus available about relative concentrations of A, U, and AU in CDCl$_3$ solution, it is of interest to examine the effect of a controlled degree of dimerization on the ultraviolet spectrum of this system. Such a study has recently been carried out by Thomas and Kyogoku [8] to see whether there is any contribution to the phenomenon of ultraviolet hypochromism by base pairing. They found that the A–U system shows considerable hypochromism, the AU dimer having some 10 to 12% smaller absorbance in the 2700-Å region than the sum of the A and U absorbances (Fig. 3). Since the infrared equilibrium studies [5] make clear that the association is dimeric and occurs through hydrogen bonding, the observed hypochromism cannot be due to parallel superposition of the bases (i.e., base stacking). Thomas and Kyogoku therefore conclude that base stacking is not the

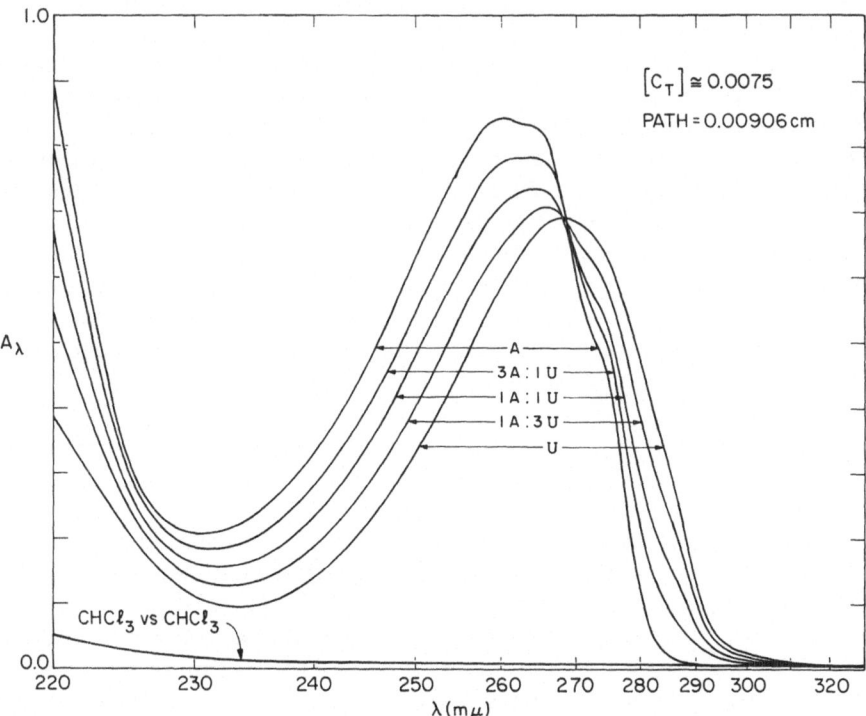

Fig. 3. Ultraviolet absorption spectra of CDCl$_3$ solutions of 9–ethyladenine and 1–cyclohexyluracil. Total solute concentration is 0.0075 M, path length is 0.00906 cm, and solvent baseline is the curve at the bottom of the figure (reproduced from Ref. 8).

only cause of hypochromism. It is not clear, however, that hydrogen-bonded pairing can cause sufficient hypochromism to account for the larger effect (about 40%) seen in the 2600-Å band of native DNA.

In summary, the infrared studies bear out quantitatively the postulate of specific base pairing by Crick and Watson and provide a free-energy basis for evaluating the part played by hydrogen bonding in stabilizing the native DNA helix. However, there are other kinds of forces involved in this stabilization—most importantly, the dispersion forces between the stacked bases and the hydration of polar sites by hydrogen bonding between water and the oxygen atoms of the phosphate and ribose groups. The quantitative evaluation of these forces is also important for an understanding of the secondary structure of DNA and the conditions under which it will unwind.

It is clear that the balance among these various forces is a delicate one. Small changes in the hydration of the DNA backbone, for example, can produce considerable instability of the double helix [9]. The instability is of course a key property of the helix, which can unwind or rewind as the biochemical environment may require. Since water is always the principal molecular constituent of the environment, infrared methods are at some disadvantage compared with other spectroscopic techniques that are less opaque to this solvent. However, the usefulness of vibrational spectra has been demonstrated, e.g., in the foregoing discussion on nonaqueous solutions, and by H. T. Miles and his associates [10] in the regions of lesser infrared opacity of H_2O and D_2O. For this reason, we have undertaken in the Massachusetts Institute of Technology Spectroscopy Laboratory the investigation of equilibria among the derivatives of nucleic acids by means of the Raman effect, in which the limitations imposed by the solvent water are considerably less severe.

RAMAN STUDIES OF NUCLEIC ACID DERIVATIVES IN AQUEOUS SOLUTION

Infrared spectra and Raman spectra give the same kind of molecular information, namely, the quantized energies of molecular vibration and rotation, but the techniques differ radically [11]. In the period immediately following the discovery of the Raman effect in 1928, it was very widely used by chemists because of the simplicity of photographic spectroscopy compared with the infrared procedures of that time. When automatically recording infrared instruments became generally available in the late 1940's, the situation was reversed and the Raman effect was relatively little used. Today, however, technical advances in spectrometers and sources, particularly the advent of the laser, have brought Raman spectroscopy once more into competition with infrared methods. In the work described here, a Cary Model 81 instrument [12] was employed. The source of excitation was the line at 4358 Å from the mercury arc, and the experimental details have been described elsewhere [13].

Summary of Results of Raman Spectral Study of Nucleic Acid Derivatives

An advantage of the Raman effect over infrared absorption results from the fact that interference from the Raman spectrum of water is negligible throughout the range of 200 to 2000 cm^{-1}. Raman spectra of aqueous solutions of moderate concentration (0.01 to 1 M for molecules of molecular weights of the order of 100) can be obtained without difficulty. The chief disadvantages are relatively large sample size (10 to 100 mg of solute) and interference from fluorescing impurities, but the laser source permits reduction of both of these by one or two orders of magnitude.

In Figs. 4 to 7, reproduced from Ref. 13, are shown the Raman spectra of aqueous and crystalline derivatives of uracil, adenine, cytosine, and guanine. From these and related spectra, a number of conclusions can be drawn:

1. Tautomerism. The Raman spectra show clearly that the purine and pyrimidine bases and their nucleoside and nucleotide derivatives exist in the keto-amino forms rather than as the enol-imino tautomers. This conclusion is not new [10a], but it follows readily from the Raman studies alone.

2. Structural effects of pH. Systematic changes in frequency are observed to follow protonation and deprotonation of the pyrimidine and purine ring systems. Cation formation raises all double-bond stretching frequencies, while anion formation lowers them. It is concluded that protonation usually occurs on the ring nitrogens, which localizes the π electrons in specific double bonds, while deprotonation results in delocalization of the π electrons.

3. Characteristics of the Raman Spectra. It is found that the intensities of Raman lines due to NH-deformation vibrations are very weak. This low intensity is presumably due to the slight effect of such motion on the electronic structure of the purine and pyrimidine rings, in contrast to the ring-stretching vibrations, whose Raman lines are strong. Likewise, the ribose residue gives very weak Raman lines,

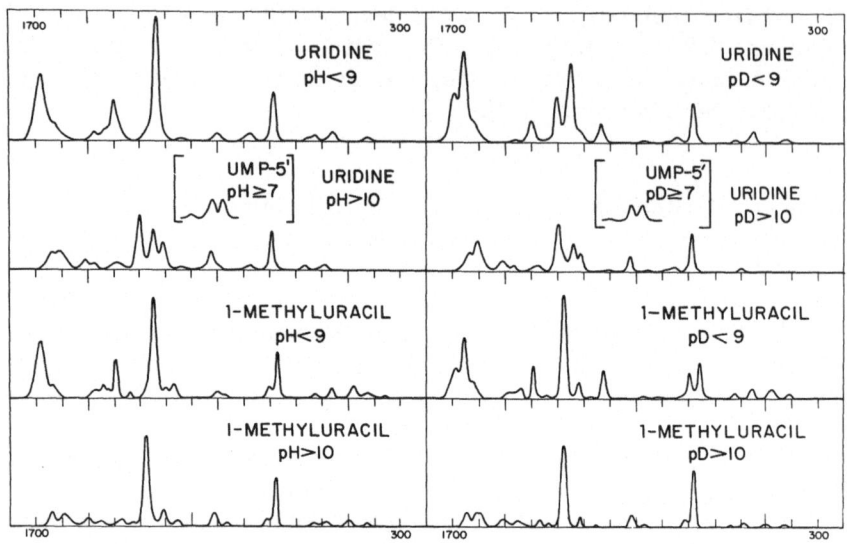

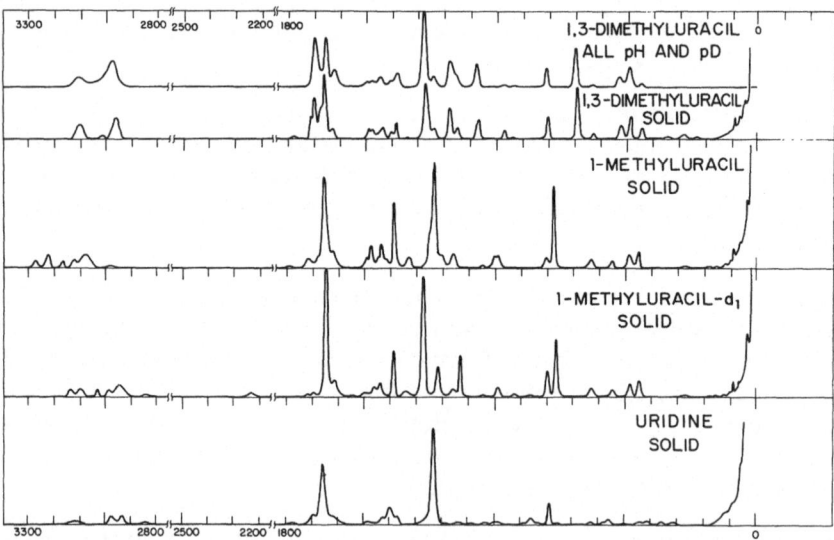

Fig. 4. Raman spectra of some uracil derivatives. The relative Raman intensity (vertical scale) as a function of wavenumber in cm^{-1} has been replotted from the original recordings. (Reproduced from Ref. 13, where experimental conditions and numerical data are given.)

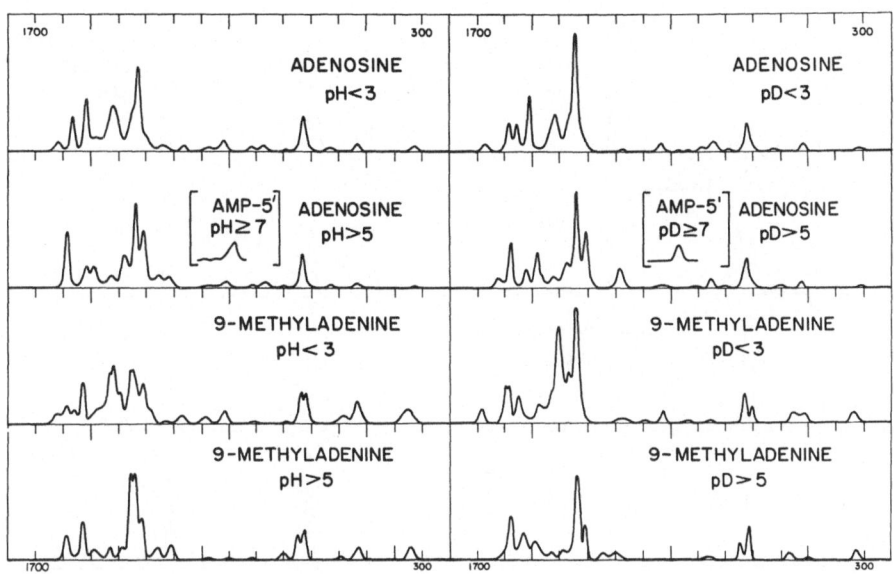

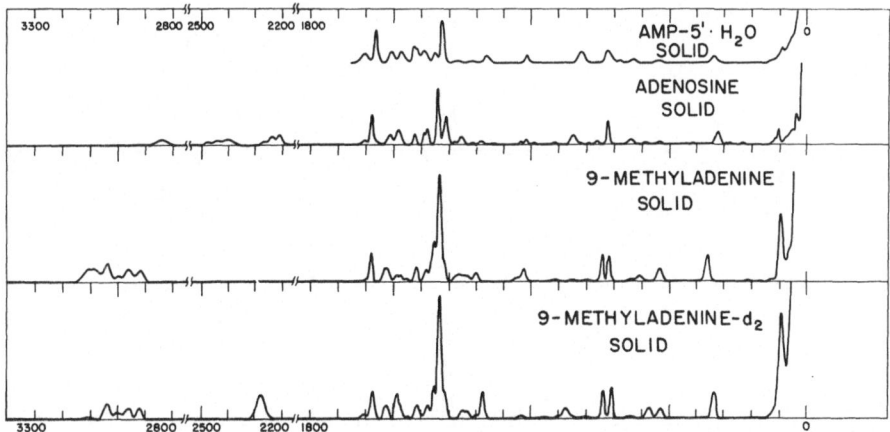

Fig. 5. Raman spectra of some adenine derivatives (reproduced from Ref. 13).

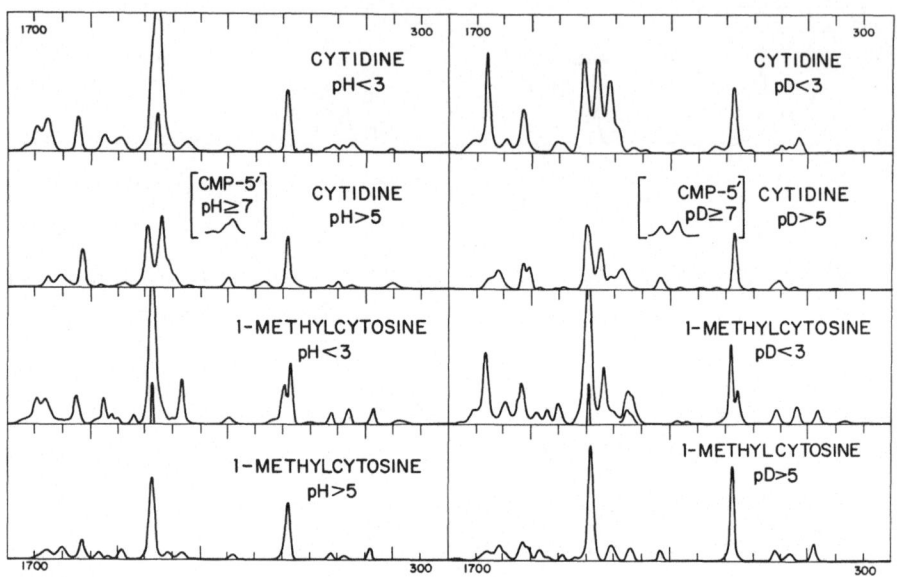

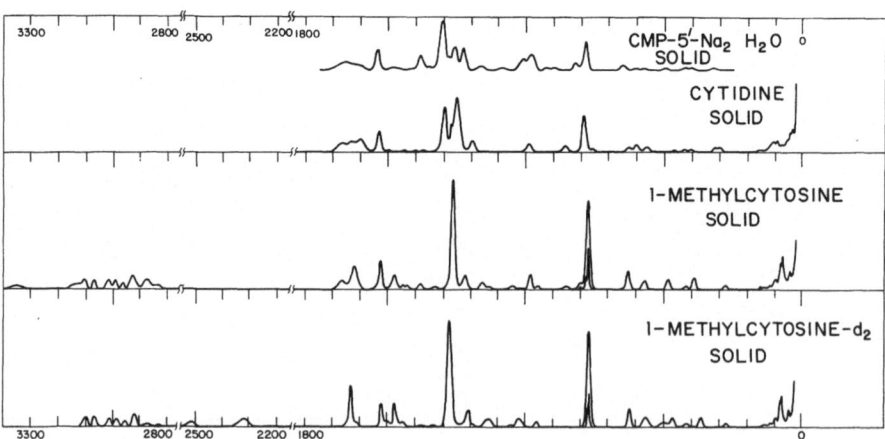

Fig. 6. Raman spectra of some cytosine derivatives (reproduced from Ref. 13).

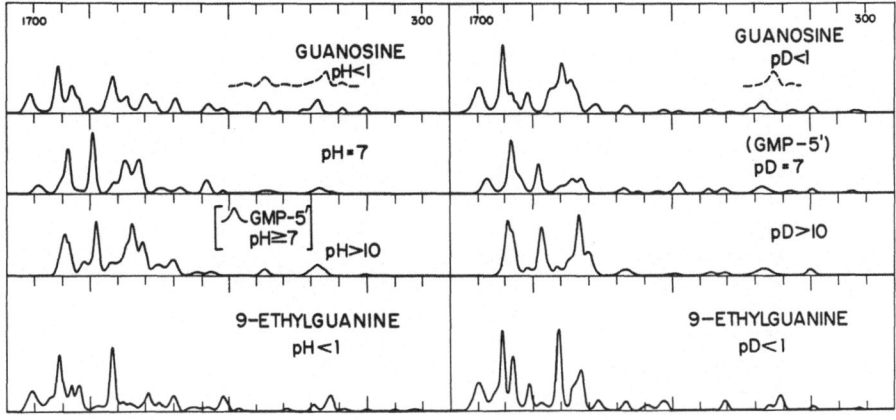

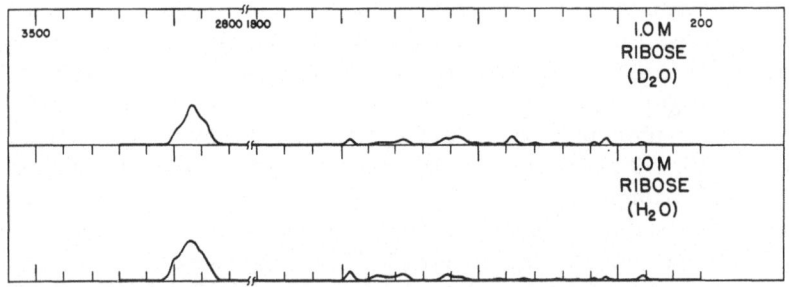

Fig. 7. Raman spectra of some guanine derivatives (reproduced from Ref. 13).

which are an order of magnitude less intense than those of the bases. The carbonyl frequencies are generally intense and are found to be considerably broader than lines arising from the ring-stretching modes, probably as a result of H bonding from the solvent to the carbonyl oxygen atoms. The carbonyl lines are also relatively strongly polarized, whereas the ring modes, which may be regarded as derived from non-totally-symmetrical vibrations of benzene, generally give rise to lines of lesser polarization.

Raman Studies of Association of Nucleotides

As an initial application of the results obtained above, the Raman spectra of a number of aqueous systems of nucleotide derivatives were examined (Table II) [14]. In all cases, it was found that the spectra of

TABLE II

Aqueous Systems of Base Pairs Studied in Raman Effect

Solutes	Solvent	pH or pD	Concentration range, M
AMP + uridine	H_2O	7.5	0.10 — 1.0
AMP + uridine	D_2O	7.5	0.10 — 0.50
AMP + UMP	H_2O	7.5	0.50
Adenosine + uridine	D_2O	2.0	0.50
GMP + cytidine	H_2O	7.5	0.20 — 0.50
GMP + cytidine	D_2O	7.5	0.10 — 1.0
GMP + CMP	D_2O	7.5	0.50
Guanosine + cytidine	D_2O	11.0	0.50

the mixtures were quantitatively equivalent to the spectra computed by adding the spectra of the individual constituents. Figure 8, taken from Ref. 14, shows the calculated and observed Raman spectra of an equimolal mixture (0.25 M each) of adenosine-5'-monophosphate and uridine. Figure 9, also from Ref. 14, gives the same information for guanosine-5'-monophosphate and cytidine. However, the agreement shown in Fig. 9 is obtained only after correction is made for a small amount of absorption of the Raman exciting radiation by guanosine-5'-monophosphate. The correction was not applied to the spectrum of GMP, since the concentration of GMP was the same, both as a single solute and mixed with cytidine. However, the Raman scattering of cytidine was reduced in the mixture compared with that of its solution by the small amount by which the exciting line was absorbed by GMP.

These results indicate that, if any association or other interaction has occurred under the conditions of Table II, the changes in Raman spectra resulting from the association were not detectable. Since dimerization due to hydrogen bonding is known to affect the carbonyl frequencies of the bases in both aqueous [10] and nonaqueous [15] solutions, it seems safe to conclude that hydrogen-bonded dimers are not formed to a detectable extent at the concentrations studied.

Raman Studies of Metal-to-Nucleotide Binding

These studies have been carried out on two systems, cytidine—mercuric chloride and monoalkylphosphate anion—magnesium (II). The former system seemed particularly suitable for study by Raman spectroscopy because the cytidine—mercury complex is sufficiently soluble to give spectra of good quality. Derivatives of adenine, guanine, and

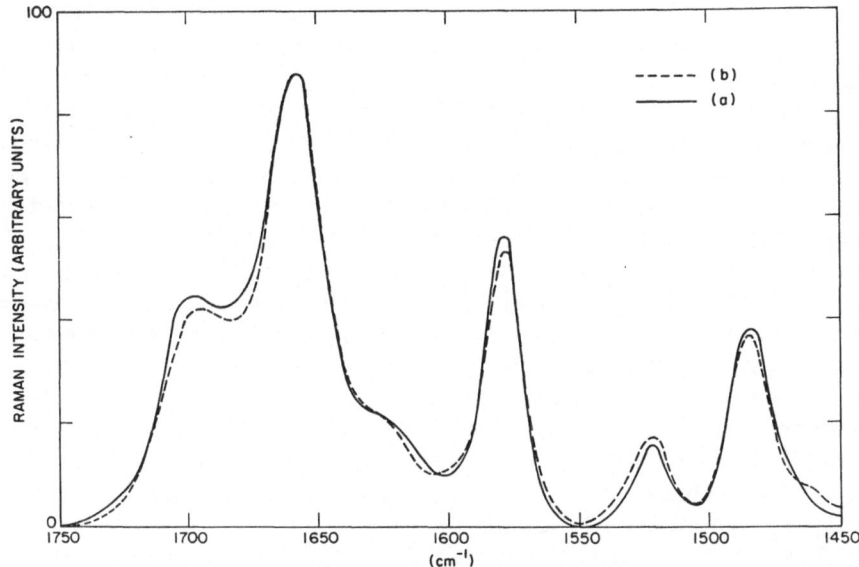

Fig. 8. Raman spectra of AMP and uridine in D_2O: (a) spectrum observed for equimolal mixture (0.25 M each) at pD = 7, t = 35°C, and spectral range as indicated by abscissa; (b) sum of separate spectra of 0.25-M AMP and uridine obtained under the same conditions (reproduced from Ref. 14).

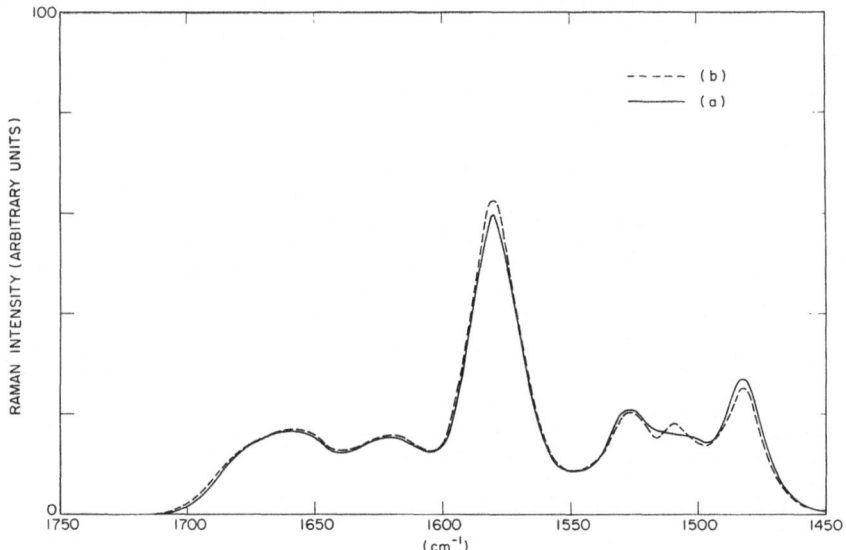

Fig. 9. Raman spectra of GMP and cytidine in D_2O: (a) spectrum observed for equimolal mixture (0.20 M each) at pD = 7.5, t = 35°C; (b) sum of separate spectra of 0.20-M GMP and cytidine, the latter corrected for absorption of exciting line (reproduced from Ref. 14).

inosine form complexes with $HgCl_2$ of lesser solubility [16, 17], and derivatives of uracil show interaction only at considerable excess of $HgCl_2$ [16].

The Raman spectra of cytidine in aqueous solution and of an equimolar mixture of cytidine and $HgCl_2$ are shown in Fig. 10 from Ref. 14. A similar set of curves has been recorded with D_2O as solvent [14]. The various differences between the two curves of Fig. 10 clearly suggest the formation of a complex. The line at 322 cm^{-1} (Fig. 10, curve b) arises from $HgCl_2$ but is much weaker than that observed in a pure solution of $HgCl_2$ at 0.25-M concentration. This decreased intensity, together with the absence of other lines in this frequency region which could be attributed to Hg–Cl stretching vibrations, is consistent with previous suggestions [16, 18] that the species bound to cytidine is Hg^{2+} rather than $HgCl^+$ or $HgCl_2$. This is also in accord with the fact that $Hg(CN)_2$, a far more tightly bound complex than $HgCl_2$, produces no change in the Raman spectrum of cytidine under conditions comparable to those of Fig. 10.

The most pronounced changes in the Raman spectrum of cytidine occur in the regions near 1550 to 1800 and 1200 to 1300 cm^{-1}, i.e., in the regions of the double-bond and single-bond stretching frequencies, as discussed earlier (cf. Fig. 6). The observed effects are best understood [14] by postulating that the binding of Hg^{2+} occurs at the 3-N ring position. Attachment at this point would be expected to have the

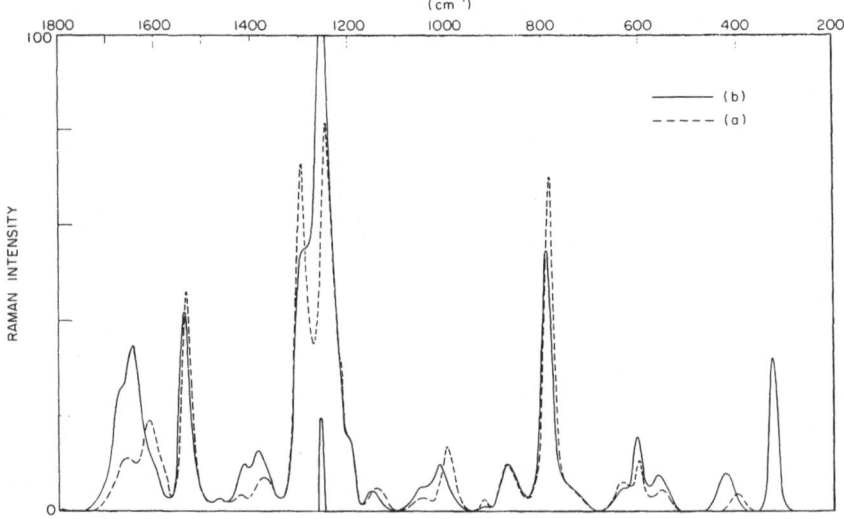

Fig. 10. Raman spectra of aqueous solutions of (a) cytidine, 0.25 M, and (b) cytidine, 0.25 M, plus $HgCl_2$, 0.25 M; pH = 5.5, t = 35°C (reproduced from Ref. 14).

effect of "fixing" the double bonds by reducing the amount of delocalization of the π electrons, with consequent increase in the double-bond frequencies. In fact, attachment of Hg^{2+} at the 3-N position produces much the same change in the Raman spectrum as protonation of cytidine, as comparison of Fig. 10 with the spectra of cytidine at low pH (Fig. 6) will show.

In addition to the cytidine—$HgCl_2$ system, the aqueous Raman spectra of cytidine and uridine in the presence of $ZnCl_2$, $ZnSO_4$, $CdCl_2$, $CuSO_4$, and $MgCl_2$ were examined and found to show no change. There was some evidence of interaction between cytidine and $AgNO_3$, which could not be taken too seriously, however, because of the possibility of photochemical action of blue exciting radiation on solutions containing silver ion. In view of the recent interest in polynucleotide—silver ion interaction [19, 20], the study of such systems by means of Raman spectra excited with the He—Ne laser is promising.

Monoalkylphosphate Anion Interaction with Mg (II)

It is well known that divalent metal cations, including Mg^{++}, Ca^{++}, and Mn^{++}, are effective in stabilizing the double helical structure of nucleic acids and polynucleotides to which they are strongly bound [21]. This stabilization has been attributed to reduced electrostatic repulsion between phosphate groups of the helix backbone when excess cations are available to shield them from one another [22]. It is therefore expected that the divalent cation Mg^{++} will be more strongly bound to phosphate anions than Na^+. This is indicated in the Raman spectrum of sodium monomethylphosphate (NaMMP)*, Fig. 11, when excess Mg^{++} ions are present. Thus the lines at 985 and 1057 cm^{-1} for the salt Na_2MMP, assigned to the symmetric $PO_3^=$ and C—O stretching vibrations, respectively [15], undergo shifts to 990 and 1050 cm^{-1}. There is also a significant broadening of the former line.

In the case of dimethylphosphate anion (DMP-), Raman spectra of Na^+, Mg^{++}, and Ba^{++} salts are indistinguishable.† These results are consistent with the expected weak effect of electrostatic binding interactions on the Raman spectrum. Apparently, only in the case of

---

*Purchased from International Chemical and Nuclear Corp. The phosphate group here ($ROPO_3^=$, where R = CH_3) is similar to that of mononucleotides, where R = sugar-phosphate substituent.

†Ba(DMP)$_2$ was isolated from a mixture of mono- and dimethylphosphate esters (City Chemical Co., N.Y.) on the basis of its greater solubility. The salts Mg(DMP)$_2$ and NaDMP were prepared by addition of the appropriate sulfate to Ba(DMP)$_2$ solutions. The phosphate group here [$(RO)_2 PO_2^-$, where R = CH_3] is similar to that of polynucleotides.

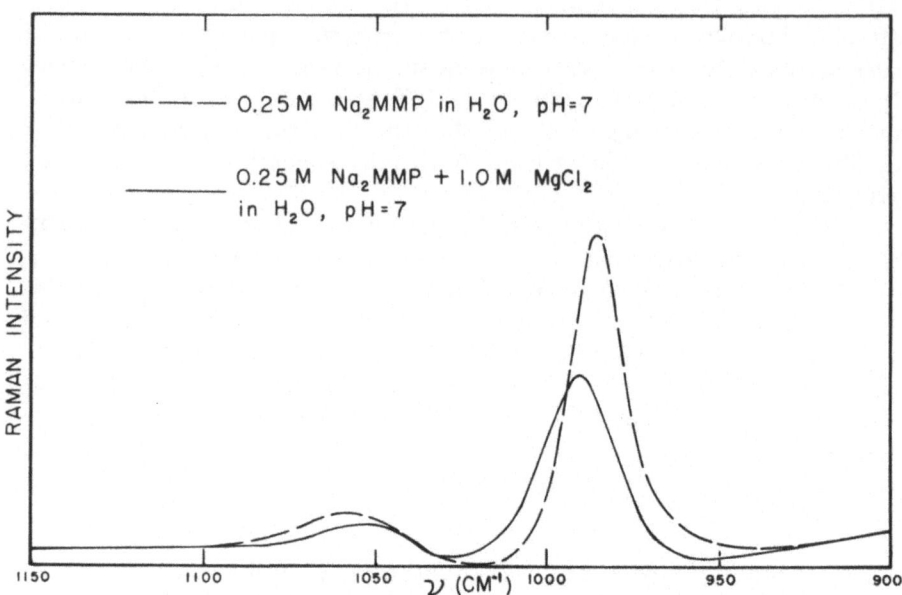

Fig. 11. Raman spectra of monomethylphosphate anion near 1000 cm^{-1} in aqueous solution; $t = 35°C$.

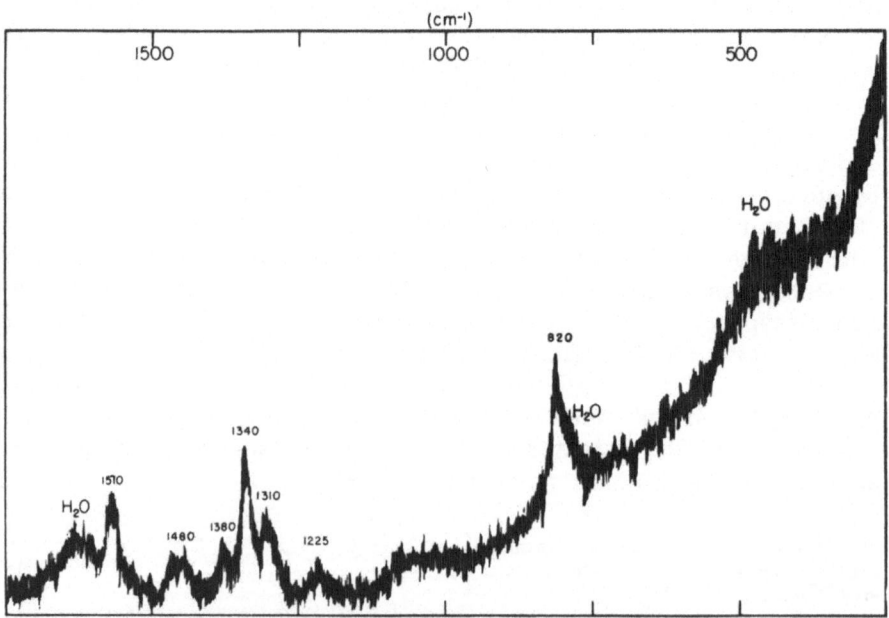

Fig. 12. Raman spectrum of polyriboádenylic acid: original spectrum, Hg-4358 excitation; spectral slit width, about 10 cm^{-1}; sample volume, 0.5 ml; solute weight, 23 mg; KCl = 0.2 M; pH = 6.1; $t = 35°C$.

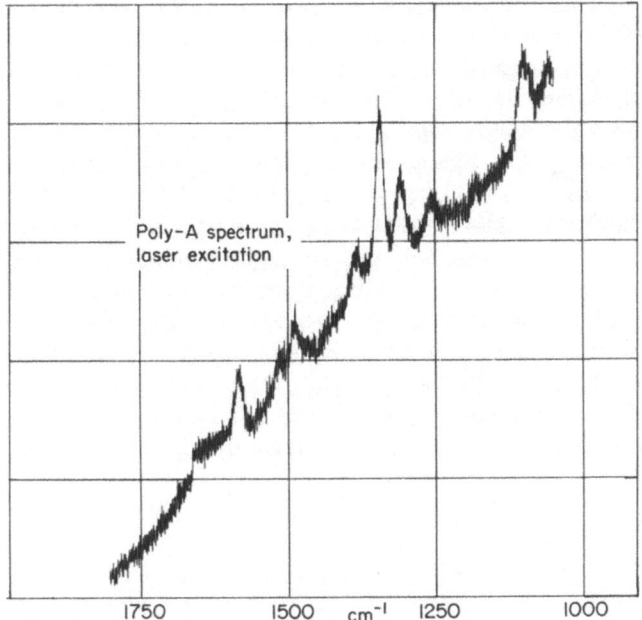

Fig. 13. Partial Raman spectrum of polyriboadenylic acid: original spectrum, Ne-6328 laser excitation; spectral slit width, about 10 cm^{-1}; sample volume, about 0.02 ml; solute weight, about 1 mg; KCl = 0.1 M; pH = 8; t = 25°C; sample, the same as in Fig. 11.

multiply charged cations and anions (Mg^{++} and MMP$^=$) can the effect of binding be detected.

A Preliminary Study of the Raman Spectrum of Polyriboadenylic Acid

Several attempts to obtain Raman spectra of aqueous DNA, RNA, and synthetic polynucleotide solutions with excitation by Hg-4358 gave unsatisfactory results. Usually the problems of sample impurity (absorption and fluorescence) and optical inhomogeneity could not be controlled at the high solute concentrations required. An exception was polyriboadenylic acid (Miles Laboratories, Inc., Control No. 12632, "poly-A") which gave nevertheless a relatively poor Raman spectrum, shown in Fig. 12 with the experimental conditions of observation. Under these conditions Poly-A is considered to have a single-stranded helical conformation [19]. The prominent Raman lines can be attributed to vibrations of the adenine ring and phosphate groups (about 820 cm^{-1}).

The problems associated with obtaining spectra of impure polymers should be largely solvable by the employment of laser sources.

As a first step in this direction, we have recently examined the Raman spectrum of Poly-A with the Cary Model 81 spectrometer modified [12] to employ a Spectraphysics He–Ne gas laser (Model 125, 85 mW). The spectrum obtained in the range of 1000 to 1750 cm^{-1} is shown in Fig. 13, where the conditions of the spectra are given. The conditions of the spectra of Figs. 12 and 13 are comparable except for the sample size, which was some 20 to 30 times smaller for excitation with the laser than with Hg-4358. The details of the spectra of Figs. 12 and 13 are quite similar except for the strongly sloping background of the latter and a few small but significant differences in the positions and intensities of the Raman lines. We are showing Figs. 12 and 13 as examples of preliminary work in Raman spectroscopy of aqueous solutions of high-molecular-weight materials; discussion of the spectra is deferred until further study is made. It is apparent, however, that laser excitation is very promising, particularly for small sample size.

ACKNOWLEDGMENTS

We are greatly indebted to our associates Prof. Alexander Rich and Dr. Yoshimasa Kyogoku for their collaboration in a large part of the work reviewed here and for much wise counsel to the authors. Dr. Kyogoku also obtained the spectrum shown in Fig. 12. One of us* is grateful to the National Institute of General Medical Sciences, U.S. Public Health Services, for the grant of a fellowship under the tenure of which he completed the Ph. D. thesis on which part of this paper is based. Acknowledgment is also made to the National Science Foundation for support by Grant GP-4923.

REFERENCES

1. R.M. Hamlin, Jr., R.C. Lord, and A.Rich, Science 148:1734 (1965).
2. J.Pitha, R.N. Jones, and P.Pithova, Can. J. Chem. 44:1044 (1966).
3. Y. Kyogoku, R.C. Lord, and A. Rich, Science 154:518 (1966).
4. E.Kuechler and J.Derkosch, Z. Naturforsch. 21b, 209 (1966).
5. Y. Kyogoku, R.C. Lord, and A.Rich, J. Am. Chem. Soc. 89:496 (1967).
6. Y. Kyogoku, R.C. Lord, and A. Rich, Proc. Natl. Acad. Sci. U.S. 57:250 (1967).
7. F.S. Mathews and A. Rich, J. Mol. Biol. 8:89 (1964).
8. G.J. Thomas, Jr., and Y. Kyogoku, J. Am. Chem. Soc., 89:4170 (1967).

---

*G.J. Thomas, Jr.

9. M. Falk, K. A. Hartman, Jr., and R. C. Lord, J. Am. Chem. Soc. 85:387, 391 (1963).
10. a. H. T. Miles, Proc. Natl. Acad. U.S. 47:791 (1961).
    b. F. B. Howard, J. Frazier, M. F. Singer, and H. T. Miles, J. Mol. Biol., 16:415 (1966) and references cited there.
11. G. R. Harrison, R. C. Lord, and J. R. Loofbourow, Practical Spectroscopy, Prentice-Hall, Inc., Englewood Cliffs, N. J. (1948), Chaps. 17 and 18.
12. R. C. Hawes, K. P. George, D. C. Nelson, and R. Beckwith, Anal. Chem. 38:1842 (1966); also various bulletins of Cary Instruments, Inc., and Applied Physics Corp.
13. R. C. Lord and G. J. Thomas, Jr., Spectrochim. Acta, 23A:2551 (1967).
14. R. C. Lord and G. J. Thomas, Jr., Arch. Biochim. Biophys., 142:1 (1967).
15. T. Shimanouchi, M. Tsuboi, and Y. Kyogoku, Advances in Chemical Physics, Vol. VII, Chap. 12, Interscience Publs. (now John Wiley and Sons, Inc.), New York (1964).
16. G. L. Eichhorn and P. Clark, J. Am. Chem. Soc. 85:4020 (1963).
17. R. B. Simpson, J. Am. Chem. Soc. 86:2059 (1964).
18. T. Yamane and N. Davidson, J. Am. Chem. Soc. 83:2599 (1961).
19. R. H. Jensen and N. Davidson, Biopolymers 4:17 (1966).
20. M. Duane, C. A. Dekker, and H. K. Schachman, Biopolymers 4:51 (1966).
21. R. F. Steiner and R. F. Beers, Polynucleotides, Elsevier Publishing Co., New York (1961), Chap. 10.
22. C. Schildkraut and S. Lifson, Biopolymers 3:195 (1965).

Coordination Properties of Magnesium in Chlorophyll from IR and NMR Spectra*

Joseph J. Katz

Argonne National Laboratory
Argonne, Illinois

The infrared (IR) and nuclear magnetic resonance (NMR) spectra of chlorophyll are best interpreted on the basis that the central magnesium atom of the chlorophyll always has a coordination number larger than 4. In electron-donor solvents, chlorophyll exists as monomer, with solvent molecules filling the axial position(s). In nonpolar solvents, the coordination unsaturation of the magnesium is relieved by coordination of the ketone oxygen function of one molecule of chlorophyll with the magnesium of another. This leads to dimer or higher aggregate formation in nonpolar media. Chlorophyll-ligand interactions can be studied by observing ring-current shielding effects on the chemical shifts of protons in ligands coordinated to the magnesium atom of chlorophyll. Chlorophyll is shown by this procedure to coordinate ethanol and lutein but not β carotene. With plant sulfolipid, chlorophyll probably dissolves in sulfolipid micelles present in CCl_4 solution as a result of phytyl long-chain fatty acid interactions rather than as a primary result of coordination interactions involving magnesium. Deuterated chlorophyll is shown to be very useful for these NMR studies.

INTRODUCTION

The structural formula of chlorophyll as usually written, Fig. 1, shows the central magnesium atom with a coordination number of 4. Recent investigations by IR [1–3] and NMR [4, 5] spectroscopy have established that magnesium is coordinatively unsaturated when only four ligands are bound. The data also suggests that the coordination number of magnesium in chlorophyll in solution is always greater than 4 and that at least one of the axial positions of the magnesium must be filled. Absorption spectroscopy in the visible region [6–9], while subject to serious ambiguities in interpretation, provides additional support for this view.

*Based on work performed under the auspices of the U.S. Atomic Energy Commission.

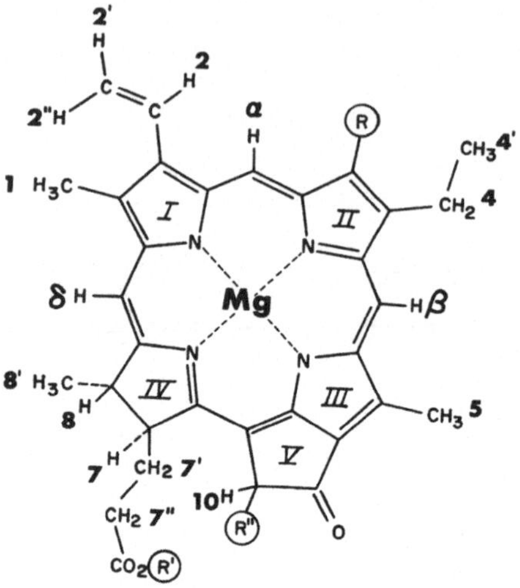

Fig. 1. Structural formula and proton numbering of chlorophyll.

	R	R'	R*	Mg
Chlorophyll a	$-CH_3$	Phytyl	$-CO_2CH_3$	x
Chlorophyll b	$-CHO$	Phytyl	$-CO_2CH_3$	x
Pheophytin a	$-CH_3$	Phytyl	$-CO_2CH_3$	-

Magnesium generally is considered to prefer a coordination number of 4 and ordinarily assumes a tetrahedral configuration in its coordination compounds. However, many compounds of magnesium halides with alcohols, ethers, ketones, esters, and amides [10] are known in which the magnesium assumes an octahedral configuration and a coordination number of 6. Magnesium phthalocyanin coordinates with water strongly [11], and magnesium-containing porphyrins form stable pyridine complexes, similar to hemichromes, in which the magnesium has a coordination number of six [12]. Chlorophyll itself has long been known to coordinate water tenaciously [13]. Strong evidence for interaction of chlorophyll with various bases was provided by Livingston [14], who showed that chlorophyll forms stable solvates with the Lewis bases water, alcohol, amines, ketones, and ethers; chlorophyll is strongly fluorescent when solvated (in solution in polar solvents) but nonfluorescent in nonpolar media free from traces of base. Originally, the postulate was made that solvation occurred by way of hydrogen bonding of the Lewis base to the ketone oxygen of ring V, but Evstigneev et al. [15] observed that the fluorescence of magnesium-free pheophytins, unlike the magnesium-containing chlorophylls, is not solvent dependent. These earlier studies thus suggested that, in some unspecified way, the coordination properties of magnesium

were implicated in important aspects of chlorophyll behavior. However, a more exact delineation of the situation and a more precise
description of the role of magnesium in chlorophyll came only from
the application of IR and NMR spectroscopy. These tools, in fact,
proved very well suited to the task. Each of these spectroscopic
techniques focuses on important but different aspects of chlorophyll
interactions. Infrared spectra provided detailed information about the
coordination behavior of the oxygen functional groups of the molecule,
whereas NMR provided direct information about ligand–magnesium
interactions. The two modes of study are thus complementary, and
the results described here illustrate benefits that accrue from a joint
application of both spectroscopic techniques. Katz, Dougherty, and
Boucher [16] have recently summarized the IR and NMR spectroscopy
of chlorophyll and describe a number of chlorophyll structural, bio-
synthetic, and physical problems to which these spectroscopic methods
are particularly adaptable.

Many of the properties of chlorophyll solutions can now be interpreted in terms of the coordination behavior of magnesium. For
example, the solvent dependence of the chlorophyll IR and NMR spectra
receive a reasonable interpretation on the premise that the coordination
number of magnesium in chlorophyll is always greater than 4. In basic
or polar (electron-donor) solvents (tetrahydrofuran, acetone, pyridine,
ether, and the like), the coordination unsaturation of the magnesium is
satisfied by coordination of solvent. One or two solvent molecules are
bound in axial magnesium positions, and, in such solutions, chlorophyll
exists predominately in monomeric form. In nonpolar solvents free
of base (e.g., carbon tetrachloride, alcohol-free chloroform, benzene,
tetrachlorethylene) the coordination unsaturation of the magnesium is
assuaged by coordination of the magnesium of one chlorophyll molecule
with an oxygen function of another chlorophyll molecule. In this kind of
coordination aggregation, it appears that it is the ketone carbonyl
oxygen of ring V that is generally implicated. In nonpolar solvents,
then, chlorophyll undergoes self-aggregation and occurs at least as
dimers [1]. It is monomeric chlorophyll that is fluorescent; chlorophyll
self-aggregated by coordination interaction is only very weakly fluorescent. Differences in the visible absorption and fluorescence spectra
between polar and nonpolar solutions of chlorophyll are accounted for
in qualitative terms on this basis. The state of aggregation of the chlorophyll is the important factor and not the solvent effects on the
relative positions of the $n-\pi$ and $\pi-\pi$ levels [4,9], as had been previously postulated [17].

The state of aggregation of chlorophyll has become a central issue in the study of the photochemical stage of photosynthesis [18]. Recent investigations [19], based mostly on absorption and fluorescence spectroscopy in the visible range provide good reason to suppose that chlorophyll *in vivo* exists in many forms and that these forms may represent different aggregation states of chlorophyll or coordination interactions of chlorophyll with various (biologically important) ligands. Studies made in defined solutions can only be extrapolated to the living plant with caution. Nevertheless, IR and NMR studies on chlorophyll can be expected to make a valuable contribution to the elucidation of these very important problems.

INFRARED SPECTRA

In nonpolar solvents, chlorophyll a shows four absorption bands in the carbonyl (1750 to 1600 cm^{-1}) region, whereas, in polar solvents,

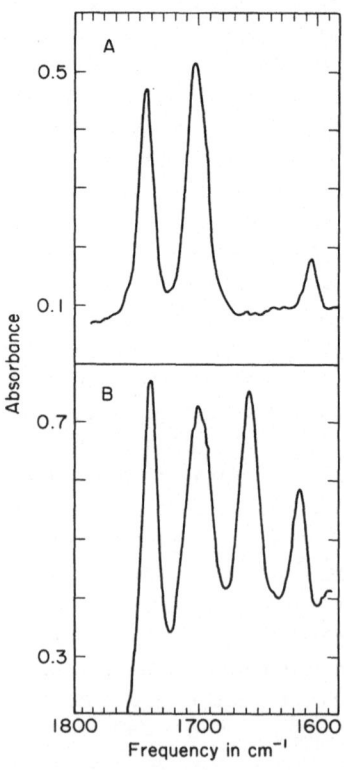

Fig. 2.   Infrared spectra of chlorophyll a in the carbonyl region in a polar and nonpolar solvent: A in tetrahydrofuran, B in CCl .

only three peaks are present, Fig. 2. From the structural formula of chlorophyll, four peaks in this region are not too surprising: the ketone $C=O$ in ring V, the two ester carbonyl groups, and the skeletal $C=C$ and $C=N$ groups could be responsible for four peaks. The three-peaked spectrum in tetrahydrofuran, however, cannot be accounted for in these terms. The situation was made additionally confusing by the report that chloroform solutions show a three-banded spectrum, whereas CCl_4 solutions are four-banded [20]. It was shown subsequently [1] that this difference arises from the small amount of alcohol usually added to chloroform to stabilize it. When the alcohol is removed from the chloroform, the spectra in $CHCl_3$ and CCl_4 are very similar in the carbonyl region. It also became apparent that the differences in the spectra taken in polar and nonpolar media must be due to the presence or absence of electron donors in the chlorophyll solution. An interpretation of the considerable body of IR data consistent with all of the observations could then be made along the following lines: the first (high-frequency) strong band at 1743 to 1732 cm^{-1} is assigned to the $C=O$ absorptions of the two ester groups at C-7 and C-10. These two ester absorption peaks generally overlap, but, in hydrogen-bonding solvents such as methanol or ethanol, definite splitting of the ester peak occurs, which verifies the composite nature of this absorption peak. The second strong peak at 1708 to 1701 cm^{-1} can be assigned to the stretching mode of the ketone carbonyl group. The relatively weak band at 1619 to 1616 cm^{-1} is best assigned to a general skeletal vibration in which the $C=C$ and $C=N$ vibrations are coupled to each other to a greater or lesser extent. This scheme, then, accounts reasonably well for the IR spectra of chlorophyll a in the carbonyl region in polar solvents or in nonpolar solvents containing bases.

In nonpolar solvents, the spectra show, in addition to the three absorption maximums described above, a strong band in the vicinity of 1650 cm^{-1}. Since no other electron donors are available in chlorophyll solutions in nonpolar media, the coordination unsaturation of the magnesium is relieved by coordination with the ketone oxygen of another chlorophyll molecule. The ketone vibration will then be split and partially shifted to lower frequency. The peak at about 1650 cm^{-1} is thus assigned to a vibration of $C=O$ coordinated with the magnesium of another chlorophyll molecule. Such coordination leads to intermolecular aggregation, and the carbonyl absorption peak that results may be referred to as an "aggregation peak." Probably for steric reasons, aggregation of chlorophyll a does not appear to proceed much beyond the dimer stage in $CHCl_3$ solution; as a consequence, about half the carbonyl groups will be coordinated and absorb at about 1650 cm^{-1}, and the other half will show normal carbonyl absorption near 1700 cm^{-1}. That magnesium is involved in the solvent dependence of the IR spectra in this region is established by the disappearance of the solvent dependence of the IR in magnesium-free derivatives of chlorophyll. Thus, pheophytin a (chlorophyll from which the magnesium has been removed) shows a three-banded spectrum in the carbonyl region in either polar or nonpolar solvents, and this provides convincing support for the thesis that self-coordination with magnesium is the important factor in the generation of the carbonyl-region spectra in nonpolar solvents.

Far IR spectra of chlorophyll are consistent with this interpretation [3]. The solution spectra of chlorophyll in the 340 to 160 cm^{-1} region are observed to be solvent dependent; as in the case of the carbonyl-region spectra, magnesium-free chlorophyll derivatives fail to show a similar solvent dependence. In benzene or cyclohexane solution as well as in the solid state, chlorophyll a shows an absorption maximum at 311 to 303 cm^{-1} that can be attributed to intermolecular coordination involving magnesium and oxygen and which leads to the

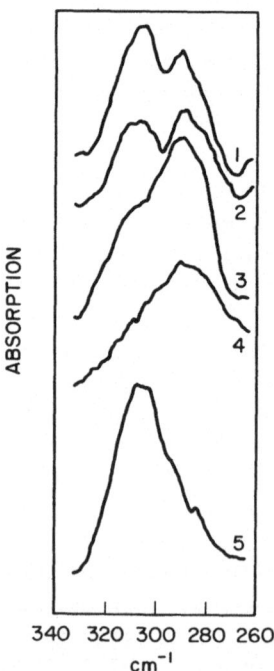

Fig. 3.   Fair infrared spectra of chlorophyll a in (1) cyclohexane, (2) Nujol mull, (3) 10% (v/v) pyridine–cyclohexane; (4) 1% (v/v) methanol–cyclohexane; (5) benzene.

formation of molecular aggregates. Strong absorption bands at 292 to 296 and 195 to 196 cm^{-1} observed in the spectra of the monomeric chlorophylls can be attributed to magnesium–nitrogen vibrational modes. The peak at 312 to 306 cm^{-1} is solvent dependent in the same way as the ketone $C = O$ stretching vibration is solvent dependent, and it is illustrated in Fig. 3.

In our experience, it is desirable to make use of solutions for chlorophyll IR studies. Mulls generally yield poorly resolved spectra that offer more than ordinary difficulties in interpretation. Obviously, working in solution makes it possible to observe the effects of the solvent, which turns out to be crucial in the interpretation of the spectra. Experimental techniques are described in [16].

NUCLEAR MAGNETIC RESONANCE

High-resolution NMR spectroscopy has demonstrated its utility in many ways for the study of the molecular structure and function of complicated organic molecules. Applied to chlorophyll, it has not only yielded new information on the self-association of chlorophyll and the structure of the chlorophyll aggregates, but it is also beginning to yield new understanding of chlorophyll-ligand interactions. The sensitivity of the Varian HA-100 spectrometer, especially when used with time-average techniques, makes it practical to secure useful data on chlorophyll solutions as dilute as 0.001 M. All of the NMR spectra shown here are taken at 100 MHz, and chemical shifts δ are given in units of parts per million (ppm) downfield from hexamethylsiloxane (HMS). Details of the special problems and techniques involved in working with the labile and reactive chlorophylls are given in [16].

The complete spectral assignment for the NMR of chlorophyll was largely the work of Prof. G. L. Closs. Dr. R. C. Dougherty has since then extended the assignments to most of the other chlorophylls [16, 21]. Figure 4 shows the NMR spectrum of chlorophyll a in tetrahydrofuran-d_8 solution, with assignments of the important proton resonances as indicated.

An unusual feature of chlorophyll NMR spectra is the extreme range of chemical shifts exhibited by its protons. The methine-proton resonances are downfield from HMS about 10 ppm, whereas the internal protons in pheophytin (chlorophyll in which the magnesium is replaced by two protons) are about 5 ppm to higher field than HMS. The spectra thus extend over a range of about 15 ppm. The internal chemical shifts of the porphyrins and chlorins are thus among the largest observed for any class of compounds.

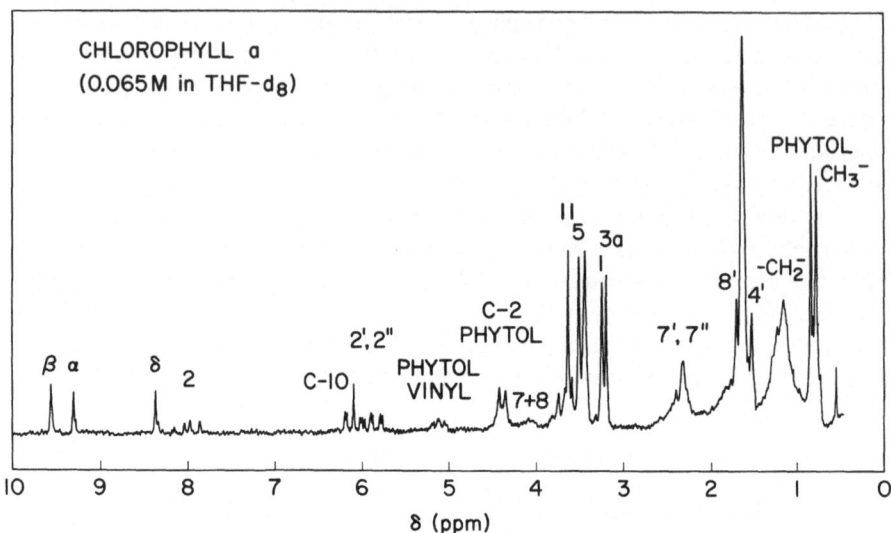

Fig. 4. Chlorophyll a NMR assignments. Chemical shifts are relative to hexamethylsiloxane (HMS) in δ, ppm.

The unusual range of the chlorophyll spectra is a consequence of
the large ring current associated with the chlorin macrocycle [22].
The highly conjugated π-electron system of the 16-atom macrocycle
exerts a strong deshielding effect on protons in the plane of the macrocycle and a strong shielding effect on protons sited above or below the
plane of the macrocycle. The methine protons, α, β, and δ, in Fig. 1,
are in the plane of the macrocycle and thus are in resonance at unusually low fields. The methyl groups at positions 1, 3, and 5, are
attached directly to the conjugated system and thus appear at rather
lower field than usual. The methyl group at position 12 likewise is at
low field because of its proximity to an ester carbonyl group. The
methyl groups at positions 4' and 8' are essentially aliphatic in nature
and appear as expected at higher field. Aside from the range of
chemical shifts, it is probably correct to say that there are few unusual features in the NMR spectrum. Indeed, the chlorophyll NMR
spectrum would seem to have little to interest the professional student
of NMR spectra, for only the hydrogen atoms in the vinyl group, the
hydrogen atoms at positions 8 and 8', the hydrogen atoms in the pro-
pionic acid side chain, and the hydrogen atoms in the ethyl group at
position 4 show spin-spin interactions. The spin-spin interactions
that are present are amenable, for the most part, to straightforward
first-order analysis. As expected, the high-field resonances associated

with the long-chain aliphatic phytyl chain at position 12 show little structure even at 100 MHz.

The ring-current effect, so prominent in the chlorophyll NMR spectra, has some other interesting consequences. A proton sited above the macrocycle ring is shifted strongly to higher field. An organic base (alcohol, amine, ether, ketone, and the like) coordinated to the central magnesium atom will have those of its protons situated close to the coordination center shifted to higher field. For example, methanol coordinated with the magnesium of chlorophyll a has a methyl resonance some 2 ppm higher than the CH_3 group in uncomplexed methanol. In ethanol, the methylene protons will be shifted to higher field to a greater extent than the protons in the methyl group because the methylene protons, for reasons of geometry, are subject to a larger ring-current effect than are the more remote methyl protons. This fortunate circumstance provides a powerful tool for the study of coordination phenomenon involving the central magnesium atom of chlorophyll. The utility of the ring-current effect for this purpose was recognized by Closs and co-workers [4], who observed the coordination of methanol with chlorophyll a, and by Storm and Corwin [5], who studied the coordination of pyridine with magnesium porphyrins by NMR.

One other circumstance considerably facilitates the use of NMR for studying chlorophyll-ligand interactions by NMR, and that is the availability of chlorophyll and other biologically important substances such as carotenoids, lipids, and other chloroplast components in fully deuterated form. It is now possible to grow many different kinds of algae autotrophically in 99.7% D_2O, which results in organisms that contain essentially no ordinary hydrogen. These serve as an excellent source of fully deuterated chlorophyll, carotenoids, lipids, and the like [23]. Where change in the NMR spectrum of chlorophyll or a ligand is being observed, it is a considerable convenience to have one or the other component of the system in fully deuterated form. In the studies described here, ligand chemical shifts resulting from coordination with magnesium are observed with fully deuterated chlorophyll providing a "transparent" partner. Alternatively, changes in the NMR spectrum of ordinary hydrogen-containing chlorophyll are observed in the presence of a fully deuterated ligand, which, in the case discussed here, is a fully deuterated sulfolipid.

As in the case of the IR spectra, the NMR spectra are solvent dependent and for much the same reasons, Fig. 5. In tetrahydrofuran solution, chlorophyll a possesses the well-defined NMR spectrum shown in Fig. 1. In nonpolar media, the situation is quite different. For

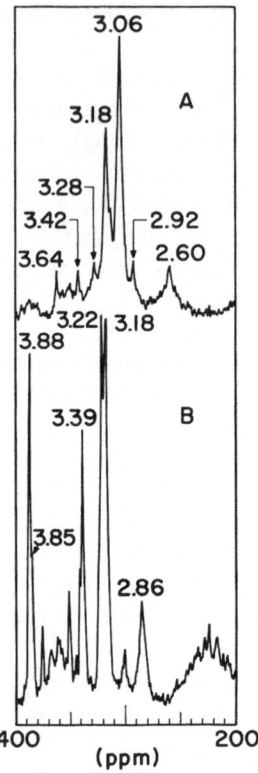

Fig. 5. Low-field methyl region of chlorophyll a in A, dry CCl_4 and B, CCl_4-CD_3OD, showing the effects of base (15 moles of CD_3OD per mole of chlorophyll a) in the NMR spectrum. Chlorophyll a concentration is 0.06 M.

example, in CCl_4 solution, the low-field methyl region is shifted, and the methyl groups at positions 1, 3, 5, and 11, and particularly those at positions 5 and 11, are upfield. The methyl-group resonances overlap, and, as can be seen from the top curve of Fig. 5, the various methyl-group resonances required by the structural formula can scarcely be distinguished. This behavior can readily be acounted for by molecular aggregation of the chlorophyll. If aggregates form in CCl_4, then the protons of one chlorophyll molecule will be brought into close proximity to the ring current of another chlorophyll molecule. These protons will then experience an upfield shift. That all the protons do not experience an equal upfield shift allows the deduction to be made that, in the aggregates, the chlorophyll molecules are not symmetrically situated with respect to each other. In fact, in the aggregates, it is mainly in the region of ring V that overlap occurs, as would be expected from the participation of the ketone carbonyl in ring V in aggregate formation. When a base capable of successful competition for the magnesium coordination positions is added to a solution of chlorophyll in a nonpolar solvent, the aggregates are gradually destroyed as the amounts of base are increased. The lower curve of Fig. 5 shows the methyl region of chlorophyll in a CCl_4 solution containing a ratio of 15 methanol molecules per chlorophyll molecule. All of the methyl groups that should be seen are now visible and are clearly differentiated. The resonances of the methyl groups at positions 5 and 11 have moved to lower field, which confirms the importance of the region in the vicinity of ring V for aggregate formation. A detailed analysis of such titration experiments permits the construction of an aggregation map, Fig. 6, that clearly defines the geometry of the aggregate [1, 16]. Experiments of this kind thus provide some rather detailed information about the self-aggregation of chlorophyll.

Figure 7 shows the result of an experiment in which the chemical shifts of the protons in ethanol are observed in the presence of deuteriochlorophyll, as a function of the chlorophyll/methanol ratio. Since the methylene protons are closer to the ring, they experience stronger upfield shifts than do the methyl protons. The upfield shift

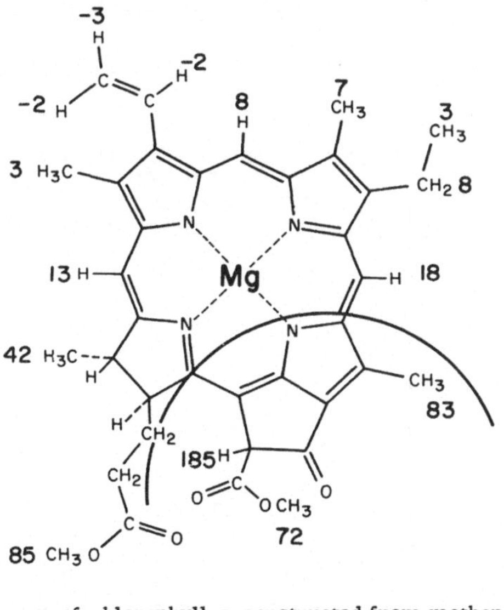

Fig. 6. Aggregation map of chlorophyll a constructed from methanol titration data. The numbers are maximum chemical—shift differences for the indicated protons (in ppm × 100) between highly aggregated chlorophyll a (dissolved in $CDCl_3$) and disaggregated chlorophyll a (in $CDCl_3 + CD_3OD$).

is large, the methylene protons of the ethanol appearing no less than 1.5 ppm to higher field than they ordinarily do in pure solution. A reasonable estimate of the ring current leads to the conclusion that coordination of ethanol with chlorophyll results in the formation of an Mg–O bond with a bond distance of about 3.4 Å. The formation of a complex with such a geometry accounts for the upfield chemical shifts actually observed. It is important to note that only one set of ethanol lines appears in the NMR spectra. This circumstance suggests that the coordination complex between ethanol and chlorophyll is highly labile on the NMR time scale. As the mole ratio of ethanol increases, the chemical shift average between the bound ethanol and the ethanol in solution is observed. The chemical shifts tend to approach the chemical shift of pure ethanol in solution at large mole ratios where most of the ethanol is not complexed.

Particularly interesting are possible coordination interactions between chlorophyll and other constituents of the chloroplast. Figure 8 shows the NMR spectrum of β-carotene, a substance universally present in chloroplasts, in the presence and absence of deuterio-chlorophyll a. It appears clear that no interaction between chlorophyll a and β-carotene occurs that involves the chlorin ring. No coordination interaction exists that would bring any of the β-carotene protons into close proximity to the chlorin ring. Possible interactions between chlorophyll and β-carotene are then presumably limited to

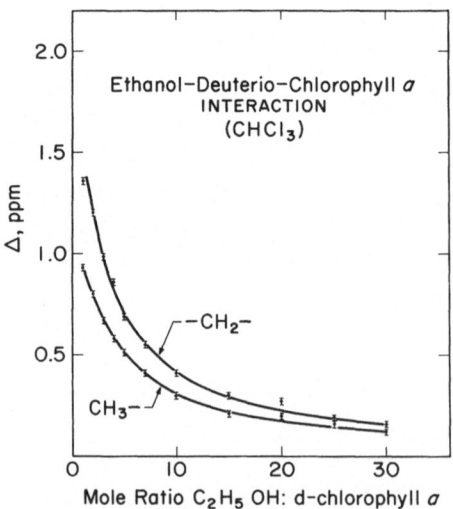

Fig. 7. Chemical shifts of methylene (—CH$_2$–) and methyl (CH$_3$–) protons of ethanol coordinated with deuteriochlorophyll a as a function of ratio of ethanol to chlorophyll [27].

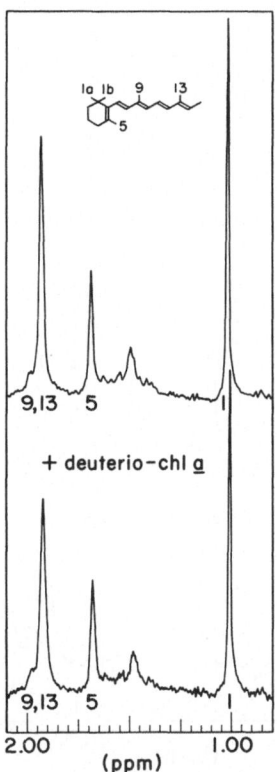

Fig. 8. Nuclear magnetic resonance spectrum of β-carotene in CDCl₃ and in the presence of deuteriochlorophyll a.

interaction of the β-carotene with the phytyl moiety of the chlorophyll. Aronoff and Kirk [24] have also inferred from visible absorption spectra that β-carotene, but not lutein, does not interact with chlorophyll.

With lutein, a dihydroxy-β-carotene, the situation is very different, Fig. 9. Here, coordination of magnesium with the oxygen function of the lutein can occur, and this brings some of the lutein protons into the deshielding zone of the chlorophyll. The methyl groups of the lutein at positions 9 and 13 are scarcely affected, being remote from the coordination center. The methyl group at position 5 experiences only a slight upfield shift for it is insulated from the ring current by the alicyclic ring of the carotenoid. The methyl groups at position 1, however, are brought directly over the chlorin ring. The magnetic nonequivalence of the two methyl groups is increased, and some of the methyl resonances are shifted considerably to higher field, the lower

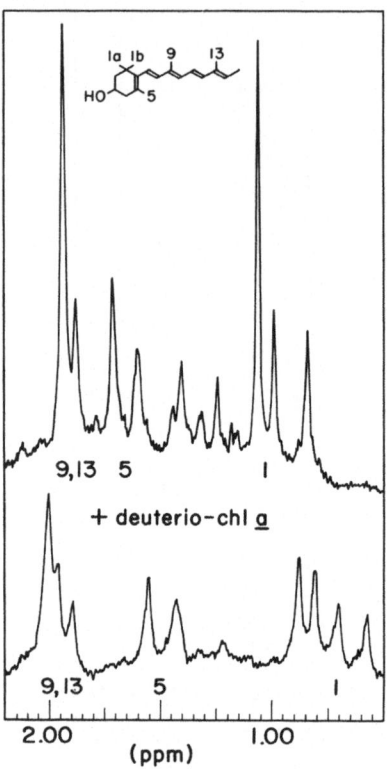

Fig. 9. Nuclear magnetic resonance spectrum of lutein in CDCl₃ and in the presence of deuteriochlorophyll a. Only one half of the lutein molecule is shown.

curve in Fig. 8. The assumption that the shifted protons belong to the methyl groups at position 1 appears to be compatible with the available assignments of the carotenoid NMR spectra [25]. In this case, also, only one set of lutein lines can be seen in the presence of deuterio-chlorophyll, and this may be taken to indicate that the lutein–chloro-phyll complex is in mobile equilibrium with its components and only the weighted average of the chemical shifts between bound and un-complexed ligand can be measured.

Another very important class of chloroplast components are the lipids, about whose function in photosynthesis relatively little is known. We have therefore carried out experiments on possible inter-actions between chlorophyll and the sulfolipid or sulfoquinovosyl diglyceride [26], Fig. 10. Coordination with magnesium could reason-ably be expected to occur by way of the ester carbonyl functions, the hydroxyl functions in the carbohydrate moiety, or with the sulfonic acid grouping. In our experiments, fully deuterated sulfolipid extracted from a fully deuterated blue-green alga, *Synechococcus lividus*, was used, and the spectral changes attendant on complex formation were followed in ordinary hydrogen-containing chlorophyll. The results of one such experiment are shown in Fig. 11. Only a preliminary analysis of the data has been made, but the indications are that the equilibrium involved in sulfolipid–chlorophyll complex formation is less mobile than is the case for any other class of complexes so far observed. By taking the data at face value, it would appear that, when 2 moles of sulfolipid per mole of chlorophyll are present, most of the chlorophyll still appears to be self-aggregated, as shown by the top curve, Fig. 11. Small peaks at higher field (3.81, 3.63, and 3.39) characteristic of methyl resonances in disaggregated chlorophyll solutions are indicative of the presence of a small amount of chlorophyll in essentially monomeric form. These may be chlorophyll molecules held in some kind of sulfolipid complex or micelle in which the chlorophyll molecules are sufficiently remote from each other to escape ring-current effects. With a ratio of 3 moles of sulfolipid per mole of chlorophyll, the relative amount of chlorophyll–sulfolipid complex appears to have increased (Fig. 11, middle curve), and, with a ratio of 10 moles of sulfolipid per mole of chlorophyll, it may be judged that the chlorophyll is quite well "disaggregated" (Fig. 11, lower curve). It appears from these data that the sulfolipid is not nearly as effective a coordinating or disaggregating agent as are the simple alcohols. This may be due to steric factors that make for inaccessibility of the oxygen functions of the sulfolipid molecule for coordination. Another and more likely possibility is that the chlorophyll–sulfolipid interaction is not primarily due to coordination forces. The sulfolipid may well form micelles in CCl_4. In these

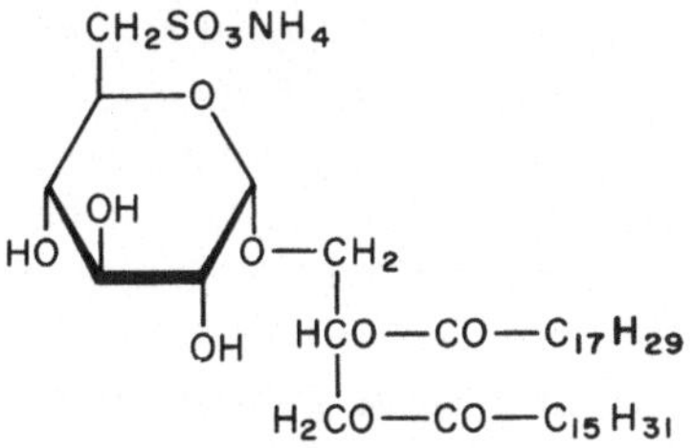

6-Sulfo-α-D-Quinovopyranosyl-
(1→1')-2, 3'-di-O-acyl-D-glycerol (SQDG)

Fig. 10. Structural formula of sulfolipid.

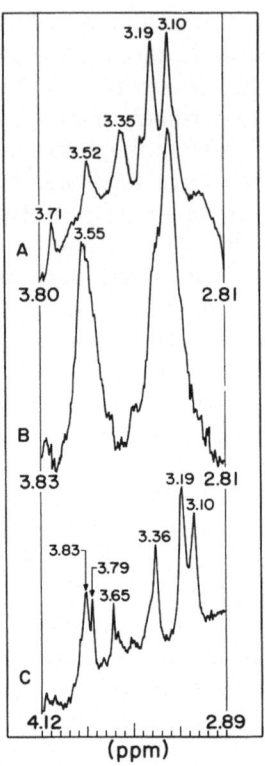

Fig. 11. Nuclear magnetic resonance spectra of chlorophyll a in CCl$_4$ in presence of 2 moles of sulfolipid per mole of chlorophyll a (A), 3 moles of sulfolipid per mole of chlorophyll a (B), and 10 moles of sulfolipid per mole of chlorophyll a (C).

micelles the hydrophilic portions of the sulfolipid would probably be in the interior, and the long fatty-acid chains would be on the outside of the micelles. The chlorophyll could be held in the micelle by the dissolution of the phytyl chain in the fatty-acid region of the sulfolipid micelle, with the chlorin rings protruding into the CCl$_4$ solvent. If the chlorin rings are sufficiently remote from each other, then they will show chemical shifts characteristic of monomeric chlorophyll. Such a model could readily account for the relative inefficiency of the hydroxyl, ester, and sulfonic acid groupings as coordinating agents.

Categorical conclusions about chlorophyll–sulfolipid interactions are certainly unjustified on the information now at hand. Chemical reaction between the sulfolipid and chlorophyll has not been entirely excluded, and other lipids and solvents must also be studied. Nevertheless, the methods described here would seem to offer good prospects for the future [27].

ACKNOWLEDGMENT

The work described here was carried out in close collaboration with Drs. Harold H. Strain and Henry L. Crespi. The work benefited greatly from the preparative skill of Mr. Walter A. Svec. Dr. Laurence J. Boucher carried out the IR studies in the far infrared. Miss Gail D. Norman recorded many of the NMR spectra, and Mrs. Therese Cotton prepared the deuterated sulfolipid. The contributions of Prof. Gerhard L. Closs and Dr. Ralph C. Dougherty in the interpretation of the NMR spectra are acknowledged in the text.

REFERENCES

1. J. J. Katz, G.L. Closs, F.C. Pennington, M. R. Thomas, and H. H. Strain, J. Am. Chem. Soc. 85:3801 (1963).
2. A.F.H. Anderson and M. Calvin, Arch. Biochem. Biophys. 107:251 (1964).
3. L. J. Boucher, H. H. Strain, and J. J. Katz, J. Am. Chem. Soc. 88:1341 (1966).
4. G.L. Closs, J. J. Katz, F.C. Pennington, M. R. Thomas, and H. H. Strain, J. Am. Chem. Soc. 85:3809 (1963).
5. a. C.B. Storm and A. H. Corwin, J. Org. Chem. 29:3700 (1964); also,
   b. C.B. Storm, A.H. Corwin, R.R. Arellano, M. Martz, and R. Weintraub, J. Am. Chem. Soc. 88:2525 (1966).
6. G.R. Seely and R.G. Jensen, Spectrochim. Acta 21:1835 (1965).
7. G.R. Seely, Spectrochim. Acta 21:1847 (1965).
8. K. Sauer, J.R.L. Smith, and A.J. Schultz, J. Am. Chem. Soc. 88 — 2681 (1966).
9. R.L. Amster and G. Porter, Proc. Roy. Soc. (London), A296:38—44 (1967).
10. a. N. V. Sidgwick, The Chemical Elements, Vol. I, Oxford Press, Oxford (1950), p. 241 ff.; also,
    b. F. Hein, Chemische Koordinationslehre, S. Hirzel Verlag, Zurich (1950), p. 380.
11. G.T. Byrne, R. P. Linstead, and A.R. Lowe, J. Chem. Soc. 1934:1017.
12. P.E. Wei, A.H. Corwin, and R. Arellano, J. Org. Chem. 27:3344 (1962).
13. a. E. Rabinowitch, Photosynthesis, Vol. I, Interscience Publishers, Inc. (John Wiley and Sons, Inc.), New York (1945), p. 450; also,
    b. E.E. Jacobs and A.S. Holt, J. Chem. Phys. 20:1326 (1962).
14. a. R. Livingston, Quart. Rev. (London) 14:174 (1960); also,
    b. R. Livingston and S. Weil, Nature 170:750 (1952).
15. V.B. Evstigneev, V.A. Gavrilova, and A.A. Krasnovskii, Dokl. Akad. Nauk. S.S.S.R. 70:261 (1950).
16. J.J. Katz, R.C. Dougherty, and L.J.Boucher, "Infrared and Nuclear Magnetic Resonance Spectroscopy of Chlorophyll," in: L. Vernon and G.R. Seely, eds., The Chlorophylls, Chap. 7, Academic Press, Inc., New York (1966), pp. 185—251.
17. a. J. Fernandez and R.S. Becker, J. Chem. Phys. 31:467 (1959); also
    b. J. Franck, J.L. Rosenberg, and C. Weiss, Jr., in: H. P. Kollmann and C.M. Spruch, eds., Luminescence of Organic and Inorganic Materials, John Wiley and Sons, Inc., New York (1962), pp. 16 ff.; also,
    c. R.S. Becker and M. Kasha, in: F.H. Johnson, ed., The Luminescence of Biological Systems, American Association for the Advancement of Science, Washington, D.C. (1955), pp. 30 ff.
18. a. E.I. Rabinowitch and Govindjee, Sci. Am. 213:74 (1965); also,
    Govindjee and E.I. Rabinowitch, J. Sci. Ind. Res. (India) 24:591 (1965).
19. Summarized by J.C. Goedheer, "Visible Absorption and Fluorescence of Chlorophyll and Its Aggregates in Solution," Chap. 6, pp. 147-184, and W. L. Butler, "Spectral Characteristics of Chlorophyll in Green Plants," Chap. 11, pp. 343—379, in: L. Vernon and G.R. Seely, eds., The Chlorophylls, Academic Press, Inc., New York (1966).

20. A.S. Holt and E.E. Jacobs, Plant Physiol. 30:553 (1955).
21. R.C. Dougherty, H.H. Strain, W.A. Svec, R.A. Uphaus, and J.J. Katz, J. Am. Chem. Soc. 88:5037 (1966).
22. R.J. Abraham, Mol. Phys. 4:145 (1961).
23. a. J.J. Katz and H.L. Crespi, Science 151:1187 (1966); also,
    b. H.A. DaBoll, H.L. Crespi, and J.J. Katz, Biotechnol. Bioeng. 4:281–297 (1962); also,
    c. A.J. Williams, A.T. Morse, and R.S. Stuart, Can. J. Microbiol. 12:1167–1173 (1966).
24. S. Aronoff and P. Kirk, Nature 213:722 (1967).
25. B.C.L. Weedon, "Chemistry of the Carotenoids," in: T.W. Goodwin, ed., Chemistry and Biochemistry of Plant Pigments, Chap. 3, Academic Press, Inc., New York (1965), pp. 75–173.
26. A.A. Benson and I.Shibuya, "Surfactant Lipids," in: R.A. Lewin, ed., Physiology and Biochemistry of Algae, Chap. 22, Academic Press, Inc., New York (1962), pp. 371–383.
27. J.J. Katz, H.H. Strain, D.L. Leussing, and R.C. Dougherty, J. Am. Chem. Soc. 90:784 (1968).

Near Infrared Spectroscopy in Structural Problems of Biochemistry

Sue Hanlon

Department of Biological Chemistry, College of Medicine
University of Illinois
Chicago, Illinois

and

Irving M. Klotz

Department of Chemistry
Northwestern University
Evanston, Illinois

Near infrared spectra have found biochemical applications in equilibrium and in kinetic problems. The region of the spectrum between 1.4 and $1.6\,\mu$ is especially useful for following transformations of —NH or —OH groups. The particular chemical state of these groups can be readily identified by the positions of the absorption maxima due to the first overtone of their stretching vibrations. Relative proportions of species can be evaluated from extinction coefficients which can be accurately determined. The equilibrium state of the peptide unit in a number of synthetic polyamino acids (poly-L-alanine, poly-L-leucine, poly-L-methionine, and polybenzyl-L-glutamate) has been followed as a function of solvent composition under conditions where the transitions in other physical properties of these polymers have been interpreted as simple peptide hydrogen-bonded helix to coil transitions. Spectral data demonstrate that these conversions involve protonated peptide species and are far more complicated than has hitherto been assumed.

Kinetic investigations of hydrogen—deuterium exchange rates have also been conducted with a variety of low- and high-molecular-weight protein models. The exchange process

$$-NH + D_2O \rightarrow -ND + HOD$$

has been studied in detail to reveal some of the factors which influence the rate. The latter may be evaluated from changes in —NH, —OH, —ND, or —OD absorption. In practice, the first two bands have proved more useful.

The spectral region referred to as the "near infrared" is generally identified as the region between 0.8 and $3\,\mu$, a region which is conveniently measured with a photoconductive-cell detector. At the low-wavelength end, a variety of low-energy-charge transfer bands occur; at the high-wavelength side, the fundamental stretching vibrations of NH, OH, and CH groups are found. Between these limits, fall a variety of stretching overtones as well as combination bands. The position and intensity of these overtone and combination bands tend to be difficult to analyze theoretically, and there has been no major systematic effort directed toward that end. Despite this absence of a theoretical basis of interpretation, this particular region of the spectrum may still be profitably employed in qualitative as well as quantitative studies.

The wavelength region between 1.4 and 1.6 μ has been particularly useful in the investigation of certain structural problems in physical biochemistry.   The solution to many of these problems requires a knowledge of the chemical state of NH and OH groups. In many cases, near infrared spectroscopy can provide such knowledge. The first overtones of the stretching vibrations of the NH and OH groups fall, for the most part, in the 1.4 to 1.6-μ region, and studies with model compounds have demonstrated that both the position and the intensity of these overtone bands are markedly influenced by the chemical state and environment in which the group finds itself. Since this is a relatively uncluttered region of the spectrum, one can usually determine qualitatively from a rapid scan in the 1.4 to 1.6-μ region which particular species of OH or NH compounds are present.

Quantitative data are also readily obtainable. The problems of temperature control and reproducibility of cell-path length, commonly encountered in the fundamental region, are nonexistent in the near infrared. Quartz is transparent in this spectral region, and extinction coefficients are low enough that conventional quartz spectrophotometer cells, normally employed in the visible and ultraviolet, can be used. Thus, extinction coefficients can be accurately determined. Furthermore, one generally finds that, as long as the chemical state of a given compound remains constant, the absorbance at a single wavelength follows Beer's law, and, hence, it is not necessary to integrate over the entire absorption band to obtain a quantity proportional to concentration.

There are, however, a few experimental precautions which should be observed if reliable data are desired. Although most solvents used in near infrared studies have low extinction coefficients in the region of the solute bands of interest, such extinction coefficients are in general not negligible. This is especially true, of course, if one is examining an amide NH in an OH solvent or vice versa. Since the concentrations of solute required for measurable absorbances are usually rather high, a spectrum obtained by placing solution in the sample beam and pure solvent in the reference beam will not reflect the actual spectrum of the solute species if either the extinction coefficient of the solvent, E_0, is appreciable or the volume fraction of the solute, φ_s, is large. This is demonstrated schematically in Fig. 1.

This problem may be obviated by conducting the experiment in the manner shown schematically in Fig. 2. The volume fraction of solvent in both the sample and the reference beam is equalized by putting a filler solute in the reference cell whose volume fraction, φ_R, is equal to that of the sample solute, φ_s. Two types of filler material may be employed: (1) One that is transparent in the wavelength region of interest, $E_R = 0$, or (2) a structural analog of the sample solute, which contains all of the component parts and functional groups of the sample except the group of interest.

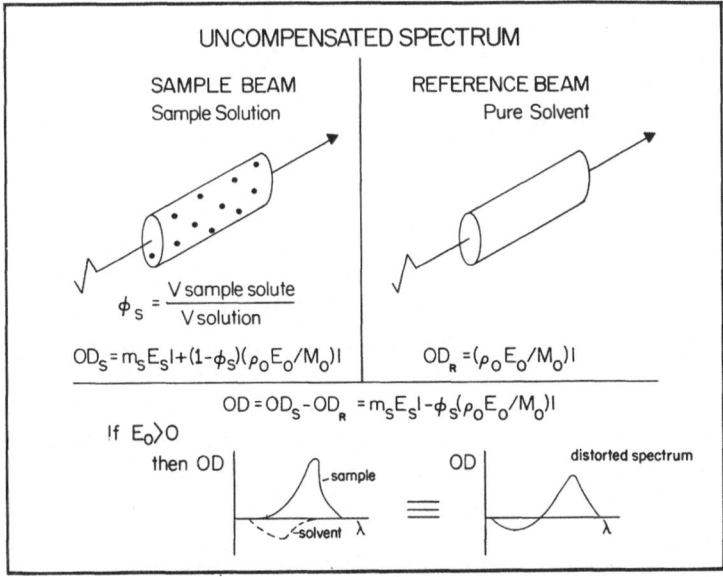

Fig. 1. Distortion of spectra in the near infrared due to solvent absorption. The observed optical density OD at a given wavelength λ in the experiment shown schematically above will be the difference between the absorbance of the sample path, OD_S, and that of the reference path, OD_R. These two quantities are in turn related, by the equation given above, to the path length l of the sample and reference cells, the volume fraction φ_S, the concentration m moles/ml, and, the extinction coefficient E_S of the sample solute of interest, as well as the extinction coefficient E_0 and the concentration of the solvent of density, ρ_0, and molecular weight, M_0, in the sample and reference cells. If the solvent is not transparent, a distorted spectrum will be obtained.

This particular technique was first employed by Klotz and Franzen [1] in their near infrared study of the self-association of N-methyl-acetamide (NMA) in a variety of solvents. The biochemist is interested in the thermodynamics of such an interaction as it serves as a model for peptide hydrogen-bond formation in polypeptides and proteins. In the case of NMA, one can easily distinguish between a monomeric amide NH (I) and the polymeric amide NH self-associated in the hydro-gen-bonded aggregate (II) on the basis of the band positions in the near IR.

In the free or solvent-associated form, present at low concentrations of amide, the first overtone of the NH stretching vibration falls any-where from 1.47 to 1.49 μ, the exact position and extinction coefficients at the maximum being dependent on the nature and interaction prop-erties of the solvent. At high concentrations of amide, the NH band shifts to higher wavelengths and becomes a doublet with maxima at 1.53 and 1.57 μ. These bands are readily resolved, as is illustrated by Fig. 3A, which shows a spectrum of N-methylacetamide in CCl_4. The band at 1.47 μ represents the monomeric NH; the characteristic doublet reflects the presence of intermolecular hydrogen-bonded NH species.

One may use absorbance data from such experiments to calculate the fractions of free and associated amide species as a function of amide concentration and temperature and, hence, evaluate equilibrium constants and free energies and enthalpies of association. The results of such experiments are shown in Table I [2]. As the data demonstrate, the free energy and enthalpy of formation of the model peptide hydrogen bond become increasingly less favorable as the solvent becomes an

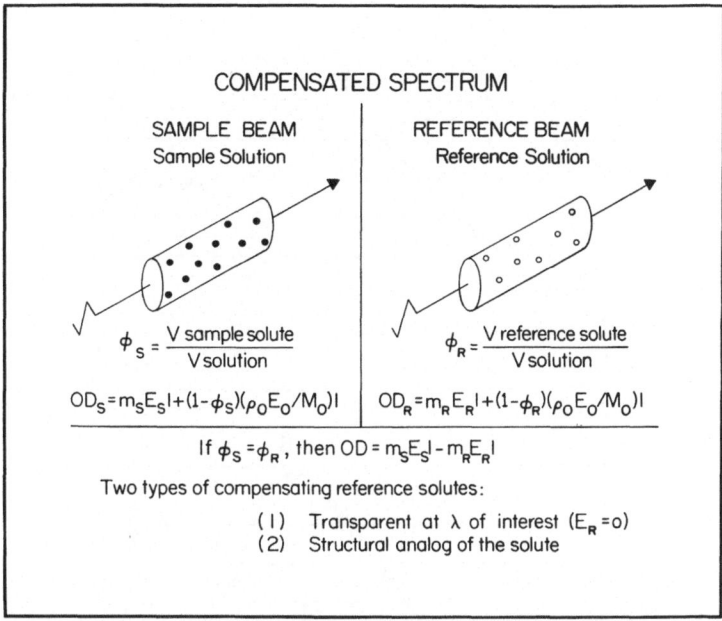

Fig. 2. Technique employed in obtaining undistorted spectra in the near infrared. A reference solution containing a compensating reference solute dissolved in the solvent at a volume fraction, φ_R, is placed in the reference beam. The sample solution contains a sample solute at a volume fraction, φ_S, in the sample beam. All other symbols used in this figure are defined in the legend for Fig. 1. If $\varphi_S = \varphi_R$, the contribution to the absorbance due to the solvent is approximately balanced in both sample and reference solutions and an undistorted spectrum of the sample solute can be obtained.

TABLE I

Thermodynamics of $C = O \cdots H-N-$Hydrogen–Bond Formation in N–Methylacetamide at 25°C*

Solvent	Association constant	$\Delta F°$, kcal/mole	$\Delta H°$, kcal/mole	$\Delta S°$, Gibbs/mole
Carbon tetrachloride	4.70	−0.92	−4.2	−11
N–methyl morpholine	1.90	−0.38		
Acetonitrile	0.75	0.17	−0.7	−3
Dioxane	0.52	0.39	−0.8	−4
Water	0.005	3.10	0.0	−10

*Taken from Ref. 2.

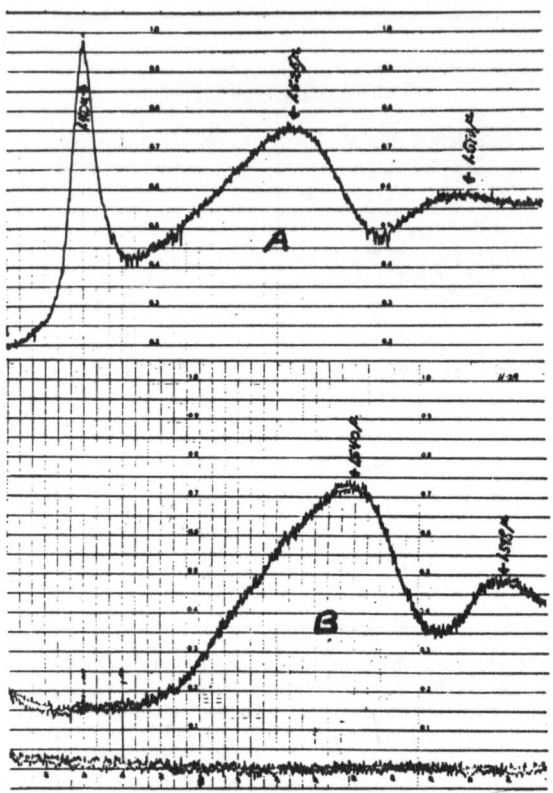

Fig. 3. Spectra of amides in a self-associated state: A, N-methylacetamide, 3.0 M, in CCl$_4$ versus CCl$_4$; B, poly-γ-benzyl-L-glutamate, 0.053 residue M, in pyridine versus CCl$_4$ in pyridine. The ordinate represents actual absorbance times 10; the abscissa wavelengths are given in μ (from Hanlon and Klotz [8]).

increasingly better hydrogen bonder. In fact, $\Delta H°$ drops to 0 in water. Extrapolating these results to the polypeptide level, one may conclude that the peptide hydrogen bond makes no appreciable contribution to the stabilization of protein or polyamino acid conformations in aqueous solutions.

These conformations, of course, have usually been thought to consist of variable fractions of α helices [3]. Denaturation of globular proteins is thought to reflect, among other things, the conversion of this variable fraction in such a helical conformation to a completely disorganized conformation, the random-coil form. Certain synthetic polymers, the polyamino acids, can be made to simulate this behavior by changes in solvent environment. One finds, for instance, that, in relatively inert solvents [CHCl$_3$, ethylene dichloride (EDC), and even pyridine], these polymers have all of the characteristics of rigid rods, whose dimensions are those predicted for the polypeptide chain wrapped in the form of the α helix [4, 5]. In strong organic acids, the random-coil form, however, predominates [4, 5]. As the character of the solvent is changed from inert to highly interacting, discontinuities in a number of physical properties are frequently observed over a very narrow range of solvent composition. Such discontinuities are thought to reflect the conversion of intact peptide hydrogen-bonded α helices to the random-coil form, and this phenomenon is correspondingly referred to as a helix–coil transition.

Until recently, the usual method of following such transitions has employed a parameter derived from optical rotatory-dispersion data. A theoretical analysis by Moffitt [6, 7] resulted in an equation which shows that the behavior of the optical rotation of an α helix as a function of wavelength is governed by three parameters, one of which, b_0, is a linear function of the fraction of residues in the helical form. The limits of b_0 at 100 and 0% helix were originally evaluated from experiments with poly-γ-benzyl-L-glutamate (PBLG) [6, 7]. In inert solvents, where PBLG is presumably completely helical, the b_0 value for PBLG is around −630. This value falls to 0 in the highly interacting solvents such as trifluoroacetic acid (TFA) or dichloroacetic acid (DCA). Thus, the fraction of residues in the helical form, α_H, of a given protein or polyamino acid is usually calculated from its b_0 value according to the relationship

$$\alpha_H = \frac{b_0}{-630}$$

Typical transformations of these values of b_0 of polyamino acids in mixed solvents are shown in Fig. 4 [8] and Table II [9]. In the case of the two polyamino acids poly-L-alanine (PLA) and poly-L-leucine (PLL) in the TFA–CHCl$_3$ solvent [8], measurements could not be made in pure CHCl$_3$ owing to the insolubility of the polyamino acids. At the lowest acid concentration required for appreciable solubility, the b_0 values of the two polymers are quite high, being about −470 for PLL and −330 for PLA. This would presumably reflect about a 75 and a 50% helix for PLL and PLA, respectively. As the acid concentration increases, the sharpest transition in b_0 is exhibited by PLL between 50 and 60% TFA. Poly-L-alanine exhibits a more gradual one over the acid concentration range of 70 to 100% TFA.

As the data in Table II demonstrate [9], the transition in b_0 exhibited by PBLG in the mixed solvent EDC–DCA, is quite marked in the 70 to 80% DCA range. The fraction of residues, α_H, in a peptide

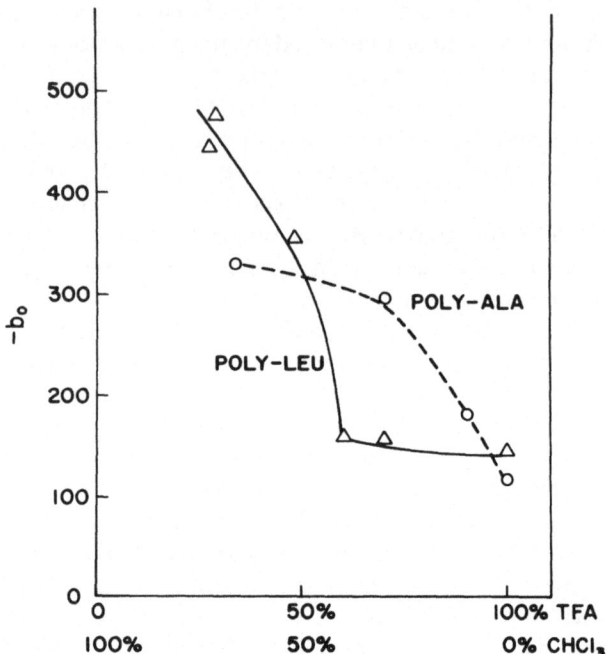

Fig. 4. Variation in the optical rotatory-dispersion parameter b_0 of polyamino acids in
$CHCl_3$—CF_3COOH: The broken line is poly-L-alanine, 0.33 residue M, and the solid line is
poly-L-leucine, 0.33 residue M (from Hanlon and Klotz [8]).

TABLE II

Calculation of the Fraction of Peptide
Hydrogen-Bonded Residues, α_H, Poly-γ-
benzyl-L-glutamate*

Component of solvent, % DCA	b_0, deg	α_H From b_0 data	From spectral data
0(100%EDC)	−667	1.06 ± 0.05	1.00 ± 0.10
1	−629	1.00 ± 0.05	0.85 ± 0.10
20	−588	0.93 ± 0.05	0.54 ± 0.10
70	−552	0.88 ± 0.05	0.38 ± 0.10
80	+67	0 ± 0.05	0 ± 0.10
100	+36	0 ± 0.05	0 ± 0.10

*Taken from Ref. 9.

hydrogen-bonded helix calculated from these b_0 values are given in the third column of this table.

If this interpretation of the optical rotatory-dispersion data is correct, characteristic shifts in the near-infrared spectra of these polyamino acids should occur in the transition region as the peptide group is converted from the intramolecular hydrogen-bonded species in the α helix to some solvated form in the random coil. The possible species that might be anticipated in mixed solvents containing strong acids are shown in the following reaction:

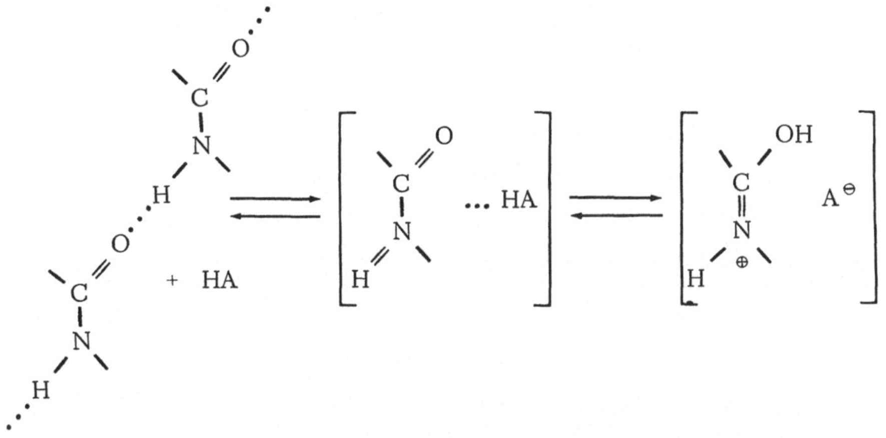

The band assignments for these polypeptide species can be readily made by returning to studies with the low-molecular-weight model N-methylacetamide [8, 10]. As was shown in Fig. 3, NMA in the self-associated state exhibits a characteristic double band with maxima at 1.53 and 1.57 μ. The spectrum of PBLG in the inert solvents exhibited similar bands with maxima at 1.54 and 1.58 μ, as is seen by the curve in the lower portion B of this same Fig. 3. The extinction coefficient at the maxima were identical with those of completely associated NMA. This 1.54- and 1.58-μ band has therefore been assigned to the peptide hydrogen-bonded species involved in maintaining the polypeptide helix.

The band positions of the protonated as well as that of the acid-solvated form were identified by the positions of the NH overtone of NMA under conditions where other data indicated that the amide was either protonated or hydrogen bonded to the acid component of the solvent. A few typical spectra from such experiments are presented in Fig. 5.

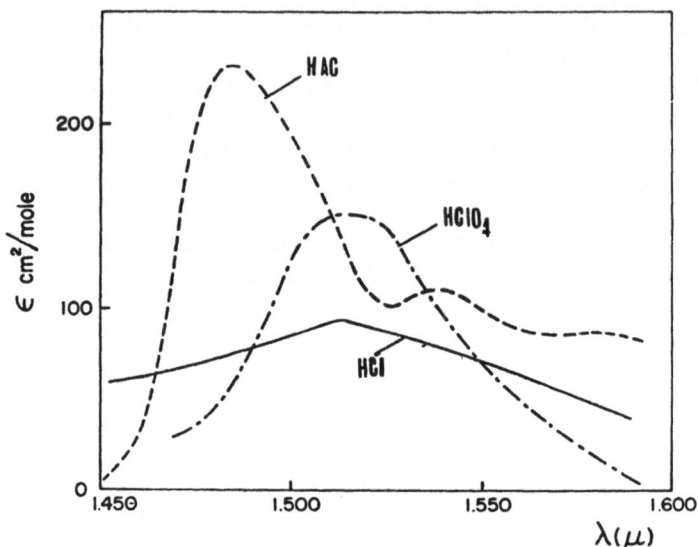

Fig. 5. Spectra of N-methylacetamide (NMA) in strong and weak acids: The dashed line is 0.419 M NMA in acetic acid; the dot-dash line is 1.10 M NMA in 8.4 M $HClO_4$, 2.5 M dioxane, and 19.2 M H_2O; and the solid line is 1.478 M NMA in 10.7 M HCl and 36.5 M H_2O (from Klotz et al. [10]).

In strong acids, HCl or $HClO_4$, where other data indicate that N-methylacetamide is protonated [11], the maximum of the NH-stretching overtone falls at 1.51 μ. The same band was also found for PLA in the perchloric acid solvent. In acetic acid, however, where there is no evidence for protonation of the low-molecular-weight amide, a maximum of 1.49 μ is observed for the stretching overtone. A summary of these band assignments for the various species, based on these experiments, is shown in Fig. 6.

On the basis of these assignments and the conclusions drawn from the optical rotatory-dispersion data, it would be expected that, in the region of the so-called helix-coil transition, the conversion of the typical spectrum showing maxima at 1.54 and 1.58 μ to one with peaks at either 1.51 or 1.49 μ should be observed. Such a prediction, however, is not supported by the spectral experimental results. For instance, the spectra of PLL in TFA, shown in Fig. 7 [8], exhibit certain changes in the solvent composition range where the b_0 values are changing. At no point, however, do these reflect the presence of or transformation of peptide hydrogen-bonded species. Even at the lowest acid concentration required for solubility (at which the b_0 values are quite high), there is only a single peak at 1.51 μ indicative of the

CHEMICAL STATE OF —NH IN POLYPEPTIDES	POSITION OF —NH OVERTONES, λ (μ)
Peptide hydrogen bond 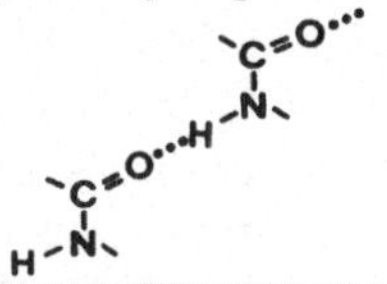 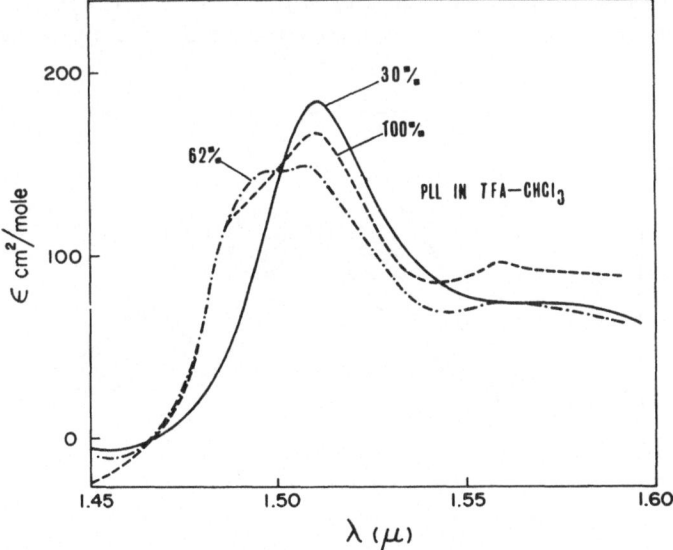	**1.54 μ, 1.58 μ**
Protonated peptide 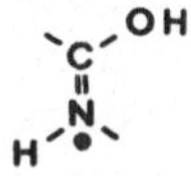	**1.51 μ**
Peptide — Acid hydrogen bond 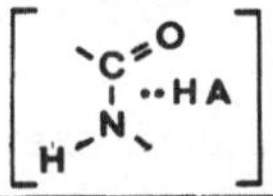	**1.49 μ**

Fig. 6. Band assignments for the first overtone of the NH-stretching vibration in polyamino acids in the near infrared.

Fig. 7. Spectra of poly-L-leucine, 0.33 residue M, in $CHCl_3$ —CF_3 COOH: The solid line is 30% CF_3 COOH; the dash-dot line is 62% CF_3 COOH; and the dashed line is 100% CF COOH (from Hanlon and Klotz [8]).

existence of the protonated peptide species. Similar results were obtained for PLA in the same solvent.

In the case of PBLG in DCA−EDC mixtures, one does indeed see the conversion of the peptide hydrogen-bonded NH to the protonated form, as is demonstrated by the spectra in Fig. 8 [9]. This conversion, however, does not parallel the changes in b_0 in this same solvent system. At 0% DCA (100% EDC), the spectrum exhibits the characteristic doublet associated with the peptide hydrogen-bonded NH. As the acid concentration increases to 80%, this doublet gradually disappears and is replaced by a band at 1.51 μ indicative of the protonated peptide species. Above 80% DCA, this is the only peak evident in the spectrum.

Since there is an isosbestic point, the fractions of peptide residues in the hydrogen-bonded and in the protonated form may be calculated from these data. Such calculations for PBLG reveal a protonation pattern of the type shown in Fig. 9. Corresponding values for the fraction of residues in the hydrogen-bonded form α_H are given in the last column in Table II.

Contrary to predictions based on the b_0 data (where only a single transition was observed between 70 and 80% acid), there are two transitions in the protonation curve, each involving roughly equal fractions of peptide residues. One occurs at fairly low concentrations of acid, between 0 and 20%; and the other, in the 70 to 80% range. The latter transition parallels the transformation in b_0. Poly-γ-benzyl-L-

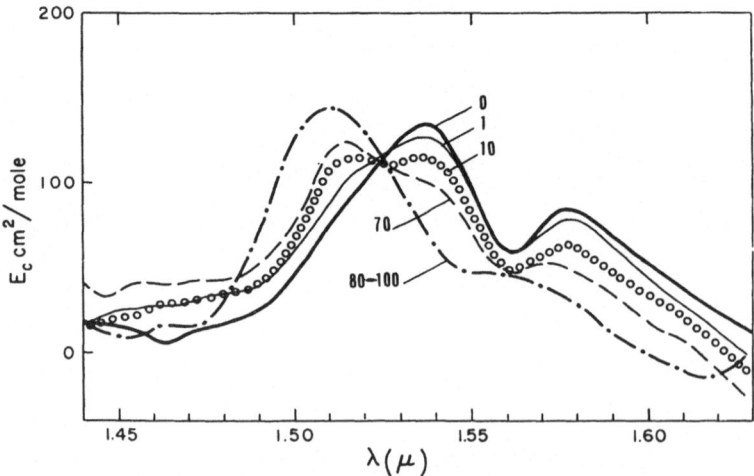

Fig. 8. Spectra of poly-γ-benzyl-L-glutamate in DCA−EDC mixtures: The heavy solid line is 100% EDC; the light solid line is 1% DCA; the open-dot line is 10% DCA; the dashed line is 70% DCA; and the dot-dash line is 80 to 100% DCA (from Hanlon [9]).

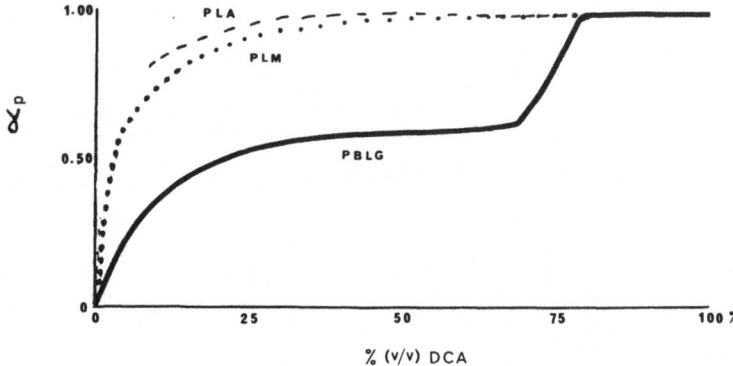

Fig. 9. Protonation of the peptide residues of polyamino acids in DCA–EDC mixtures: The fraction a_p of protonated residues at 25°C is plotted as a function of solvent composition in per cent (v/v) of DCA for poly-γ-benzyl-L-glutamate, solid line; poly-L-methionine, dotted line; and poly-L-alanine, dashed line (from Hanlon [9]).

glutamate is unique in this regard, as neither poly-L-alanine nor poly-L-methionine in this same solvent system exhibits this two-step transition.

The particular form of the protonation pattern of each polyamino acid depends on the strength of the acid as well as the steric properties of the side chains. All the titration curves of the polyamino acids examined thus far are explainable by a common reaction scheme dependent upon the formation and disassociation of certain types of ion pairs [8, 9]. Such a reaction scheme is primarily of interest to the chemist. The important conclusions for the biochemist are: (1) The so-called helix-coil transitions of polyamino acids in mixed solvents are more complex chemically than the optically rotatory dispersion data indicate, and (2) a high value of b_0 does not necessarily reflect the existence of peptide hydrogen-bonded residues.

In addition to equilibrium studies, near infrared methods can also be employed in kinetic experiments. Examples of such applications are the extensive series of studies by Klotz and co-workers [12–15] on deuterium exchange in low- and high-molecular-weight amides and polyamino acids. The assumption was made some years ago by Linderstrom-Lang [16] that the exchange of hydrogen for deuterium in a peptide group that was hydrogen bonded in a protein would be slow compared with the rate of exchange of a peptide group in a random-coil section of the polypeptide chain. As a consequence, one could in principle measure the fraction of peptide residues involved in these hydrogen-bonded sections by determining the number of slowly exchangeable hydrogens. Presumably, these were the same peptide residues which were involved in maintaining a unique protein conformation in solution.

In order to evaluate this assumption, however, one must know something about the general factors which influence the rate of such an exchange process.   Near infrared spectroscopy is particularly useful for such studies.   As can be seen from the exchange reaction for NMA

$$
\underset{\substack{\|\\O}}{CH_3C}-\overset{H}{\underset{|}{N}}-CH_3 + DOD \longrightarrow \underset{\substack{\|\\O}}{CH_3C}-\overset{D}{\underset{|}{N}}-CH_3 + HOD
$$

this process can be conveniently followed by measuring either the appearance of the OH band or the disappearance of the NH band. (The ND and OD overtones fall considerably out of this wavelength range and do not interfere.)

For all the amides examined so far [12–15], a plot of the optical density at the maxima of these bands as a function of time yields apparent first-order rate constants.   Experiments conducted under a variety of pH and temperature conditions yield typical rate-profile curves of the type shown in Fig. 10 [13].

This particular figure illustrates the exchange behavior of poly-L-glutamic acid under conditions where it is helical.   The parabolic form of this curve, found for all amides, is characteristic of an acid and base-catalyzed reaction.   To compare the D—H exchange in different amides and polymers, it is necessary to compare the rates at the minima of the curves.   When such rates for polyglutamic acid are compared with the exchange rate of a low-molecular-weight analog, it is found that the helical form of the polyamino acid exchanges at a rate almost a thousandfold slower [13].

This might at first be interpreted as a clear-cut demonstration that α-helix formation, as such, is responsible for the striking decrease in rate.   There are, however, a number of other factors which could also slow down the exchange.   The fact that the peptide residues are incorporated in a polymer may in itself have an appreciable effect on the exchange rate.   In fact, when the rates of exchange of NH of free and bonded amide species are examined in poly-N-isopropyl acrylamide, the exchange of both NH species is slowed by over one hundredfold [15] compared with the exchange of the monomer.   This particular polymer has about one-third of its NH groups in the free form and the other remaining two-thirds in —NH···O=C—hydrogen bonds.   The exchange of each of these species may be followed separately by the disappearance in the spectrum of the free NH as well as the bonded

NH. These two amide species can be clearly differentiated from each other, as is shown from the spectrum presented in Fig. 11 [15].

Shown in Fig. 12 [15] is a plot of the rate of disappearance of each of the two peaks. It can be seen that the rate constant for the D—H exchange of either form is identical. Hence, it seems likely that the lowered rate of exchange of the peptide species in polyglutamic acid is not so much a matter of α -helical hydrogen bonding as it is due to the fact that the peptide residue is embedded in a polymeric structure.

These are some of the problems to which near infrared spectro-photometry has been applied with advantage. There are a number of others, of course, which immediately suggest themselves. Nucleic acid derivatives have overtones in this region, and their association and exchange rates can in principle be followed by near infrared techniques. It seems likely, therefore, that the technique will find increasingly wider application in the future.

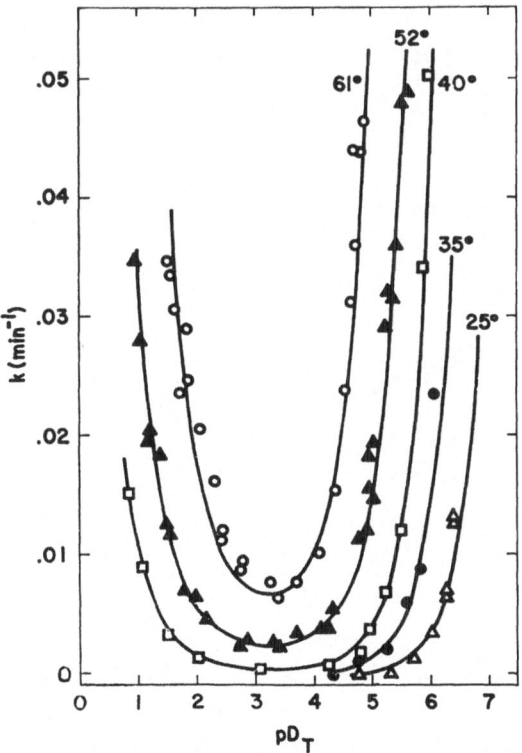

Fig. 10. Hydrogen—deuterium exchange-rate constants versus pD_T at the respective temperatures shown for 2% poly-L-glutamic in D_2O-dioxane, 1:1 (from Leichtling and Klotz [13]).

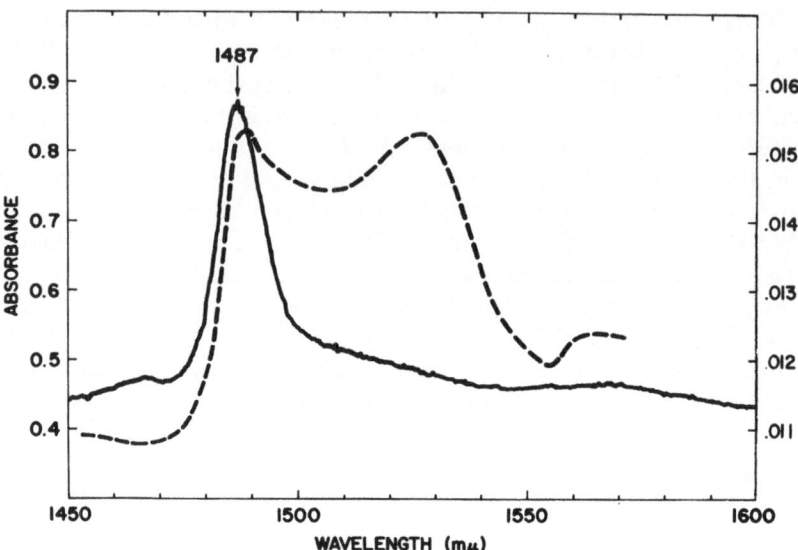

Fig. 11. Near infrared spectra of poly-N-isopropylacrylamide: The solid line is in CHCl₃ (left, ordinate scale) showing free NH band; the dashed line is in D_2O (right, ordinate scale) (from Scarpa et al. [15]).

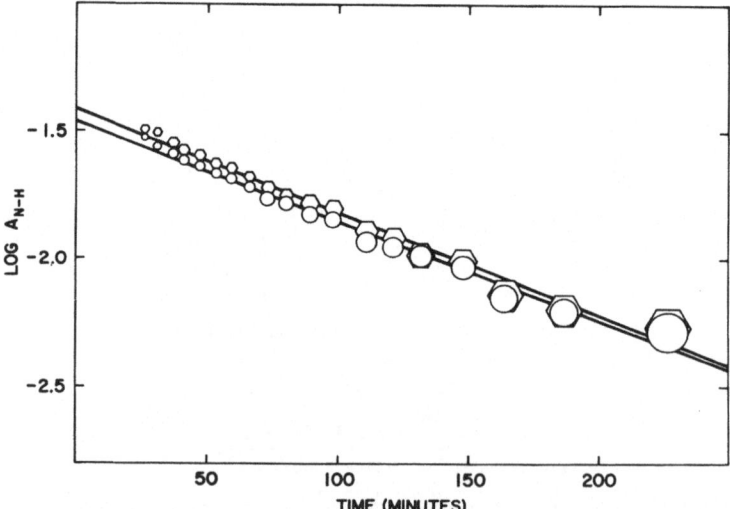

Fig. 12. Hydrogen–deuterium exchange of poly-N-isopropylacrylamide: The circles mark the absorbance change with time at 1.489 μ of free NH; the hexagons mark the absorbance change with time at 1.525 μ of hydrogen-bonded NH (from Scarpa et al. [15]).

REFERENCES

1. I.M. Klotz and J.M. Franzen, J. Am. Chem. Soc. 84:3461 (1962).
2. I.M. Klotz, Federation Proc. 24 III:S-24 (1965).
3. L. Pauling and R.B. Corey, Proc. Natl. Acad. Sci. U.S. 37:729 (1951).
4. P. Doty, J.H. Bradbury, and A.M. Holtzer, J. Am. Chem. Soc. 78:947 (1956).
5. P. Doty and J.T. Yang, J. Am. Chem. Soc. 78:499 (1956).
6. W. Moffitt, J. Chem. Phys. 25:467 (1956).
7. W. Moffitt and J.T. Yang, Proc. Natl. Acad. Sci. U.S. 42:596 (1956).
8. S. Hanlon and I.M. Klotz, Biochem. 4:37 (1965).
9. S. Hanlon, Biochem. 5:2049 (1966).
10. I.M. Klotz, S.F. Russo, S. Hanlon, and M.A. Stake, J. Am. Chem. Soc. 86:4774 (1964).
11. A. Berger, A. Loewenstein, and S. Meiboom, J. Am. Chem. Soc. 81:62 (1959).
12. I.M. Klotz and B. H. Frank, J. Am. Chem. Soc. 87:2721 (1965).
13. B.H. Leichtling and I. M. Klotz, Biochem. 5:4026 (1966).
14. I.M. Klotz and P.L. Feidelseit, J. Am. Chem. Soc. 88:5103 (1966).
15. J.S. Scarpa, D.D. Mueller, and I.M. Klotz, J. Am. Chem. Soc. 89:6024 (1967).
16. A. Hvidt and K. Linderstrom-Lang, Biochim. Biophys. Acta 14:574 (1954).

Infrared Spectroscopy of Carbohydrates in Water (1600 to 900 cm⁻¹)*

Frank S. Parker†

New York Medical College
New York, New York

Coblentz [1] used water as a solvent for some infrared spectro-scopic work as early as 1905. Gore, Barnes, and Petersen [2] published IR spectra of aqueous solutions of some amino acids in 1949. Plyler and Acquista [3] in 1954 published spectra of pure water, in which the region from about 6.5 to 10 μ had sufficient transmittance to suggest the use of water for quantitative analytical work.

Several years ago, a program was begun in the laboratory of the author to examine IR spectra of aqueous solutions of various substances of biological interest, including compounds of the tricarboxylic acid cycle [4], amino acids [5], amines [6], and carbohydrates [7].

The kinetics of the mutarotations of α-D-glucose, β-D-glucose, and β-D-mannose were measured in BaF_2 cells in H_2O solutions by IR spectroscopy [7]. The calculated mutarotation constants were in good agreement with values that had previously been determined polarimetrically [8, 9]. The IR method specifically avoided the use of D_2O since D_2O lowers the values of the rates of mutarotation [10, 11]. Besides measuring the mutarotation values, the spectra of several other carbohydrates in water were also recorded, namely, D-levulose, maltose, D-galactose, D-ribose, L-arabinose, and dipotassium glucose 1-phosphate, which showed that these carbohydrates have characteristic spectra in water as well as in the solid state. Goulden has also published spectra of some sugars in H_2O [12].

*Supported by grant AM 08417-01-02 from the National Institutes of Health.
†Recipient of Career Scientist Award of the Health Research Council of the City of New York, under contract I-323.

The IR experiments with mutarotation kinetics and the spontaneous hydrolysis of gluconolactone to form gluconic acid [13] led to the study of enzymic systems by IR spectroscopy, specifically, systems in which carbohydrate was the substrate acted upon by the enzyme. Carbohydrates have intense absorption bands in water as well as in the solid state and would be expected to lend themselves easily to measurements of changes of absorbance with time at low concentrations. The following enzymic reactions were studied: glucose oxidase on glucose [13], invertase on sucrose [14, 15], and β-glucosidase on phenyl-β-D-glucoside [16]. In all these reactions, the solvent was a buffer system containing H_2O, not D_2O. It is known that D_2O decreases the rate of hydrolysis by invertase by 25% [17].

Also recorded have been the spectra (700 to 250 cm^{-1}) of many carbohydrates in the solid state by transmission and ATR (attenuated total reflection) spectroscopy [18]. By both techniques, a- and β-anomers are readily distinguishable as are other sugars [18, 19].

REFERENCES

1. W. W. Coblentz, Investigations of Infrared Spectra, Carnegie Institution of Washington, Washington, D.C. (1905), p. 56.
2. R. C. Gore, R. B. Barnes, and E. Petersen, Anal. Chem. 21:382 (1949).
3. E. K. Plyler and N. Acquista, J. Opt. Soc. Amer. 44:505 (1954).
4. F. S. Parker, Appl. Spectr. 12:163 (1958).
5. F. S. Parker and D. M. Kirschenbaum, Nature 187:386 (1960); also Spectrochim. Acta 16:910 (1960).
6. D. M. Kirschenbaum and F. S. Parker, Spectrochim. Acta 17:785 (1961).
7. F. S. Parker, Biochim. Biophys. Acta 42:513 (1960).
8. C. S. Hudson and J. K. Dale, J. Am. Chem. Soc. 39:320 (1917).
9. C. S. Hudson and H. L. Sawyer, J. Am. Chem. Soc. 39:470 (1917).
10. J. Nicolle and F. Weisbuch, Compt. Rend. 240:1340 (1955).
11. E. L. Purlee, J. Am. Chem. Soc. 81:263 (1959).
12. J. D. S. Goulden, Spectrochim. Acta 15:657 (1959).
13. F. S. Parker, Perkin-Elmer Instrument News 13:No. 4, 1 (1962).
14. F. S. Parker, Fed. Proc. 21:243 (1962).
15. F. S. Parker, Progress in Infrared Spectroscopy, Vol. III, Plenum Press, New York (1967).
16. F. S. Parker, Nature 203:975 (1964).
17. E. W. R. Steacie, Z. Physik. Chem. B27:6 (1934).
18. F. S. Parker and R. Ans, Appl. Spectry. 20:384 (1966).
19. a. R. S. Tipson, H. S. Isbell, and J. E. Stewart, J. Res. Natl. Bur. Std. 62:257 (1959); also,
    b. R. S. Tipson and H. S. Isbell, J. Res. Natl. Bur. Std. 64A:239, 405 (1960); 65A:31, 249 (1961); and 66A:31 (1962).

Structure Studies of Ice, Water, and Aqueous Solutions

Nuclear Magnetic Resonance Studies of Water Structure

Jay A. Glasel

Department of Biochemistry
College of Physicians and Surgeons of Columbia University
New York, New York

The use of spin-lattice relaxation time measurements in the study of water structure is discussed. Particular emphasis is laid upon the problem of determining correct correlation functions and upon the information derived from quadrupole-induced relaxation in H_2^2O and H_2O^{17}. The derivation and importance of quadrupole-coupling constants is discussed. An attempt is made to fit data together from various sources.

Interest in magnetic-resonance relaxation studies in aqueous systems stems from the experimental data that have been obtained bearing on rotational correlation times and molecular electronic structure. These data both complement and supplement other spectroscopic and relaxational studies. The magnetic resonance work is aided by the existence of isotopic water molecules, which yield different sorts of magnetic-resonance information. These are normal water H_2O; heavy water D_2O, and oxygen-17 water, H_2O^{17}. Whenever the decay of the macroscopic Z component of nuclear magnetism is exponential, and spin–rotation interactions are unimportant, the equations describing the observed macroscopic spin-lattice relaxation times T_1 are:

Protons:

$$\left(\frac{1}{T_1}\right)_{obs} = \left(\frac{1}{T_1}\right)_{intra,\,D} + \left(\frac{1}{T_1}\right)_{inter,\,D} \tag{1}$$

Deuterons and oxygen-17:

$$\left(\frac{1}{T_1}\right)_{obs} = \left(\frac{1}{T_1}\right)_Q \tag{2}$$

Thus, for protons the relaxation time observed is governed by two terms which describe nuclear-magnetic relaxation induced by nuclear-magnetic dipole—dipole interactions from nuclei in the same molecule and in neighboring molecules, respectively. For deuterons and oxygen-17, on the other hand, magnetic relaxation is induced by interactions between the nuclear electric-quadrupole moment, which these isotopes possess, and the molecular electronic structure surrounding them. Brownian motion in the liquid makes these interactions time-dependent at the positions of the relaxing nuclei.

For deuterons and oxygen-17, the observed spin-lattice relaxation time is governed by an equation of the form [1]

$$\left(\frac{1}{T_1}\right)_Q = \frac{3}{80}\frac{2I+3}{I^2(2I-1)}\left(\frac{e^2qQ}{\hbar}\right)^2\left[J_l^m(\omega)\right]_Q \tag{3}$$

where I is the nuclear-spin quantum number, $e^2qQ/\hbar$ is the quadrupole-coupling constant, and $J_l^m(\omega)$ is the Fourier transform of a normalized angular autocorrelation function [2]

$$\left[J_l^m(\omega)\right]_Q = \int_{-\infty}^{\infty} \left[G_l^m(t)\right]_Q e^{-i\omega t}dt$$

$$G_l^m(t) = < Y_l^m[\vartheta(t),\ \varphi(t)]Y_l^m[\vartheta(0),\ \varphi(0)] > \tag{4}$$

The motional narrowing approximation $J_l^m(\omega) = J_l^m(l\omega)$ is assumed in equation (3), and in the rest of this discussion. Also, $Y_l^m(\vartheta,\ \varphi)$ are spherical harmonics as functions of ϑ and φ, the polar and azimuthal angles of the interaction vector with respect to fixed laboratory coordinates. The quadrupole-coupling constant is a function of q, which is the molecular electric-field gradient at the nucleus in question. Changes in this constant reflect changes in molecular electronic structure. These constants are known for individual, noninteracting, molecules from microwave spectroscopic studies on gaseous D_2O and H_2O^{17}, where they are 319 kHz [3], and 9.8 MHz [4], respectively, and from NMR wide-line studies on polycrystalline solid D_2O, where it is 192 kHz [5]. The subscript on the correlation function in equation (3) emphasizes that it relates to the direction of the electric-field gradient at the nucleus.

For protons, the general equation for dipole—dipole induced relaxation between two nuclei with $I = 1/2$, magnetogyric ratio γ, and internuclear separation r, is [6]

$$\left(\frac{1}{T_1}\right)_D = \frac{3}{4}\frac{\gamma^4\hbar}{r^6}\left[J_l^m(\omega)\right]_D \tag{5}$$

where the subscript emphasizes that the correlation function refers to the dipole–dipole interaction and its direction. This equation must be used to find the intra- and inter-molecular contributions by taking into account the correlation functions for molecular angular reorientation and molecular translational motion.

Deriving the correlation functions on the basis of theoretical models for the liquid structure is not an easy matter. It has been more popular to assume a form for the entire function or to assume that the liquid is continuous and that a molecule undergoes Brownian rotational or translational motion as a classical sphere. Upon this latter assumption, the correlation functions for angular reorientation are simple exponentials of the form

$$G_l^m(t) = e^{-t/T_c} \tag{6}$$

where T_c is the correlation time. The following equations may then be derived for the relaxation times of the various nuclei:

Protons:

$$\left(\frac{1}{T_1}\right)_{D,\,\text{intra}} = \frac{3}{2}\frac{\gamma^4 \hbar^2}{r^6} T_c$$

$$\left(\frac{1}{T_1}\right)_{D,\,\text{inter}} = \frac{\pi}{5}\frac{N\gamma^4 \hbar^2}{aD} \tag{7}$$

where N is the number density of the molecules, a is the "radius" of a molecule, and D is the macroscopic diffusion coefficient.

Deuterons:

$$\left(\frac{1}{T_1}\right)_{Q,\,\text{H}^2} = \frac{3}{8}\left(\frac{e^2 qQ}{\hbar}\right)^2_{\text{H}^2} T_c \tag{8}$$

Oxygen-17:

$$\left(\frac{1}{T_1}\right)_{Q,\,\text{O}^{17}} = \frac{3}{125}\left(\frac{e^2 qQ}{\hbar}\right)^2_{\text{O}^{17}} T_c \tag{9}$$

and T_c is related to the microscopic dielectric-relaxation time by,

$$T_{\text{diel, mic}} = 3T_c \tag{10}$$

It is worthwhile asking how close these predictions come to the experimental results. Approximate numerical evaluation of equation (7) indicates that, for water,

$$\left(\frac{1}{T_1}\right)_{D,\text{intra}} \approx \left(\frac{1}{T_1}\right)_{D,\text{inter}} \tag{11}$$

The usual method for deriving the values of the correlation time from observables involves the assumption that the intercontribution is given accurately by an equation similar to (7) but corrected for changing number density. This contribution is then subtracted from the observed $1/T_1$ to find $(1/T_1)_{D,\text{intra}}$. From a knowledge of r ($r = 1.58$ Å for a free water molecule), T_c may be derived. It should be noted that it is there-

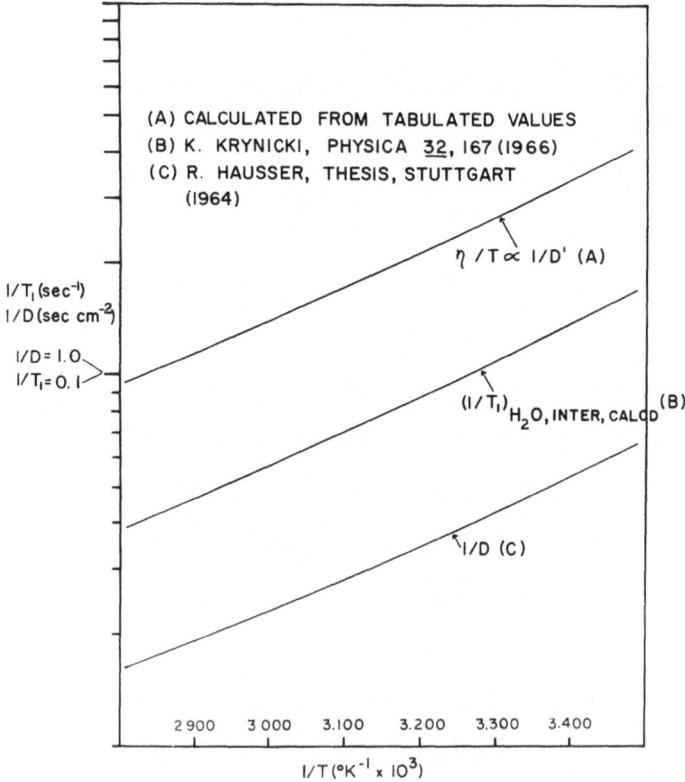

Fig. 1. Calculated and measured temperature dependences of reciprocal diffusion coefficients and spin-lattice relaxation time (intermolecular contribution) viscosity taken from standard tables).

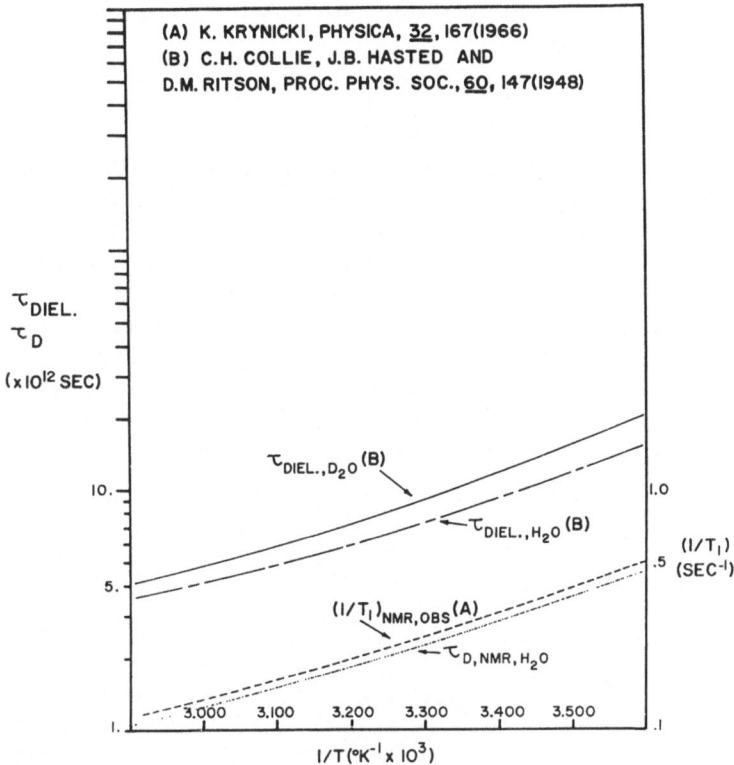

Fig. 2. Measured macroscopic dielectric relaxation times for H_2O and D_2O, measured reciprocal spin-lattice relaxation time for H_2O, and calculated dipole correlation time for H_2O.

fore implicitly assumed that the relaxing units are individual, unbonded water molecules. There is reasonable evidence to suggest that the values of $(1/T_1)_{D,\,inter}$ obtained are not sensitive to the liquid model chosen [7] and that the values of T_c are therefore reasonably accurate provided that the relaxing units are single molecules.

Figure 1 shows the temperature dependence of $(1/T_1)_{D,\,inter}$ calculated by such a procedure [8], contrasted with the reciprocal of the measured diffusion coefficient [9] and the viscosity (which, classically, should follow the relation $\eta/T \propto 1/D$. The agreement is quite good. Figure 2 shows the temperature dependence of the macroscopic dielectric relaxation time [10] and the correlation time derived as described above [8]. The ratio of $T_{diel,\,mac}$ to T_c is 3.7 and is found to be nearly temperature independent. If the Powles [11] correction is used to convert macroscopic to microscopic dielectric relaxation times, then, experimentally,

$$T_{\text{diel, mic, exptl}} = 2.5T_c \tag{12}$$

which compares favorably with the theoretical value of 3.0 from equation (10) for a Debye classical diffusion model. Figure 3 shows the results obtained from various NMR experiments [8, 12–14]. If the classical diffusion picture is retained, then the curvature of the $1/T_1$ plots for D_2O and H_2O^{17} must represent changes in the quadrupole-coupling constants as functions of temperature. As calculated from equations (8) and (9) [12, 14], this behavior is illustrated in Fig. 4. The curves show changes of electronic-field gradient at the oxygen and hydrogen nuclei of water as a function of temperature. Data of this type may fit into a model which explains the continuous changes

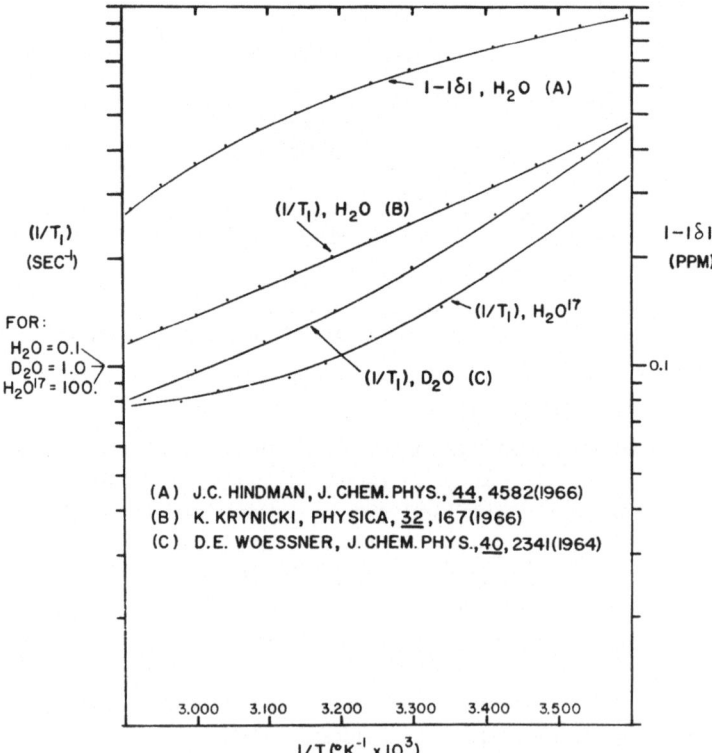

Fig. 3. Measured NMR parameters in isotopic waters, including the chemical shift, and $1/T_1$ for H_2O, D_2O, and H_2O^{17}.

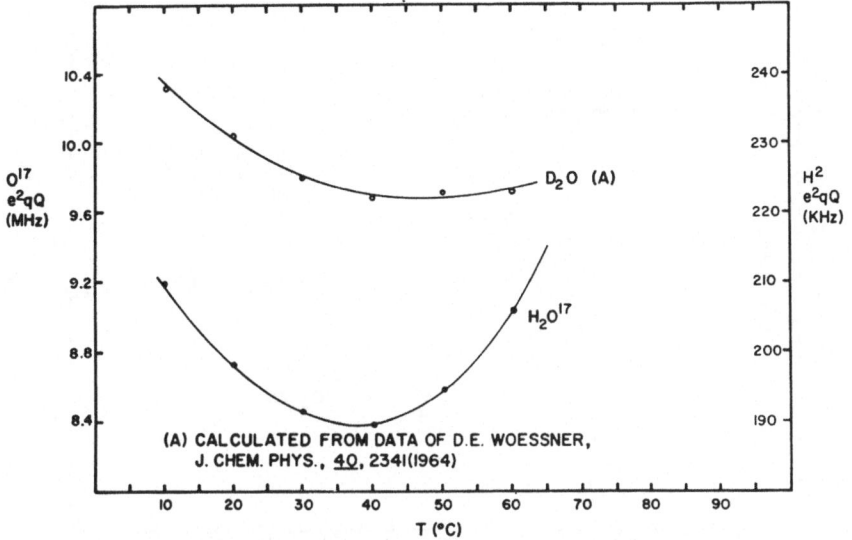

Fig. 4. Calculated temperature dependences for quadrupole-coupling constants of liquid D_2O and H_2O^{17}.

of some other related parameters with temperature, as shown in Fig. 5 [15–17].

On a more complex level, it must be realized that, for a molecule which does not reorient randomly, as a classical sphere, the correlation functions will not be single exponentials but more generally will be sums of exponentials with different sums entering into the dipole-dipole and quadrupole expressions [1, 2]. That is,

$$[G_l^m(t)]_D \neq [G_l^m(t)]_Q \tag{13}$$

and, also, in general,

$$[G_l^m(t)]_{Q, O^{17}} \neq [G_l^m(t)]_{Q, H^2} \tag{14}$$

Furthermore, the molecular electric-dipole moment angular-auto-correlation function $[G_1^o(t)]_m$ is different from all of the above. The theoretical problem cannot be solved rigorously in closed form for an asymmetric top molecule such as H_2O. Therefore, continuing experimental work is necessary to attempt to work out the nature and form of the various correlation functions and the relationships between them. Some information may possibly be derived from investigations of the shapes of IR and Raman bands, which, at least

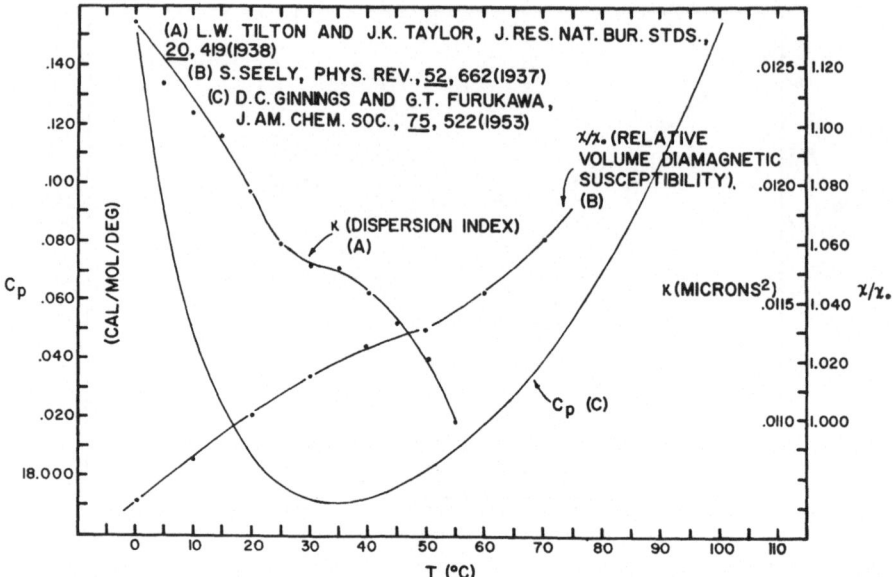

Fig. 5. Measured temperature dependences of heat capacity at constant pressure, relative volume diamagnetic susceptibility, and optical-dispersion index.

for symmetric top molecules, may in some cases be related to the correlation functions $G_i^\circ(t)$ and $G_i^m(t)$, respectively [18]. It is, in fact, hard to see how the Gaussian shapes usually assumed for Raman bands may be rigorously reconciled with the simple exponential decay which usually is found in NMR relaxation work and as discussed above, gives rather good correspondence with experiment, even on the basis of a very naïve theory.

Up to the present time all of the experimental work on bulk water indicates:

1. The macroscopic-dielectric and NMR-relaxation phenomena are described by exponential decays to thermal equilibrium.

2. The main features of relaxation of all types are described rather well by classical rotational and translational diffusion equations, so well, in fact, that it becomes problematical how much better more complex calculations will be.

3. The apparent activation energies of the transport properties such as macroscopic viscosity and diffusion are very close to the corresponding values for dielectric and NMR relaxational rotational reorientation times.

4. The molecular electronic structure of the liquid, as revealed by the quadrupole-coupling constants derived from D_2O and H_2O^{17} measurements, indicates a surprisingly small change from the gas-phase structure for O^{17} and a surprisingly small change from solid state structure for H^2.

5. The identity of the relaxing units has not yet been definitely established, either by experiment or theory, nor has the temperature dependence of their concentration. All of the above observations are, however, consistent with relaxation by single molecules which are not fully hydrogen bonded.

REFERENCES

1. H. Shimizu, J. Chem. Phys. 40:754 (1964).
2. W.A. Steele, J. Chem. Phys. 38:2411 (1963).
3. P. Thaddeus, L.C. Krisher, and J.H.N. Loubser, J. Chem. Phys. 40:257 (1964).
4. M.J. Stevenson and C.H. Townes, Phys. Rev. 107:635 (1957).
5. J.A. Jackson and S.W. Rabideau, J. Chem. Phys. 41:4008 (1964).
6. W.B. Moniz, W.A. Steele, and J.A. Dixon, J. Chem. Phys. 38:2418 (1963).
7. P.K. Sharma and S.K. Joshi, Physiol. Rev. 132:1431 (1963).
8. K. Krynicki, Physica 32:167 (1966).
9. R. Hausser, Thesis, Stuttgart (1964). (Values are quoted in Ref. 8.)
10. C.H. Collie, J.B. Hasted, and D.M. Ritson, Proc. Phys. Soc. (London) 60:147 (1948).
11. J.G. Powles, J. Chem. Phys. 21:633 (1953).
12. J.A. Glasel, unpublished experiments. (T_1 was measured by pulsed magnetic field technique.)
13. J.C. Hindman, J. Chem. Phys. 44:4582 (1966).
14. D.E. Woessner, J. Chem. Phys. 40:2341 (1964).
15. L.W. Tilton and J.K. Taylor, J. Res. Natl. Bur. Std. 20:419 (1938).
16. S. Seely, Phys. Rev. 52:662 (1937).
17. D.C. Ginnings and G.T. Furukawa, J. Am. Chem. Soc. 75:522 (1953).
18. R.G. Gordon, J. Chem. Phys. 42:3658 (1965).

Proton Magnetic Resonance Studies of Water Structure*

J. C. Hindman

Chemistry Division
Argonne National Laboratory
Argonne, Illinois

The problem of interpreting the proton chemical-shift data for water and its variation with temperature is discussed. By using a derived value for the chemical shift of fully hydrogen-bonded water (ice) and experimental data for the chemical shift of monomeric water molecules, it is shown that the gross features of the data can be accounted for by a model involving an equilibrium between a hydrogen-bonded "icelike" component and a monomeric species interacting with the lattice or with itself only by dispersion forces. Different kinds of experimental data do not, however, yield results completely consistent with this model, which indicates that there are inadequacies in the model. Discrepancies are also found when various other two-component models are compared. The possibility is considered that changes other than the breaking of hydrogen bonds must be taken into account. It is noted that the proton resonance is sensitive to the stretching and bending of hydrogen bonds. The implications of this observation for the interpretation of data on water structure is discussed.

INTRODUCTION

Without question the proton chemical shift in water and its variation with temperature are connected with and reflect changes in the hydrogen-bonded structure of the liquid [1–4]. The nature of these changes are not, however, determinable from the experimental data. It is necessary, therefore, to consider alternative interpretations of the data. We have considered two general possibilities. The first is that the chemical shift and its temperature variation are direct measures of the number of hydrogen bonds present in the liquid. The second is that other structural changes such as the stretching or bending of hydrogen bonds are responsible for the shielding changes.

*Based on work performed under the auspices of the U.S. Atomic Energy Commission.

251

If the first possibility is correct, the proton-resonance data might provide a definitive answer to the question of which of the many models for water structure where the breaking of hydrogen bonds is considered the primary process, with their widely different predictions as to the number of bonds broken in the ice-water transition and in the liquid as a function of temperature, is most nearly correct. There are, however, certain difficulties in making the comparison. First, we cannot determine the shielding parameters for either the fully hydrogen-bonded species or for unbonded water molecules from the data on water alone. We have not, e.g., been able to determine the chemical shift of ice because of the breadth of the proton resonance in the solid, nor have we a chemical shift for liquid water under conditions where we know we have only monomeric water molecules. Despite these difficulties, there are several methods of treating the data that provide a comparison of the proton-resonance results with the predictions of the different models.

A procedure that has been used at Argonne is to calculate the fraction of hydrogen-bonded species at 0°C from thermal data upon assuming an equilibrium between ice and monomeric water molecules [2]. The shielding properties of the monomeric water molecules are derived from chemical-shift measurements on water molecules in organic solvents [2]. The shielding parameter for the hydrogen-bonded fraction is then calculated from the equilibrium relationship. The temperature coefficient for the equilibrium can then be obtained from the experimental observations on the chemical shift. An alternative procedure, used by Muller [3], is to calculate the shielding constants of the two species by using the values for the fraction of hydrogen-bonded species for a given model at two temperatures and then to compare these values with, e.g., the shielding constant of a nonbonded water molecule obtained from observations on water in an organic solvent. Another procedure, used by Ruterjans and Scheraga [4], is to calculate the chemical shift for different models as a function of temperature. They start with the assumption that the temperature coefficient of the shift is a constant determined by the mole fraction of the hydrogen-bonded species given by a specific model and the chemical shift at 0°C. The shift as a function of temperature is then obtained by multiplying this constant by the integral over the temperature range of the change in mole fraction with temperature. Variations of these different procedures are used in the present paper for comparison of the different models.

To investigate the alternative possibility that the shielding changes can be accounted for by the bending or stretching of the hydrogen bonds in the water structure, the procedure was as follows [2]: It is assumed that the hydrogen bond can be represented by an electrostatic model [5, 6]. Use is made of the theory for the electric-field effect on the shielding of a hydrogen in a bond [7, 8] to calculate the polarization contribution to the shielding as a function of the intermolecular separation of the bonded water molecules and as a function of the angle of bend of the hydrogen bonds. Since it is expected that this polar contribution is the predominant contribution to the shielding of the hydrogen-bonded species, the calculated values can be directly compared with the experimental observations. The results of such comparisons are discussed in the text.

EXPERIMENTAL

A detailed discussion of the experimental procedures has been given elsewhere [2]. Essentially it consisted of measuring the chemical shift of either gaseous or liquid water as a function of temperature (and, for the gas, as a function of pressure) relative to gaseous methane by using a Varian Associates V-4300B high-resolution spectrometer operating at 40 to 60 Mc. The chemical shifts were measured by a sideband technique. Temperature control was maintained with a gas-flow cryostat. Temperature calibration was obtained by a substitution technique with a copper-constantan thermocouple as reference.

Samples for the measurement of the chemical shift of methane gas in water were made by condensing a given volume of gas at known pressure into tubes containing water and then sealing the tubes. The "solvent"-shift measurements were made by using samples of dried and water-containing solvents sealed in tubes containing water-filled reference capillaries. For cyclohexane and cyclopentane, the solvent shifts were obtained by extrapolation to unit solvent concentration of measured shifts in mixtures of these solvents with benzene [9]. In nitromethane, the shifts were measured as a function of water concentration and extrapolated to zero water concentration.

By convention, signals at higher field than the reference have more-positive values of the chemical shielding parameter σ. The experimental chemical shift for the gas-to-liquid transformation at 0°C is −6.16 ppm. After correction for bulk susceptibility, the shielding change for the transformation of water monomer to water at 0°C is −4.656 ppm. Values at other temperatures are given in Table I [2]. Values for the solvent shifts of water at infinite dilution in

TABLE I

Chemical Shifts of Water
and Methane in Selected Solvents

Solvent	Temp., °C	"Solvent shifts," ppm	
		σ_{CH_4}	σ_{H_2O}
H_2O	0	−0.41	−4.66
	25	−0.37	−4.37
	50	−0.39	−4.12
	75	−0.39	−3.88
	100	−0.39	−3.66
C_6H_{12}	25	−0.24	−0.33*
C_5H_{10}	25	−0.16	−0.35
CH_3NO_2	0	. . . .	−1.72
	25	−0.20	−1.62
	50	. . . .	−1.41
	75	. . . .	−1.03
	100	. . . .	−0.90

*Independent of temperature, 25–75°C.

selected organic solvents are also given in Table I. These data, together with the solvent shift of methane gas in water (Table I) and the solvent shifts of methane gas in these same organic solvents [10, 11], provide an experimental basis for the ·estimation of the shielding for "monomeric" water molecules in various environments.

TREATMENT OF THE DATA

One of the simplest models that results from the concept of water as a broken-down ice structure is that of a two-component system composed, on one hand, of a hydrogen-bonded "icelike" material and, on the other, of a non-hydrogen-bonded disordered fluid. Because such a model, modified in various ways, has proved adequate as a basis for explaining many of the static and dynamic properties of water, it has been considered that it provides a suitable initial framework for treatment of the chemical-shift data.

For the equilibrium between the states of such a two-component system, we can write

$$\text{Non-hydrogen-bonded water} \underset{\rightleftharpoons}{K} \text{Hydrogen-bonded water} \tag{1}$$

with

$$K = X_{HB}/(1 - X_{HB}) = \frac{\text{Fraction hydrogen-bonded water}}{\text{Fraction non-hydrogen-bonded water}} \tag{2}$$

from which

$$X_{HB} = \frac{\sigma_E - \sigma_f}{\sigma_{HB} - \sigma_f} = \frac{\sigma_E - \sigma_f}{\sigma_T} = \frac{E_s - E_f}{E_{ice} - E_f} \tag{3}$$

where E_s and σ_E are, respectively, the experimental sublimation energy and shielding constant of liquid water, and E_f and σ_f are the corresponding quantities for the non-hydrogen-bonded water fraction.

Since there was no direct measurement of either σ_{HB} or σ_f, these quantities have been evaluated in a semiempirical manner. One procedure used [2] is to calculate the quantity X_{HB} at 0°C by using thermodynamic arguments. Experimental observations on the shielding of methane in water and in organic solvents are then used to calculate σ_f for non-hydrogen-bonded water. These data are next combined in equation (3) to yield a value for σ_{HB}. Alternatively, a procedure used in our earlier paper has been reversed and the theory for the electric field effect [8] has been used to calculate σ_{HB} from the dipole moment of a water molecule in ice.

In making the thermodynamic calculations, two idealized cases have been considered. In both, it is thought that the hydrogen-bonded fraction has the properties of ice. It is thought that, in one (model 1), the nonbonded molecules interact with the lattice or with one another only by dispersion forces, and that, in the second (model 2), these molecules form a separate phase and interact by dipole forces [12]. In both cases, it is assumed that the separations between unbonded water molecules and their nearest neighbors are between 3.2 and 3.4 Å. This near-neighbor separation corresponds to that derived for such unbonded water molecules for X-ray radial-distribution data [12–14]. Calculations then indicate that, when the structural breakdown yields unbonded water molecules interacting only by dispersion forces, the fraction of unbroken hydrogen bonds X_{HB} is 0.845 at 0°C, in agreement

with earlier calculations by Pauling [15]. Where the water is assumed
to form a dipolar liquid, the results are not significantly different,
with X_{HB} equal to 0.82 at 0°C. It can also be noted that the orders of
magnitude of X_{HB} obtained from these calculations are very close to
those obtained by using other data, e.g., dielectric [2] or thermodynamic [16, 17], where it is assumed that the nonbonded water
molecules are monomeric species. It can also be noted that, although
the calculations are for the formation of the nonpolar water (model 1)
as a separate phase, the model calculations [2] indicate that they
should be reasonable approximations for the case in which the water
is in an interstitial site [16–18].

To calculate σ_f for the two kinds of water, use is made of the
experimental solvent-shift data in the following way: The experimental shielding or solvent shift of water in an organic solvent reflects contributions from both dispersion and dipole interactions, i.e.,

$$\sigma_{exp} = \sigma_f = \sigma_L + \sigma_R \tag{4}$$

In a low-dielectric solvent, the dipole or reaction-field contribution
is minimized, and it is convenient to estimate its order of magnitude
by using the procedure of Buckingham [8], where

$$\sigma_R = -A(R) - B(R^2) \tag{5}$$

with

$$R = [2(\varepsilon_1 - 1)(n_2^2 - 1)/3(2\varepsilon_1 + n_2^2)][\mu_2 \cos \theta_2 / a^2] \tag{6}$$

With $A = 2.26 \times 10^{-12}$ (see later text), $B = 0.74 \times 10^{-18}$, $n_{H_2O}^2 = 1.79$,
$\mu_{H_2O} = 1.84$ D, $a_{H_2O} = 1.44 \quad 10^{-24}$, $\cos \theta_2 = 0.6115$, we find, that for
H_2O in cyclohexane with $\varepsilon = 2.013$, $\sigma_R = -0.17 \times 10^{-6}$ [2] yield a value
of $\sigma_R = -0.17 \times 10^{-6}$. From equation (4) and the experimental solvent
shifts in Table I, the dispersion contributions are then σ_L (H_2O in cyclohexane) $= -0.16 \times 10^{-6}$ and σ_L(H_2O in cyclopentane) $= -0.18 \times 10^{-6}$.
To obtain the quantity of interest, i.e., the dispersion contribution to
the shielding of a nonbonded water molecule in water, use is made
of a relationship derived from a theory [19] for the dispersion effect,
i.e.,

$$\sigma_L(H_2O \text{ in } H_2O) = [\sigma_{exp} - \sigma_R]_{H_2O \text{ in } C_6H_{12}} \times \frac{\sigma_{exp}(CH_4 - H_2O)}{\sigma_{exp}(CH_4 - C_6H_{12})} \tag{7}$$

together with the experimental values for the solvent shifts for methane in water and in organic solvents. From the data for cyclohexane and water, $\sigma_L(H_2O$ in $H_2O) = -0.27 \times 10^{-6}$, and, from the data on cyclopentane and water, $\sigma_L(H_2O$ in $H_2O) = -0.45 \times 10^{-6}$. Comparable values have been obtained by similar procedures [2].

For the dipolar water, consideration of the theoretical relationships for the dispersion and reaction-field contributions suggest that a more direct estimate of the solvent shift could be obtained from measurements of the chemical shift of water in a noncomplexing polar solvent of selected dielectric properties. Nitromethane is one possibility. Its dielectric behavior indicates that it is a nonassociated polar solvent [20]. Use of such a solvent, however, requires that it not form a complex with water. The literature is contradictory on this question [21]. The possibility that a weak water—nitromethane complex is formed is indicated by two aspects of the NMR data. At low temperatures, the chemical shift of water in nitromethane is strongly dependent on the water concentration. It is also observed that the temperature coefficient of the chemical shift is significantly different than would be expected if only dispersion and dipole forces are important. Fortunately, however, at higher temperatures, both of these effects are markedly reduced so that, at 100°C, where the dielectric constant of the nitromethane is estimated to approximate closely that of our dipolar water, the experimentally observed shift is probably a good approximation of the value that we are seeking. Some confidence in this conclusion is suggested by examination of the data in Table II, where it is noted that the calculated and observed values for the chemical shift of water in nitromethane are of the same order of magnitude at this temperature. When the data are combined, we obtain, for model 1, $\sigma_{HB} = -5.44$ ppm and $\sigma_f = -0.36$ ppm and, for model 2, $\sigma_{HB} = -5.48$ ppm and $\sigma_f = -0.90$ ppm.

These σ_{HB} values can be compared with those calculated from the dipole moment of the water molecule in ice. These calculations are made as follows: It is assumed that the hydrogen bond can be represented by an electrostatic model [5, 6] in which the hydrogen atom is in an internal field λ/R^2 and an external field arising from the polarizing molecule in the hydrogen bond [2]. The shielding change induced in the hydrogen atom owing to this external field E can be written

$$\sigma_p \cong \sigma_{HB} = -AE_z - BE^2 \tag{5a}$$

where E_z is the component of the external field along the O—H bond, and the constant A is dependent upon the internal field λ/R^2. In our

TABLE II

Comparative Properties of Nitromethane and Water at 0°C

Solvent	n^2	ε	μ_{gas}, debyes	σ_R(calc),* ppm	σ_L(calc)*	σ_{H_2O}, ppm
CH_3NO_2	1.80	39.35	3.44	-0.47	-0.20	-1.72[†]
H_2O (d)[‡]	1.79	26.5	1.84	-0.46	-0.36	(-0.90)[§]
H_2O (n)[¶]	1.79	88.15	1.84	-0.48	-0.36	-4.66

*For water as solute, see text.
[†]Observe for $H_2O-CH_3NO_2$ at 0°C.
[‡]Model 2 water, dipolar.
[§]Observed for $H_2O-CH_3NO_2$ at 100°C, $\varepsilon_{CH_3NO_2} \sim 28$.
[¶]Normal water.

original treatment, a value of $A = 2.26 \times 10^{-12}$ was deduced from a consideration of the internal charge distribution in a polarized water molecule. With a value of 2.45 debyes for the dipole moment for ice derived from dielectric [2] or a model calculation [22], $\sigma_{HB} = 5.50$ ppm. In order to get some idea of the possible limits for σ_{HB}, further calculations have been made as follows: From perturbation theory [23], it has been calculated that A should be approximately 0.7 of the value for a C—H bond. With the experimental [24] and theoretical [23] values for A for C—H bonds, this indicates that A for the O—H bond should be between 1.8 and 2.2×10^{-12}. If we also take into account the recent calculation of $\mu = 2.60$ debyes for the dipole moment [25], the calculated values of σ_{HB} lie within the limits of −4.64 and −6.02 ppm. In actuality, the experimental value for water at −15°C, $\sigma_{exp} = -4.85$ ppm, indicates a lower limit for σ_{HB}. The σ_{HB} values for the two models are midrange of these limits.

DISCUSSION

It is, at this point, of interest to compare the deductions about the properties of the component species made in the present paper with those obtained from other two-component models. One method of doing this is to use the estimates of the fractions of the two states at some temperature reported for the different models to calculate the shielding properties for the hydrogen-bonded and -unbonded species and to compare them with those obtained in the present work. In particular, we can compare the results for the unbonded water molecules, whose shielding properties have been defined experimentally. A second method is to compare the temperature coefficient for the equilibrium between the components as deduced for the specific model with that obtained from the chemical shift. The procedure adopted in the present paper for making this comparison is to use the X_{HB} value at 0°C for the model under consideration, together with the experimentally derived values for the shielding of the unbonded species, to calculate a σ_{HB} value. The σ_f value to be used is selected on the basis of the original author's statement as to the kind of water molecules produced by the breakdown of the lattice structure. The shielding values for the two species are then combined with the experimental chemical-shift data to calculate X_{HB} for 100°C. These X_{HB} values are then compared with those given by the original authors for their models.

The data for various representative models are summarized in Table III. In addition to the idealized models treated in the present paper, these include: (1) the Nemethy—Scheraga model, involving a hydrogen-bonded cluster and a monomeric dipolar liquid phase; (2) the Marchi—Eyring model, with an "icelike" component and a monomeric "gaslike" component; (3) the Davis—Litovitz model, with an icelike fraction and a partially hydrogen-bonded close-packed fraction. A number of comments can be made with respect to these data. First, we can note that an extension of the thermodynamic

TABLE III

Comparative Values of Hydrogen Bonding and Shielding Parameters

Model	X_{HB} 0°C	X_{HB} 100°C	σ_{HB} (calc), ppm	σ_f (calc), ppm	σ_f (exp), ppm	σ_{HB} (calc), ppm	X_{HB} (calc) 100°C
H bond			-4.84 to -6.02				
Nonpolar water					-0.36		
Dipolar water					-0.90		
Ice nonpolar water	0.845				-0.36	-5.44	0.65
Thermodynamic*	0.845	0.71	-5.22	-1.58	-0.36	-5.44	0.65
Marchi—Eyring	0.975	0.60	-4.72	-2.07	-0.36	-4.77	0.75
Davis—Litovitz	0.82	0.64	-5.66	-0.07	-0.36	-5.60	0.63
Ice dipolar water	0.82				-0.90	-5.48	0.54
Nemethy—Scheraga	0.528	0.325	-6.97	-2.07	-0.90	-8.01	0.39

*From energy of sublimation data at 0 and 100°C. Dielectric data [2] yield similar results.

calculations [equation (3)] used to derive the value of σ_{HB} for the system of ice–nonpolar water does not yield a value of σ_t or a temperature coefficient for the equilibrium in agreement with that obtained by using the chemical-shift data in conjunction with the value of X_{HB} at $0°C$. This is most certainly an indication that this very simple model is inadequate. Similar comments can be made concerning both the Marchi–Eyring and Nemethy–Scheraga models, where even greater discrepancies are noted. For the Marchi–Eyring model, the temperature coefficient of the equilibrium, in particular, cannot be reconciled with the shielding data. Both the σ_{HB} and σ_t values calculated from the Nemethy–Scheraga model appear to be outside the limits that we should expect for these quantities. Further, the discrepancy between the σ_{HB} value calculated for this model and the limits within which we expect σ_{HB} to lie is increased if this quantity is calculated by using X_{HB} for the model at $0°C$ and the appropriate value for σ_t. As noted by Muller [3], the Davis–Litovitz model appears to be most concordant with the chemical-shift data. The reason for this concordance, in spite of the differences in details for the structures of the two components in the models, is, of course, the agreement between the two models in the estimate of the number of hydrogen bonds broken at a given temperature and the assumption in both that nonbonded molecules interact only by dispersion forces. The apparent agreement is, however, somewhat misleading since a closer examination reveals that the two models should not in fact yield the same results. For example, in the Davis–Litovitz model, the hydrogen-bonded near–neighbor distance is thought to be between 2.80 and 2.82 Å, significantly greater than the 2.76-Å value for ice. As indicated later in the text, we should expect this amount of bond stretching to reduce significantly the number of hydrogen bonds broken in the ice-water transition.

It is apparent that the various data cannot be reconciled on the basis of the simplest kind of model, involving only an ice fraction and monomeric fraction. In considering various modifications that might be made to improve the fit of the data, it must be recognized that it is by no means certain that monomeric water molecules exist in appreciable concentration in liquid water [26]. Although X-ray data indicate the existence of short-range order in liquid water, there is, in fact, no conclusive evidence for the existence of either ice fragments or monomeric water molecules, or, indeed, for any kind of discretely structured regions in the liquid. Consequently, much of the experimental data can be interpreted equally well with a variety of models, including the continuum model of Pople [6], where it is

assumed that few hydrogen bonds are broken but there is a considerable amount of stretching and bending of the bonds. In particular, the
increase in the average intermolecular distance of near-neighbor
water molecules, observed in melting ice and in liquid water with increasing temperature, indicates that the hydrogen bonds are stretched.
It seems desirable, therefore, to consider how the interpretation of
the chemical-shift data might be modified if the hydrogen bonds are
bent or stretched.

To examine this question, the idea was again used that the hydrogen bond can be represented by the electrostatic equivalent of a
quantum mechanical model. With this model, values of E_z, E, and
σ_p [equation (5a)] are readily calculated as a function of the intermolecular separation of the water molecules or the angle of bend of
the hydrogen bond. Since $\sigma_p \cong \sigma_{HB}$, we can, by comparing the calculated σ_p values with the experimental shielding values σ_E, determine
how much bending or stretching would have to occur to account for
the whole of the shielding changes. Alternatively, we could calculate
the fraction of bonds broken or an assumed amount of stretching or
bending [equation (3)]. Examples of such calculations are given in
Table IV. Several aspects of these calculations are of interest. For
example, consideration of the data in this table indicates that:

1. We could essentially account for the experimental chemical-
shift data on the assumption that the change in average intermolecular
distance as deduced from X-ray data is due to the stretching of hydrogen bonds.

TABLE IV

Calculation of Fraction O-Bonded Water as Function
of O—O Distance or Angle of Bend in Hydrogen-
Bonded Molecules at 0°C*

O—O, Å	σ_p(calc), ppm	f_0	Angle, deg	σ_p(calc)	f_0
2.76	−5.42	(0.155)	0	−5.42	(0.155)
2.80	−5.12	0.10	10	−4.96	(0.068)
2.83	−4.64	0	13	−4.66	0
2.90	−4.02	....	23	−3.84	
3.00	−3.32	....	26	−3.11	

*σ_{exp} = −4.66 ppm at 0°C and −3.66 ppm at 100°C; A [equation (5a)] taken
as 2.26×10^{-12}.

2. The chemical shift is markedly sensitive to the bending of hydrogen bonds, and a relatively small amount of bending could account for all of the observed shielding change.

3. All three processes, bond breaking, bond stretching, and bond bending, affect the chemical shift in a similar fashion with the result that, if some bonds are broken, as must be the case since water is a liquid, then the amount of stretching or bending that can occur simultaneously will be limited.

As examples, we can consider the Davis—Litovitz [14] and Pople models [6]. As noted before, in the Davis—Litovitz model, the hydrogen-bond distance is considered between 2.80 and 2.82 Å. The data in Table IV indicate that, for this intermolecular separation, the amount of bond breaking at 0°C would be limited from 0 to 10%. Similarly, in the Pople model, excluding the requirement that some bonds must be broken, calculations similar to those in the table indicate that the assumption that the O—O distance is 2.80 Å would limit the amount of bond bending to 7° compared with the 26° calculated by the author.

In summary, we can state that, although proton-resonance data cannot be interpreted by a unique model for the water structure, it does appear that the data can be used in several ways to test the consistency of suggested models. This usefulness should increase as further data become available, particularly if we are able to determine within closer limits the shielding constant for fully hydrogen-bonded water and better define the electric-field effect on the hydrogen in the O—H bond.

REFERENCES

1. W. G. Schneider, H. J. Bernstein, and J. A. Pople, J. Chem. Phys. 28:601 (1953).
2. J. C. Hindman, J. Chem. Phys. 44:4582 (1966).
3. N. Muller, J. Chem. Phys. 43:2555 (1965).
4. H. H. Rüterjans and H. A. Scheraga, J. Chem. Phys. 45:3296 (1966).
5. J. Lennard-Jones and J. A. Pople, Proc. Roy. Soc. (London) A205:155 (1951).
6. J. A. Pople, Proc. Roy. Soc. (London) A205:163 (1951).
7. T. W. Marshall and J. A. Pople, Mol. Phys. 1:199 (1958).
8. A. D. Buckingham, Can. J. Chem. 38:300 (1960).
9. W. B. Dixon and J. C. Hindman, unpublished observations.
10. A. D. Buckingham, T. Schaefer, and W. G. Schneider, J. Chem. Phys. 32:1227 (1960).
11. N. Lumbroso, T. K. Wu, and B. P. Dailey, J. Phys. Chem. 67:2469 (1963).
12. G. Nemethy and H. A. Scheraga, J. Chem. Phys. 36:3382 (1962).
13. J. Morgan and B. E. Warren, J. Chem. Phys. 6:666 (1938).
14. C. M. Davis and T. A. Litovitz, J. Chem. Phys. 42:2563 (1965).
15. L. Pauling, The Nature of the Chemical Bond, 2nd ed., Cornell University Press, Ithaca, N. Y. (1944), p. 304.

16. E. Försland, Acta Polytech. Chem. Met. Ser. 3, 12:42 (1952).
17. H.S. Frank and A.S. Quist, J. Chem. Phys. 34, 604 (1961).
18. L. Pauling, Hydrogen Bonding, Symp. Papers, Ljubljana, 1957, 1 (1959).
19. B.B. Howard, B. Linder, and M.T. Emerson, J. Chem. Phys. 36:485 (1962).
20. C.P. Smythe and W.S. Walls, J. Chem. Phys. 3:557 (1935).
21. H.E. Ungnade, E.M. Roberts, and L.W. Kissinger, J. Phys. Chem. 68:3225 (1964).
22. E. J.W. Verwey, Rec. Trav. Chim. 60:887 (1941).
23. J.T. Musher, J. Chem. Phys. 37:34 (1962).
24. T. Schaefer and W.G. Schneider, Can. J. Chem. 41:966 (1963).
25. C.A. Coulson and D. Eisenberg, Proc. Roy. Soc. (London) A291:445 (1966).
26. D.P. Stevenson, J. Phys. Chem. 69:2145 (1965).

The Frequency Distribution of Ice by Neutron Scattering

Henry Prask* and Henri Boutin†

Picatinny Arsenal
Dover, New Jersey

and

Sidney Yip

Massachusetts Institute of Technology
Cambridge, Massachusetts

The frequency distribution of phonons in a solid known to be proportional to the one-phonon neutron incoherent-scattering cross section when the crystal is monatomic and cubic. In the present work, it is shown that, for molecular crystals, an "effective" frequency distribution, which is simply related to the true frequency distribution, can be derived directly from a measurement of the energy and angle differential neutron cross section if translation-rotation couplings can be ignored. This approach is applied to ice, for which the differential neutron cross section at 150°K has been measured from 20 to 1000 cm⁻¹ with the use of a beryllium-filter time-of-flight spectrometer. The number of degrees of freedom of rotational and translational motions is assumed equal from which an effective rotational mass of 1.3 amu is obtained. The "true" frequency distribution determined from the neutron data with this value of effective mass is used to calculate absolute values of thermodynamic quantities as a function of temperature (20 to 270°K), in reasonably good agreement with directly measured values. The application of this approach to liquid water and H_2O in crystal hydrates is currently being investigated.

INTRODUCTION

In recent years, inelastic neutron scattering has become an important tool for studying thermal vibrations in solids [1]. Perhaps the most significant contribution thus far has been the measurement of phonon dispersion relations in simple crystalline solids by coherent scattering. Neutron spectra of molecular solids, especially substances

*Guest at the U. S. Army Materials and Mechanics Research Center, Watertown, Mass.
†Presently at Research Laboratories Division, Bendix Corporation, Southfield, Mich.

containing hydrogen, have also been obtained, but these measurements have not been systematically analyzed because of the formidable computational problem involved. With the exception of polyethylene [2], very few normal-mode calculations have been carried out for molecular systems which have not been limited to optically active modes.

It is well known that in one-phonon incoherent scattering by monatomic cubic crystals, the energy spectrum of scattered neutrons is directly proportional to the density of phonon states [3]. This relation makes it possible to measure the phonon frequency-distribution function $g(\omega)$ in certain crystals (e.g., vanadium). For crystals with more than one atom per unit cell, the neutron spectrum is determined by a frequency function $g'(\omega)$ which differs fundamentally from $g(\omega)$. Whereas $g(\omega)$ can be defined in terms of the dispersion relation, $g'(\omega)$ involves both the dispersion relation and the polarization vector of normal modes.

In this paper we propose a method for determining the low-frequency portion of $g(\omega)$ for molecular solids. It is assumed that, in this frequency region, $g(\omega)$ has contributions from motions associated with center-of-mass translations and molecular rotations (or librations). We show that the neutron-scattering data depend directly on an effective frequency distribution $g_{eff}(\omega)$ which is quite simply related to $g(\omega)$. In this theoretical approach, intra- and intermolecular modes are assumed separable since intramolecular interactions are in general much stronger than intermolecular forces. To make the calculation tractable, it is also assumed that translation—rotation coupling is negligible. This constitutes the most probable source of error in the theoretical model. Using the same assumptions, Hahn [4] has derived an expression for the incoherent cross section for molecular crystals and has introduced different frequency functions for translations and librations. His analysis is quite formal and bears close resemblance to lattice dynamical calculations. We have used more physical arguments but arrived at comparable results. In both cases, an enhancement factor for the librational effects in the neutron spectrum was found. It is then to be concluded that neutron data alone are not sufficient to derive $g(\omega)$. However, this difficulty can be avoided if either the translational or librational component is known a priori or if they can be assumed to occur in distinct frequency regions.

A test of the theoretical model is made by application to inelastic neutron scattering data obtained for H_2O ice. The effective and true frequency distributions are determined from the data and the specific heat and other thermodynamic quantities calculated in the harmonic approximation. There is reasonable agreement between calculated and measured thermodynamic quantities, which suggests that the model proposed may have useful application for determining frequency distributions of other molecular solids.

NEUTRON SCATTERING BY HYDROGENOUS SOLIDS

The scattering of low-energy neutrons by a hydrogenous sample is known to be predominantly incoherent because the bound-atom cross section σ for hydrogen is 81.2 barns as compared with a few barns or less for other nuclei. It is therefore usually a good approximation to interpret the scattering data by only the hydrogen contributions. For an assembly of molecules each containing N hydrogens, the energy and angular-differential cross section can be written as

$$\left(\frac{d^2\sigma}{d\Omega\,d\omega}\right)_{molecule} = \left(\frac{E}{E_0}\right)^{1/2} \frac{\sigma_H}{8\pi^2} \sum_{s=1}^{N} \int_{-\infty}^{\infty} dt\, e^{-i\omega t} \chi_s(\kappa, t) \tag{1}$$

where

$$\chi_s(\kappa, t) = \left\langle e^{i\kappa \cdot r_s(t)} e^{-i\kappa \cdot r_s(0)} \right\rangle \tag{2}$$

In equation (1), $\omega = E - E_0$, $\kappa = k_0 = k$, where E, k and E_0, k_0 are, respectively, the final and initial neutron energies and wave vectors, and σ_H refers to hydrogen. The position of the sth hydrogen at time t is denoted as $r_s(t)$, and, in (2), the angular bracket indicates an average over a thermal equilibrium distribution.

The intermediate scattering function $\chi_s(\kappa_1, t)$ contains all the relevant dynamical information for the collision process. By considering external and internal modes separately, we are effectively taking χ to be a product of two scattering functions. Since the incident neutron energies in inelastic scattering experiments are always below the threshold for excitation of internal modes, we can safely replace χ_{int} by unity (or its zero-point vibration expression) and consider the molecules rigid bodies. The assumption of no translation—rotation couplings is less justified, but it is necessary in the present approach, as it is in all other existing calculations [5]. To introduce this approximation, we write

$$r_s(t) = R(t) + b_s(t) \tag{3}$$

and consider the center-of-mass position $R(t)$ and internal position $b_s(t)$ dynamically independent. Then χ_s becomes a product of translation and rotation scattering functions,

$$\chi_T(\kappa, t) = \left\langle e^{i\kappa \cdot R(t)} e^{-i\kappa \cdot R(0)} \right\rangle \tag{4}$$

$$\chi_R(\kappa, t) = \left\langle e^{i\kappa \cdot b_s(t)} e^{-i\kappa \cdot b_s(t)} \right\rangle \tag{5}$$

The calculation of χ_T offers no difficulty because the problem of monatomic crystals has been discussed extensively in the literature. We have the well-known expression [3, 6]

$$\chi_T(\kappa, t) = \exp\left\{\frac{\kappa^2}{2M}\left[\gamma_T(t) - \gamma_T(0)\right]\right\} \tag{6}$$

$$\gamma_T(t) = \int_0^\infty d\omega\ g_T(\omega)\Lambda(\omega, t) \tag{7}$$

$$\Lambda(\omega, t) = \omega^{-1}\left[\coth\left(\frac{\omega}{2T}\right)\cos\ \omega t - i\sin\ \omega t\right] \tag{8}$$

where M is the molecular mass and $g_T(\omega)$ is the translational frequency-distribution function and the measured neutron spectrum.

In view of our familiarity with an expression like equation (6), it is reasonable to express χ_R in a similar form. Such a procedure would be quite appropriate if the rotational (librational) degrees of freedom of the molecules are expected to be strongly hindered by intermolecular forces. Some care, however, is required in formulating an oscillator type of approximation for χ_R because differences in center-of-mass and librational motions cannot be altogether ignored. As a first-order approximation we have considered a model description of hindered rotation and examined its behavior in the limit of small angle torsions. For a linear molecule, it can be shown that [7]

$$\chi_{\text{hin rot}} \cong \exp\left[\frac{\kappa^2}{2M_r}\Lambda(\omega_r, t)\right] \tag{9}$$

where ω_r is the torsion frequency and M_r is an effective rotation mass. Then,

$$M_r = \frac{3I}{2b_2} = \frac{3m_H}{2}\ \frac{M}{M - m_H} \tag{10}$$

where I is the moment of inertia, $b = |b|$, and m_H is the mass of the scatterer, i.e., hydrogen.

It is interesting to compare our model result with the familiar expression for a simple oscillator of mass M_0 and frequency ω_0,

$$\chi_{\text{osc}} = \exp\left\{\frac{\kappa^2}{2M_0}\left[\Lambda(\omega_0 t) - \Lambda(\omega_0, 0)\right]\right\} \tag{11}$$

Aside from the time-independent part $\exp[-(\kappa^2/2M)\Lambda(\omega_0, 0)]$, which is the Debye–Waller factor, the expressions differ only in the mass factor.

In analogy with equation (6), we can therefore introduce a torsional frequency-distribution function $g_R(\omega)$ and obtain

$$\chi_R \cong \exp\left[\frac{\kappa^2}{2M_r} \int_0^\infty d\omega \, g_R(\omega) \Lambda(\omega, t)\right] \tag{12}$$

The combination of (12) and (6) gives the total scattering function as

$$\chi(\kappa, t) \cong \exp\left\{-\frac{\kappa^2}{2M}\left[\gamma_T(0) - \int_0^\infty d\omega \, g_{\text{eff}}(\omega) \Lambda(\omega, t)\right]\right\} \tag{13}$$

with

$$g_{\text{eff}}(\omega) = g_T(\omega) + \frac{M}{M_r} g_R(\omega) \tag{14}$$

This result indicates clearly that translational and librational modes do not influence the scattered neutron spectrum to the same extent, the libration effects being enhanced by a relative factor of M/M_r. Equation (10) gives the rotation mass for a linear molecule; however, this expression probably will not lead to an optimum value of M_r for molecules of more-complicated structure. In specific cases, an intuitive generalization of (10) is possible. For example, if all N hydrogens are isotropically distributed on the surface of a sphere and the nonscattering atoms assumed to be at the center of the sphere, then $I = 2Nm_H b^2/3$ and $M_r = Nm_H$, a result in agreement with Hahn's analysis.

Equation (13) can be numerically integrated to give the scattered neutron spectrum provided g_{eff} is known. Often the experimental conditions are such (low sample temperature and scattering at small momentum transfer) that the one-phonon approximation can be used,

$$\chi(\kappa, t) \cong e^{-2W}\left[1 + \frac{\kappa^2}{2M} \int_0^\infty d\omega \, g_{\text{eff}}(\omega) \Lambda(\omega, t)\right] \tag{15}$$

with $2W = \kappa^2 \gamma_T(0)/2M$. If we ignore the possibility that different hydrogens in the molecule may not be dynamically equivalent, then the differential cross section becomes

$$\left(\frac{d^2\sigma}{d\Omega \, d\omega}\right)_{\text{molecule}} = N\frac{\sigma_H}{8\pi^2}\left(\frac{E}{E_0}\right)^{1/2} e^{-2W}\left[\delta(\omega) + \frac{\kappa^2 e^{-\omega/2T}}{2M\omega \sinh(\omega/2T)} g_{\text{eff}}(\omega)\right] \tag{16}$$

Equation (16), when applicable, enables an experimental determination of the effective frequency-distribution function g_{eff}. A quantity of general interest is the frequency function $g(\omega)$, which, in the case of a lattice, is the distribution of normal mode or phonon frequencies. In the present treatment of molecular solids, it is consistent to regard the low-frequency portion of $g(\omega)$ as the sum of translation and libration frequencies. Hence,

$$g(\omega) = g_T(\omega) + g_R(\omega) \tag{17}$$

It should be noted that $g(\omega)$ cannot be directly derived from neutron-scattering experiments without additional information. Our need, in general, is an *a priori* knowledge of g_T or g_R, or, alternatively, under favorable conditions, one can assume that g_T and g_R contribute to distinct regions in $g(\omega)$.

It is interesting to compare equation (16) formally with the result one would obtain if the atomic motions were treated as lattice normal modes. Suppose the solid is polycrystalline and there are N hydrogens per unit cell. Ignoring the scattering by other atoms as before, we find in the one-phonon approximation that $(d^2\sigma/d\Omega\,d\omega)$ cell aside from a slight difference in the Debye−Waller factor is also given by equation (16), except that g_{eff} is replaced with

$$g'(\omega) = \frac{M}{m_H} \frac{v_a}{(2\pi)^3} \frac{1}{3} \sum_j \int_{\mathcal{S}(\omega)} dS\, [A_j^H(q)]^2 |\nabla \omega_j(q)|^{-1} \tag{18}$$

In this expression, v_a is the volume per unit cell, $\omega_j(q)$ is the phonon dispersion relation, where j denotes the branch, and $A_j^H(q)$ is the hydrogen polarization vector. The integral is to be carried out in phonon momentum (q) space over $\mathcal{S}(\omega)$, a surface of constant frequency. Since the normal-mode frequency-distribution function is defined as

$$g(\omega) = \frac{v_a}{(2\pi)^3} \frac{1}{3} \sum_j \int_{\mathcal{S}(\omega)} dS\, |\nabla \omega_j(q)|^{-1} \tag{19}$$

it is evident that g and g' are not simply related. Therefore, in the lattice approach, a knowledge of $g(\omega)$ is not sufficient to analyze incoherent-scattering experiments on solids with more than one atom per unit cell.

RESULTS

A test of the theoretical model presented in the previous section was made by obtaining the frequency distribution $g(\omega)$ of ice from inelastic-neutron-scattering data and comparing thermodynamic quantities calculated from the "neutron"-frequency distribution with measured values. Ice was chosen because it has been the subject of considerable study and is of fundamental importance in understanding the structure of water. The frequencies of vibrations have been tentatively assigned from IR (infrared) [8, 9], Raman [8], and neutron measurements of D_2O and H_2O ice [10], and these results show a clear separation between external ($\hbar\omega \lesssim 1000$ cm^{-1}) and internal ($\hbar\omega \gtrsim 1500$

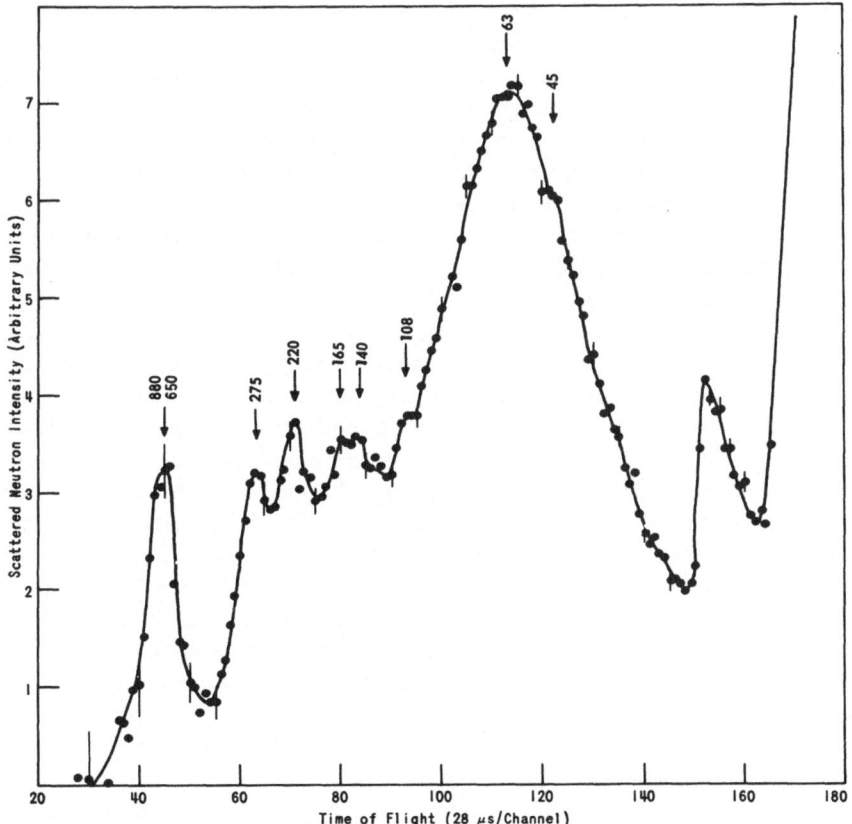

Fig. 1. The time-of-flight spectrum of neutrons scattered from H_2O ice at 150°K and a scattering angle of 65°. The indicated vibrational frequencies are in wave numbers.

cm^{-1}) and also between hindered rotational and translational modes. In addition, the neutron scattering from ice is almost entirely (> 95%) from the hydrogen and therefore incoherent, so that the theory is applicable.

The experimental apparatus was the "cold" neutron time-of-flight spectrometer at the Union Carbide Research Reactor, Tuxedo, New York, and has been described in detail elsewhere [11]. In this apparatus, a quasimonochromatic beam of neutrons ($E < 42$ cm^{-1}) is obtained by passing the beam of neutrons from the reactor through a length of refrigerated beryllium, which passes only those neutrons below the cutoff energy. The filtered neutrons are scattered either elastically (no change in energy) or inelastically (gaining energy from thermally populated levels) by the sample. The energy spectrum is determined by chopping the scattered neutron beam, measuring the time-of-flight, and correcting the observed spectrum for background, chopper transmission, counter efficiency, and window scattering.

The time-of-flight spectrum of a 0.3-mm-thick sample of H_2O ice at 150°K and a scattering angle of 65° is shown in Fig. 1. The very large peak at approximately channel 170 corresponds to elastically scattered neutrons, as does the smaller secondary peak at channel 155. The indicated frequencies (in wave numbers) were determined after instrumental resolution, thermal population, and incident-beam width were taken into account.

The effective frequency distribution $g_{eff}(\omega)$ is obtained directly from the data shown in Fig. 1 by using equation (16) of the previous section and is the curve labeled $M_R = 18$ in Fig. 2. The $g_{eff}(\omega)$ shown has been corrected for instrumental resolution and incident-beam width, the effects of which were found to be appreciable both in peak positions and spectral shape. The curve labeled $M_R = 1.3$ in Fig. 2 is the frequency distribution for ice obtained from the effective frequency distribution after separation into translational ($E < 450$ cm^{-1}) and rotational ($E > 350$ cm^{-1}) intermolecular modes. The separation is indicated by dashed lines and is based on the over-all shape of the spectrum and isotope-shift measurements of H_2O and D_2O ice, mentioned earlier. The translational part of the frequency distribution can also be compared with dispersion curves for the a and c axes calculated for hexagonal ice by Forslind [12]. This lattice dynamical calculation does not include hydrogen motions explicitly, so that dispersion curves for hindered rotational modes were not obtained. The observed maxima in the translational part of the measured frequency distribution are in reasonable agreement with those expected from the calculated dispersion curves, with the exception of the intense maximum we observe at 280 cm^{-1}.

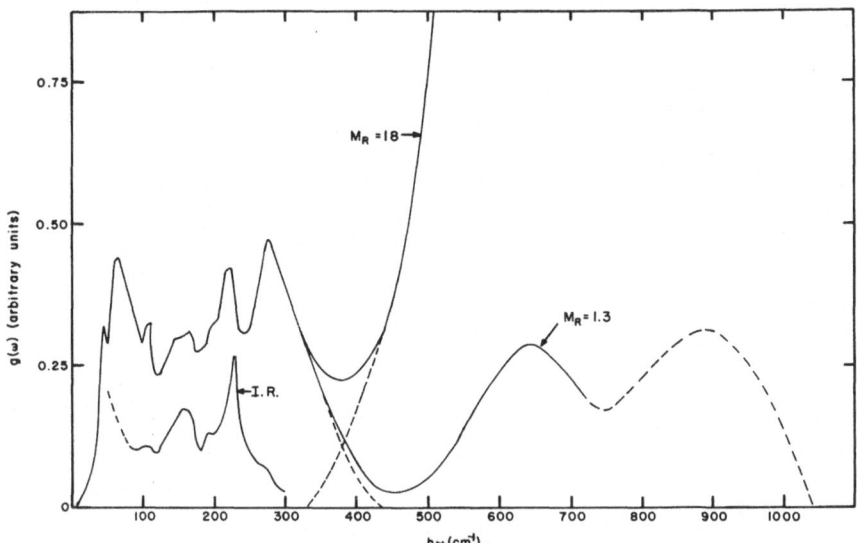

Fig. 2.   The frequency distribution of hexagonal ice obtained from the inelastic neutron-scattering data (M_R = 1.3). The M_R = 18 curve is $g_{eff}(\omega)$. The curve labeled "I.R." is infrared optical density per ν^2 [9].

The normalization of translational and rotational parts of the frequency distribution was performed by assuming that there are three degrees of freedom per molecule for each mode of vibration and that all external degrees of freedom are observed in the spectrum; quantitatively, $\int g_T(\omega)\,d\omega = \int g_R(\omega)\,d\omega$.   This fixes the effective mass and allows no adjustable parameters.   In this way, a value of M_R = 1.3 amu is obtained.   Approximating the H_2O molecule as linear, a value of M_R = 1.59 amu is obtained from equation (10).

Also shown in Fig. 2 is the IR optical density per ν^2 curve recently obtained by Bertie and Whalley [9].   These authors have derived a theory which shows that, for disordered crystals, all vibrations should be optically active and that optical density per ν^2 versus ν should show features of the density of states or frequency distribution [13].   A later IR study of the region below $100\,\mathrm{cm}^{-1}$ is in much better agreement with the neutron data, and, over-all, the IR data do show the features of the frequency distribution obtained in this work [9, 14].

The specific heat C_v, calculated from the frequency distribution by means of the harmonic approximation [15], is shown in Fig. 3. Curve 1 was calculated from a Debye frequency distribution with θ_D = 160°K obtained by fitting the measured C_p ($C_p \cong C_v$), curve 5 [16]. Curve 2 is calculated from a frequency distribution which includes the Debye spectrum and also Einstein terms for optical translations and

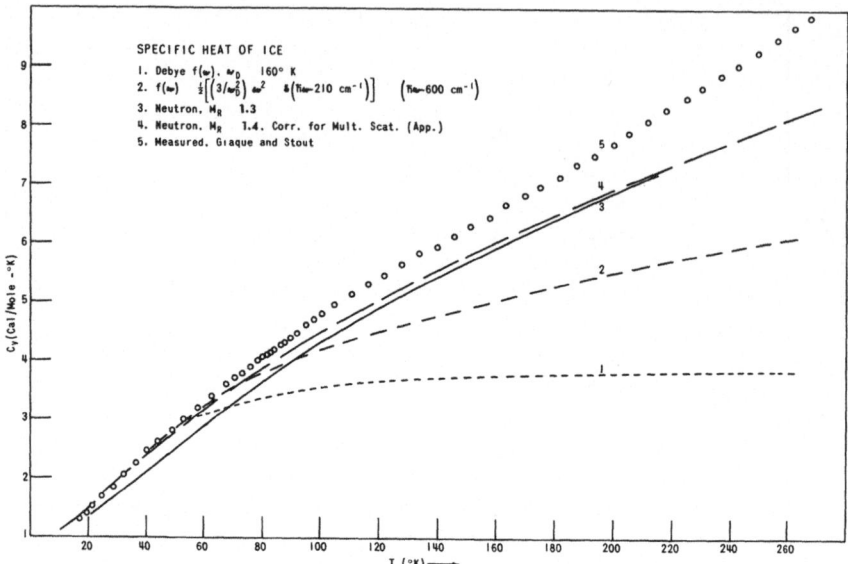

Fig. 3. Comparison of measured $C_p(T)$ (curve 5) [16] and calculated $C_v(T)$ (see text).

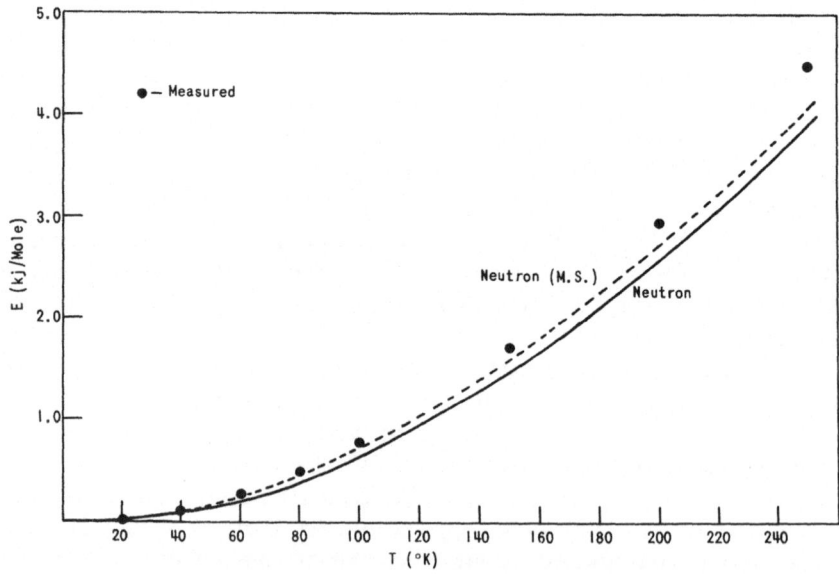

Fig. 4. Comparison of measured internal energy with T [17] and internal energy calculated from $g(\omega)$ and $g(\omega)$ approximately corrected for multiple scattering (M.S.).

hindered rotations. Curve 3 is obtained directly from the frequency distribution ($M_{\dot{R}} = 1.3$) of Fig. 2, and curve 4 is $C_v(T)$ calculated from the neutron-frequency distribution after applying an approximate correction for multiple scattering in the sample. Figure 4 shows the internal energy versus temperature, calculated from the frequency distribution of Fig. 2, and the internal energy calculated from our frequency distribution corrected for multiple scattering (labeled "M.S."). The dots are values of internal energy obtained from measured C_p values [17]. Entropy and free energy calculated from our frequency distribution show comparable agreement with measured values.

There are three corrections to the frequency distribution which, when taken into account, should lead to improved agreement between measured and calculated thermodynamic quantities for ice. A preliminary calculation of the multiphonon contribution showed that, at 150°K, the two-phonon scattering is less than 10% of the one-phonon scattering. However, the multiphonon scattering has not been corrected for exactly. The approximate correction made for multiple scattering is based on corrections calculated for liquid H_2O [18], the vibrational spectrum of which is only qualitatively similar to that of ice. A calculation for ice is being done. The calculated multiple-scattering correction can also be checked experimentally by studying samples of different thicknesses, and these experiments are in progress. Finally, the calculation of thermodynamic quantities from the frequency distribution has been made by assuming the lattice vibrations of ice to be purely harmonic. Leadbetter [19] has analyzed the thermodynamic properties of both H_2O and D_2O ice and finds appreciable anharmonicity. Correcting for anharmonicity on the basis of Leadbetter's results increases the calculated specific heat by about 4% at 150°K and 11% at 270°K which brings our values into much better agreement with the measured values of Giaque and Stout [16].

DISCUSSION

In this work, an intuitive theoretical approach has been presented, which shows that the relation that exists between incoherent-neutron spectra of monatomic cubic crystals and the normal-mode frequency-distribution function is also approximately applicable to molecular solids. The method differs from the earlier normal-mode treatment by Hahn [4]; however, comparable results are obtained in both cases under similar approximations, which indicates that the main arguments are basically equivalent.

The most important result of the theory is that a good approximation of the true frequency distribution for a molecular solid can be obtained directly from incoherent inelastic neutron-scattering data when
the enhancement factor for librations is taken into account. The determination of the exact frequency distribution for molecular solids is,
in principle, possible from spectroscopic data but requires a complete
lattice dynamical calculation, which has up to now not been accomplished.

The results for ice indicate that reasonable values are obtained
for thermodynamic quantities calculated from the frequency distribution obtained using the theoretical model. Although the calculation of
thermodynamic quantities is not the most sensitive test of the cor-
rectness of the frequency distribution, the $g(\omega)$ for ice obtained in this
work is the most detailed over the 0 to 1000 cm^{-1} frequency region
presented thus far.

The theoretical model should also be useful in determining $g(\omega)$
for other hydrogenous solids and liquids, where $g_{trans}(\omega)$ and $g_{rot}(\omega)$ are
separable or where one of the two distributions is known.

REFERENCES

1. P. A. Egelstaff, Thermal Neutron Scattering, Academic Press, Inc., New York (1965).
2. a. M. Tasumi and T. Schimanouchi, J. Chem. Phys. 43:1245 (1965); also, b. T. Kitagowa
   and T. Miyazawa, Rept. Progr. Polymer Phys. (Japan) 8:53 (1965).
3. G. Placzek and L. van Hove, Phys. Rev. 93:1207 (1954).
4. H. Hahn, in: Inelastic Scattering of Neutrons, Intern. At. Energy Agency, Vienna, 1965,
   Vol. II, p. 279.
5. P. A. Egelstaff, ed., Thermal Neutron Scattering, Chap. 9, Academic Press, Inc., New
   York (1965).
6. A. Sjolander, Arkiv. Fysik. 14:315 (1958).
7. S. Yip, to be published.
8. N. Ockman, Advan. Phys. 7:199 (1958).
9. J. E. Bertie and E. Whalley, J. Chem. Phys. 46:1271 (1967).
10. K. E. Larsson and U. Dahlborg, in: Inelastic Scattering of Neutrons, I.A.E.A., Vienna,
    Vol. I (1963), p. 317.
11. G. J. Safford, Cryobiology (1966), in press.
12. E. Forslind, Proc. Swed. Cement Concrete Res. Inst., Stockholm, No. 21 (1954).
13. E. Whalley and J. E. Bertie, J. Chem. Phys. 46:1264 (1967).
14. W. Bagdade, Mid-America Symposium on Spectroscopy, Chicago, May 1967.
15. A. A. Maradudin, E. W. Montroll, and G. H. Weiss, Solid State Physics, Suppl. 3 (1963).
16. W. F. Giaque and J. W. Stout, J. Am. Chem. Soc. 58:1144 (1936).
17. F. Simon, Hand. d. Physik 10:363 (1929).
18. L. Slaggie, private communication.
19. A. J. Leadbetter, Proc. Roy. Soc. (London), Ser. A 287:403 (1965).

Structures of Ice and Water as Investigated by Infrared Spectroscopy*

E. Whalley

Division of Applied Chemistry
National Research Council of Canada
Ottawa, Canada

The infrared spectra of ice and liquid water and the interpretation of them is reviewed. The best understood band is the translational lattice band in ice below 350 cm^{-1}, largely because the translational lattice vibrations can be taken to be mechanically regular in a first approximation, but electrically irregular owing to the orientational disorder. This allows a relatively simple dynamical and electrical theory to be worked out that is necessary for the interpretation. The rotational vibrational and the O—H stretching bands are less well understood, and there is essentially no detailed interpretation that can be reasonably maintained. The spectrum of the liquid is even less well understood, except in the most general terms, and, contrary to the conclusions of a number of papers, there is little reasonably well based structural information to be obtained at present.

The importance of intermolecular coupling is emphasized. Little progress will be possible until detailed dynamical calculations including the effects of intermolecular coupling are made.

INTRODUCTION

Many attempts have been made to learn something about the structure of ice and water from experiments on the optical absorption and scattering caused by nuclear translational motions. Contradictory conclusions have been drawn. It is obvious that the absorption and scattering experiments measure the properties of the normal vibrations, such as frequency and lifetime, and the electrical properties like dipole moment and polarizability derivatives. It is this information that must be used to deduce anything about the structure. Unless we have a reasonable understanding of the normal vibrations, little structural information can be obtained. I should like to review the under-

---

*National Research Council of Canada No. 10083.

TABLE I

Vibration Frequencies of Isolated Water Molecules

	ν_1	ν_2	ν_3
H_2O	3657	1595	3756
D_2O	2671	1178	2788
HDO	2727	1402	3707

standing we have at present of the detailed origin of the optical spectrum.

My main points will be that there are effects which are often neglected that should not be, namely, the orientational disorder of the crystal, the disorder of the liquid, and the strong intermolecular coupling, and that little information can be obtained without a knowledge of the lattice dynamics of the system.

There has been a large literature on this topic in recent years, and I do not intend to review it in detail. A full bibliography is not included; other references can be obtained from the papers cited.

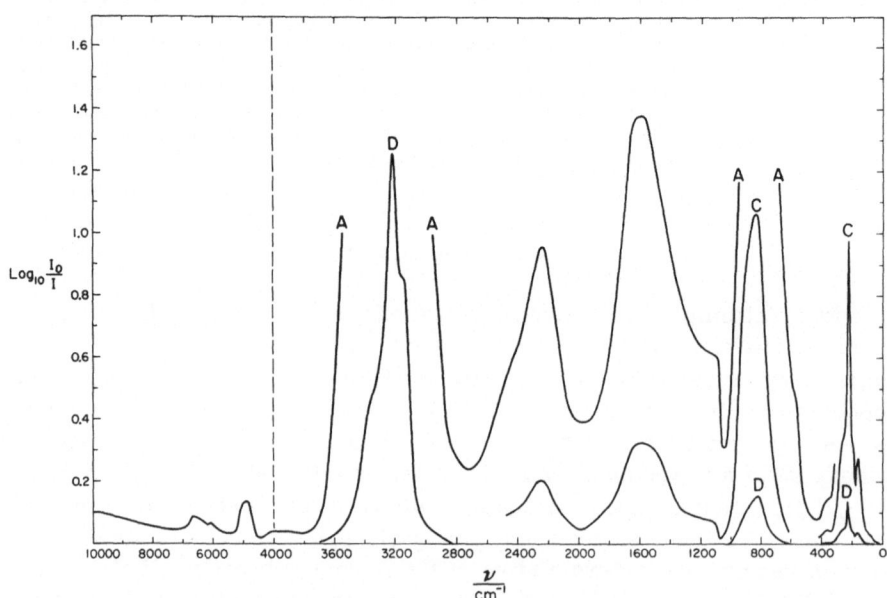

Fig. 1. Infrared spectrum of ice in the range of 10,000 to 30 cm^{-1}.

An interpretation of the spectrum of the solid and the liquid must start with the spectrum of the vapor, and, accordingly, the fundamental frequencies in the vapor [1] to the nearest cm^{-1} are summarized in Table I. Each of these intramolecular vibrations gives rise to a band of vibrations in the solid owing to the intermolecular coupling. In addition, the three translations and the three rotations of the vapor give rise to translational and rotational vibrations.

The interpretation of the spectrum is much simpler for the solid than for the liquid because the long-range order makes the dynamical problems simpler. Accordingly, we start with the solid. A preliminary description of the infrared spectrum, which is shown in Fig. 1 in the range 10,000 to 30 cm^{-1} for several thicknesses, is relatively simple because there are essentially four reasonably distinct regions, as follows: (1) the intramolecular O—H stretching vibrations centered about 3300 cm^{-1}, which are derived from ν_1 and ν_3 of the isolated molecule; (2) the intramolecular bending vibrations centered at about 1600 cm^{-1}, which are derived from ν_2; (3) the intermolecular rotational vibrations centered at about 840 cm^{-1}; and (4) the intramolecular translational vibrations with peaks at 229 and 164 cm^{-1}. It is worth noting that the absorption is continuous all the way from the lowest frequency to over 4000 cm^{-1} and, by and large, the bands just miss overlapping. In this respect, ice is quite different from most crystals, which have relatively sharp bands, and any detailed interpretation must take account of this. There are corresponding bands in the liquid, which are broader, have less fine structure, and overlap more.

It is useful first to review the structure of ice. According to diffraction experiments [2], the space group is $P6_3/mmc$ (D_{6h}^4) and there

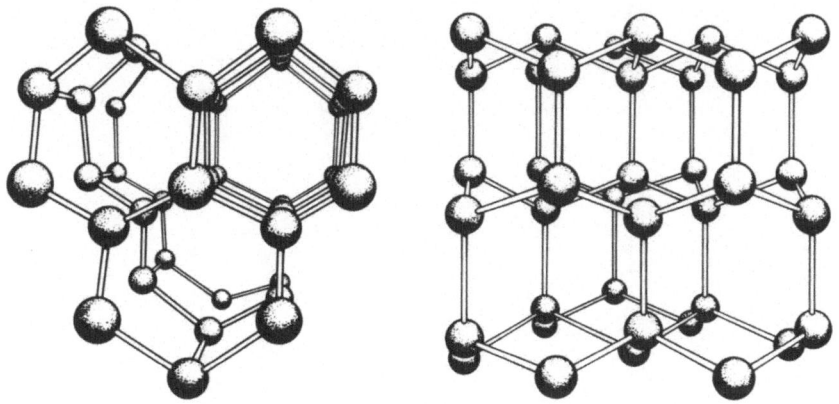

Fig. 2. Position of water molecules in ice.

are four molecules in the unit cell on sites of point symmetry $3m$ (C_{3v}). The structure is shown in Fig. 2. Water molecules of symmetry C_{2v} can fit into this structure only if they are orientationally disordered. The hydrogen atoms are not shown in Fig. 2, but there is one near each O — — —O line close to one or other oxygen atom, and two near each oxygen atom to form discrete water molecules. Each molecule takes on one of six possible orientations shown in Fig. 3. The arrows show the direction of the molecular axis and hence of the molecular dipole.

This orientational disorder plays an important role in determining the forms of the normal vibrations and their infrared activity. In a perfect crystal, the normal vibrations can be described as standing (or running) waves that have constant amplitude through the crystal and can be described by a wave vector. In orientationally disordered crystals such as ice I, this will not in general be possible because the intermolecular coupling constants are not periodic in space and so neither, in general, are the amplitudes of vibration. In a perfect crystal, the only vibrations that are potentially active in infrared or Raman spectra (they may be forbidden by the unit-cell symmetry) are those whose wave vector equals the wave vector of the light in the crystal. Since orientationally disordered crystals have no unit cell, no such selection rule applies and all vibrations are spectroscopically active. Different vibrations will have different intensities, which can be determined only from a detailed analysis.

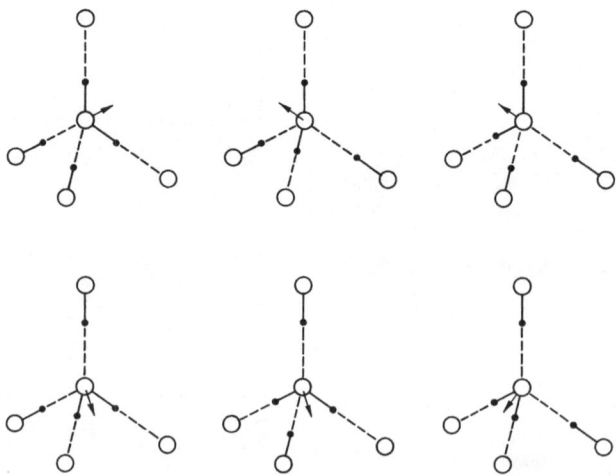

Fig. 3. The six possible orientations of a water molecule in ice [after P.G. Owston, Quart. Rev. 5: 344 (1951)].

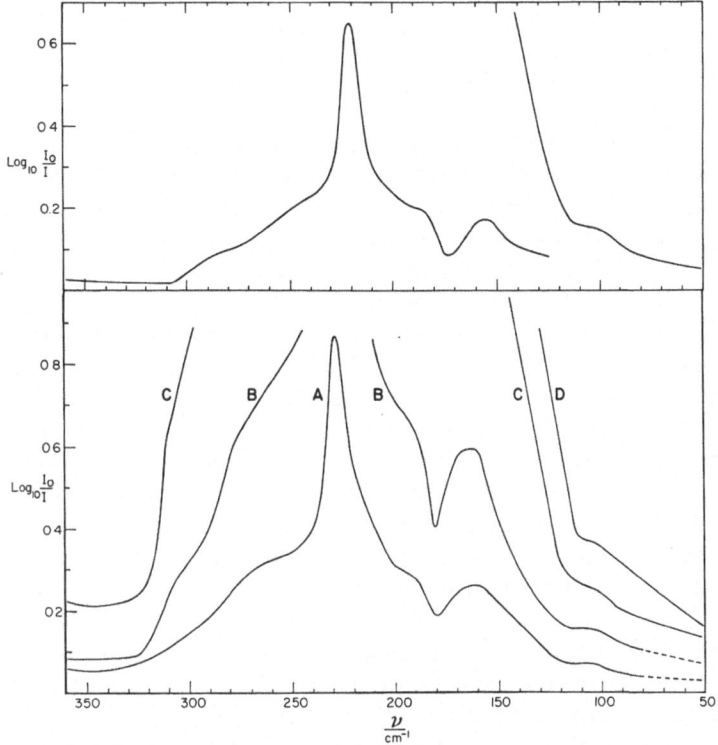

Fig. 4. Infrared spectrum of ice in the range 350 to 50 cm^{-1}. The upper field is D_2O and the lower field is several thicknesses of H_2O (reprinted with permission from Ref. 3).

Structural information which is firmly based in theory can be obtained from the spectrum only if the bands can be interpreted in detail. Accordingly, present progress toward this goal will be briefly reviewed. The translational band is probably the one understood in most detail, and we shall start with this.

THE TRANSLATIONAL BANDS

The translational IR and Raman bands of ice I have been discussed [3] in detail recently. The IR spectrum [3] in the range 350 to 50 cm^{-1} is shown in Fig. 4. The upper frame is of D_2O, and the lower is of several thicknesses of H_2O. The absorption extends downwards in frequency from at least 318 cm^{-1} and is still measurable at 17 cm^{-1} [4] and probably is continuous down to zero frequency. The disorder of the orientations causes the molecules to be only slightly displaced from their mean position (see the section, "The O — H Stretching Bands of Ice"). This suggests that, to a first approximation, the vibrations can be considered those of a mechanically regular crystal, only slightly perturbed by the disorder. They can therefore be described by a wave vector in the usual way. The electrical properties of the vibrations, on the other hand, strongly depend on the disorder. If the crystal structure is taken literally as that obtained by diffraction experiments, i.e., $P6_3/mmc$ (D_{6h}^4) with the disorder averaged out, there are nine zero-wave-vector vibrations of species $A_{1g}, B_{1g}, B_{2u}, E_{1g}, E_{2g},$ and E_{2u}. All are inactive in absorption; so there are no infrared-active translational lattice vibrations. The fundamental bands occur therefore only because of the local departure from the diffraction symmetry. Consider, for example, a particular normal vibration in which a particular O—H— — —O bond is stretched. The stretching will cause a particular change of dipole moment. In another part of the crystal, a crystallographically similar bond will also be stretched, but it might have its hydrogen atom attached to the other oxygen, thus O— — —H—O. The change of dipole moment will then be opposite in sign to that of the first bond. It is easily seen that the total effect of the stretching of all the randomly oriented O—H— — —O bonds will be to cause a most probable dipole-moment derivative for any normal vibration to be zero. The intensity of absorption, however, depends on the most probable square of the dipole-moment derivative, and this is finite, although small. All normal vibrations are in consequence slightly active, and a theory [5] based on a simple force field suggests that the integrated intensity of a normal vibration is proportional to the square of its frequency. Consequently, the absorptivity divided by the frequency squared should be closely related to the density of vibrational states.

Ice Ih and Ic have the same infrared spectrum within the experimental accuracy, and, since the discussion of ice Ic is the simpler, it only will be treated. The discussion of ice Ih is then only a matter of translation. Ice Ic has the face-centered-cubic diamond structure [2], space group $Fd3m$ (O_h^7) with two molecules in the primitive cell on sites of symmetry 23(T). The orientations are disordered as in ice Ih. The frequency—wave-vector dispersion curves in the [100] and [111] directions expected for diamond crystals are shown in Fig. 5, and similar curves are expected in other directions. Maxima in the density of states are expected in the TO (transverse optic), the LO (longitudinal optic), and the LA (longitudinal acoustic) branches near the boundary L in the [111] direction and in the TA (transverse acoustic) branch near the boundary in the [111] direction. A minimum is expected near where the LO and LA come together at the boundary X in the [100] direction. The features in the spectrum can be identified as follows: The TO

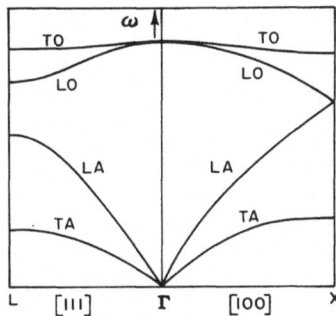

Fig. 5. Frequency–wave-vector dispersion curves for diamond type of crystals with short-range forces. Γ is the center of the Brillouin zone, X the boundary in the [100] direction, and L the boundary in the [111] direction.

maximum = 229.2 cm^{-1}, LO maximum = 190 cm^{-1}, LA maximum = 164 cm^{-1}, LO–LA minimum = 180.5 cm^{-1}, and TA = about 65 cm^{-1} maximum in optical density divided by ν^2. These assignments fit a nearest-neighbor bond-stretching angle-bending force field well enough to confirm the general validity of the interpretation. The Raman spectrum should show similar features with, in addition, the zero-wave-vector A_{1g}, E_{1g}, and E_{2g} vibrations active independently of the disorder. The observed spectrum agrees with this as far as it is known. More work, however, is needed.

At low frequencies the normal vibrations are short acoustic waves, and the density of states is nearly proportional to the frequency squared, and so the absorbtivity should, on the basis of the theory, be proportional to the fourth power of the frequency. This prediction has been approximately verified [4].

There is, however, still much to be learned, especially the origin of the absorption near 300 cm^{-1}. This is not explicable on the basis of a short-range force field, and clearly long-range forces are at work [6].

Although the infrared and Raman translational lattice band adds little to our knowledge of the structure of ice, it adds considerably to our knowledge of the lattice dynamics. Furthermore, there is a similar band in the liquid, which is sometimes used to extract information about the structure of the liquid. An interpretation of the liquid band must surely start with the crystal band.

In a perfect crystal there is long-range order in both the molecular positions and orientations. In crystalline ice there is long-range order in the molecular positions but not in the orientations. In vitreous ice, all long-range order has disappeared, and, in the liquid

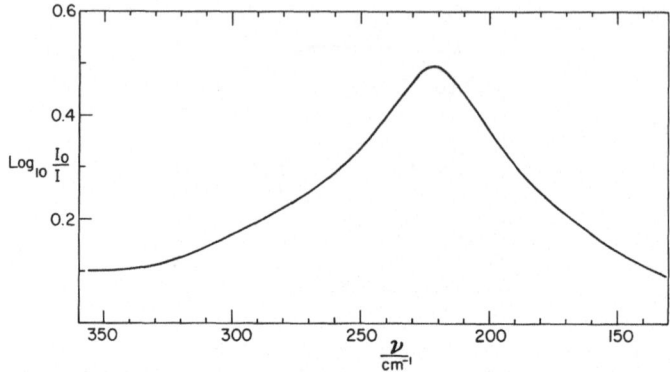

Fig. 6. Infrared spectrum of vitreous ice (reprinted with permission from Ref. 3).

in addition, the short-range order has further decreased and orientational and translational diffusion occurs. The far-infrared spectrum [3] of vitreous ice is shown in Fig. 6. It resembles that of the crystal, except that the fine structure has disappeared, presumably because the lack of long-range order broadens the peaks, minima, critical points, etc., in the density of states so much that none is visible. The peak is at 220 cm^{-1}, slightly lower than the 229 cm^{-1} in the crystal.

The spectrum of the liquid is further broadened, presumably by further reduction of the short-range order and by the occurrence of rotational and translational diffusion. So far, no understanding of these effects in detail has been achieved, and any structural information deduced from the spectrum is pure speculation. Variables such as temperature, pressure, dissolved materials, etc., will change the density of vibrational states and will modify the effects of rotational and translational diffusion. They will also affect the intensity of particular normal vibrations. None of these effects is at all understood at present; until they are, there is no structural information to be obtained.

THE ROTATIONAL VIBRATIONS

There is a band (Fig. 7) in the infrared spectrum of ice with its peak at about 840 cm^{-1} in H_2O and 640 cm^{-1} in D_2O that was first directly observed by Giguère and Harvey [7] and assigned by them to rotational vibrations ν_R. Little structural information has been obtained from this band. It was pointed out by Blue [8] that, if the hydrogen bond is cylindrically symmetrical so that the two bending vibrations

of the group O–H– – –O are doubly degenerate, then the three rotational vibrations of a water molecule in the static field of its neighbors
are nearly triply degenerate.   They are exactly degenerate if the
rotation axes pass through the oxygen nucleus.   The infrared band,
however, covers $1\frac{1}{2}$ octaves, from at least 1050 to 400 cm^{-1}. This
can only be explained if the rotational vibrations are strongly coupled
intermolecularly to form a broad band of vibrational frequencies.
Thus, further evidence is provided of the importance of intermolecular coupling in determining the density of vibrational states and
hence the optical spectrum of ice.

A puzzling feature of the ν_R band has been that the heat capacity
[9] indicates that the density of states is approximately symmetrical
about the frequency of 650 cm^{-1}, but, at this frequency, the infrared
absorption is only a small fraction of that at the maximum at 850
cm^{-1}.   It appears therefore that the high–frequency components of the
band of vibrational states are much more optically active than the low-
frequency components.

The cause of the high intensity in the high–frequency and low intensity in the low–frequency parts of the band is undoubtedly connected in
part with the orientational disorder. Since the ν_R vibration of a molecule in the static field of its neighbors is nearly triply degenerate, the
potential energy for a small rotation is nearly independent of the position of the molecular axis and, for a first approximation, the center of
rotation can be taken as the mean position of the oxygen atom.   Since

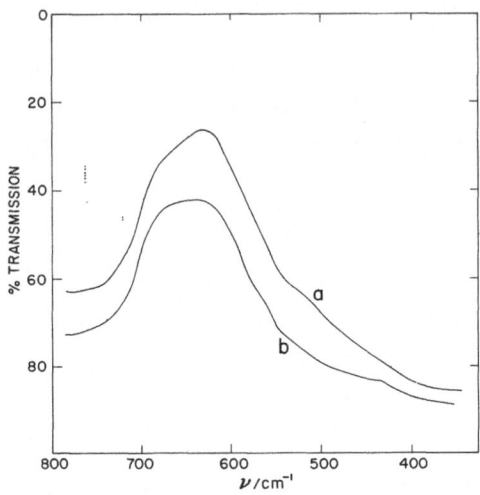

Fig. 7. Rotational vibrational ν_R band of ice (reprinted with permission from Ref. 13).

the frequency in the static field is nearly the mean frequency of the ν_R vibrations, 650 cm^{-1}, the force constant k_1 defined by the potential energy V

$$2V = a^2 k_1 \theta^2$$

where a is the O–H distance and θ the angle of rotation, is 0.50 mdyn/Å. The stronger assumption is also made that the coupling force constant for the motion of two neighboring molecules is also independent of their relative orientation. On this basis, the ν_R vibrations are mechanically regular and can be described by three optical branches added to Fig. 5 for each molecule in the unit cell. There would therefore be six branches for ice Ic and twelve for ice Ih.

The change of dipole moment during a normal vibration is composed of two parts, that due to the motion of a single molecule, which is dependent on its own vector dipole moment, and that due to the relative motion of neighbors, which depends on the moment induced in one molecule by another. A discussion [10] along these lines suggests that the infrared intensity is weak in the low-frequency part of the spectrum and strong in the high-frequency, as is observed.

THE O – H STRETCHING BANDS OF ICE

When molecules are placed in a crystal, the intramolecular vibrations are perturbed by two main effects: (1) The static field of the crystal perturbs the molecules and changes the intramolecular force constants and dipole-moment derivatives so that the frequencies, amplitudes, intensities, etc., are changed; (2) the intermolecular coupling of the motion of neighboring molecules causes the frequencies of the coupled vibrations to occupy bands whose widths depend upon the strength of the coupling.

Uncoupled O – H and O – D Stretching Vibrations

In general, it is difficult to determine from experimental vapor and crystal bands the separate contributions of the static field and the correlation field. The static field can, however, be investigated in ice and water by using HDO; the mass ratio of H and D is enough to ensure that coupling of their motions can be neglected in a first approximation. The first study of this kind was made by Hornig, White, and Reding [11] in the early 1950's, although it was not published until 1958. More

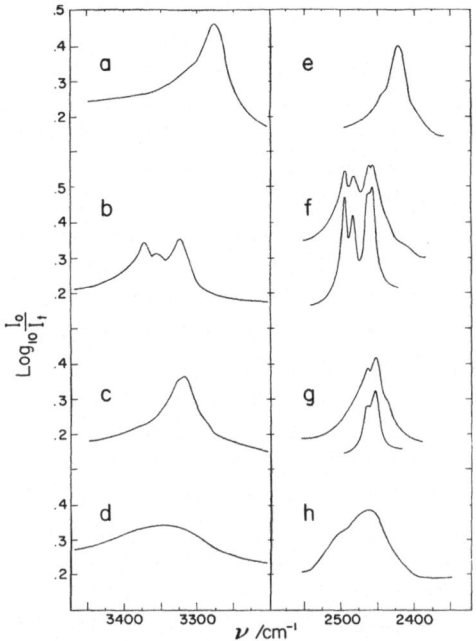

Fig. 8.   Uncoupled (approximately) O—H and O—D stretching vibrations of HDO in dilute (1 or 5%) solution in H_2O, left frame, and D_2O, right frame (reprinted with permission from Ref. 14). The phases of ice in order from the top are I, II, IX, and V.

extensive measurements were reported by Haas and Hornig [12] and by Bertie and Whalley [13]. The latter work had the major advantage that it could be compared with similar work [14] on the high-pressure phases of ice, namely II, IX*, and V, and the comparison of the spectra of the different phases threw much light on the interpretation of the ice I spectrum.

The uncoupled O—H and O—D stretching infrared bands of ice Ih, II, IX, and V are shown in Fig. 8. The peaks of the uncoupled O—H and O—D bands in ice I are at 3277 and 2421 cm^{-1}, which gives a crystal field shift of 430 and 306 cm^{-1}. The important point to notice is that, in ice II and IX, the bands are multipeaked. This shows more clearly in the O—D bands because the relative change of width on deuteration is greater than the relative change of frequency, and so

*The high-pressure phases were examined after cooling them under pressure to 77°K and removing them to atmospheric pressure. Later work, as yet unpublished, has shown that ice III transforms to ice IX over the temperature range of –75 to –108°C so the spectra attributed to III in Ref. 14 should be attributed to ice IX.

peaks not resolved in the O–H bands because of their width are resolved in the O–D bands.

In both ice II and IX, the water molecules are orientationally
ordered, and there is no doubt that the component bands are due to
crystallographically nonequivalent hydrogen (or deuterium) atoms and
the bands can be correlated with the crystal structure. The frequencies of the four bands in D_2O ice II are 2493, 2481, 2460, and 2455
cm^{-1}. The bands at 2460 and 2455 cm^{-1} are just resolved, so their half
width is less than 5 cm^{-1}. The cause of this half width is not certainly
known, but it is probably partly anharmonic interaction between the
fundamental and the translational lattice vibrations.

The uncoupled O–H and O–D bands in ice I (Ih and Ic have the
same spectrum), on the other hand, are appreciably wider—about 50
and 30 cm^{-1} respectively. According to diffraction experiments, there
are only two nonequivalent O– – –O bonds in ice Ih, those parallel to
and those at an angle to the c axis, and one in ice Ic. The half widths
cannot therefore be explained by environments nonequivalent to diffraction. However, the local environments of the O– – –O bonds are
not equivalent because of the orientational disorder. There are three
different ways of arranging the three hydrogen atoms attached to a
given O–H– – –O pair, as shown in Fig. 9, and each of these pairs
has itself many possible environments described by the positions of
the hydrogen atoms attached to the six oxygen atoms bonded to it. The
observed band is the sum of the bands from all the different environments, and the width is largely caused by the variation in environment.
If the variation in O–H frequency is all ascribed to variation in O– – –O
distance and the dependence is taken to be about 3000 $cm^{-1}/\text{Å}$, a half
width of the distribution of O– – –O distances of about 0.01 Å is obtained. It is a measure of the sensitivity of vibrational spectra to
some structural features that no other evidence of this variation appears to have been reported.

Fig. 9. Arrangements of hydrogen atoms about a bonded O–H– – –O pair of water molecules.

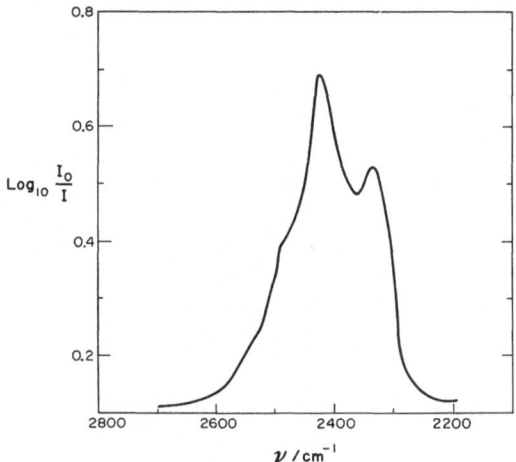

Fig. 10. Spectrum of D_2O ice I near 2400 cm^{-1} (adapted from Ref. 13).

Coupled O — H and O — D Stretching Vibrations

The spectrum [13] of the coupled O—D stretching vibrations of pure D_2O is shown in Fig. 10. D_2O is used rather than H_2O because the component bands are narrower than in H_2O by more than the ratio of the frequencies and, consequently, more structure shows. There are two peaks, the stronger at 2425 cm^{-1} and the weaker at 2332 cm^{-1}, two shoulders at 2545 and 2485 cm^{-1} on the high-frequency side of the main peak, and another shoulder at 2395 between the two peaks. There also appears to be a weak shoulder near 2240 cm^{-1}.

In the Raman spectrum at the same temperature [15], the main peak is at 2283 cm^{-1}, about 50 cm^{-1} to the low-frequency side of the weaker infrared peak. There are also two weaker peaks at 2416 and 2489 cm^{-1} that coincide with bands in the infrared spectrum. There are therefore at least six bands to explain.

The coupling of the uncoupled O—H stretching vibrations to form the band of vibrations in the crystal can be thought of as occurring in two steps, as depicted schematically in Fig. 11, although the separate steps have not been investigated experimentally. In the first, the two O—H vibrations in the same water molecule couple to form the ν_1 and ν_3 vibrations of a molecule that is uncoupled (approximately) from its neighbors. The splitting in the crystal is not known, but in the vapor it is 99 cm^{-1}, and the coupled bands are symmetrical about the uncoupled band (see Table I). In the absence of further information, we can only

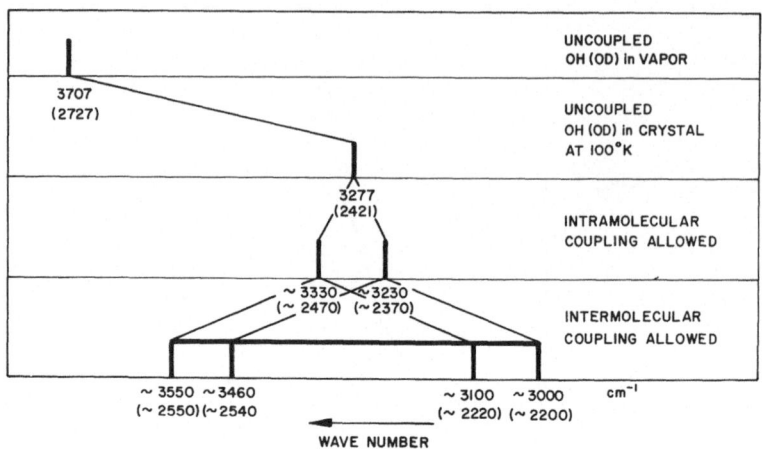

Fig. 11. Schematic description of the evolution of the O–H (O–D) stretching band.

assume that the splitting in the crystal is also nearly symmetrical and of the same magnitude.

In the second step, the ν_1 and ν_3 vibrations themselves couple by intermolecular interaction to form bands of normal vibrations, but the extent of these bands and the density of vibrational states within them is quite unknown except for information obtainable from the infrared and Raman spectra. Two neighboring O–D pairs in H_2O, i.e., O–D– – –O–D, couple to form two vibrations about 70 cm^{-1} apart [12] and symmetrical about the main peak. The coupled band of vibrations in the crystal will be much wider than this. Some of the vibrations may be mainly pure ν_1 or ν_3 motions of the various molecules, but, as no symmetry prevents coupling of ν_1 and ν_3 in part of the band, a designation as ν_1 or ν_3 may not be appropriate.

The translational vibrations in ice can be taken to be regular because it is the whole molecule that moves during the vibrations and the molecules are arranged very nearly periodically in space. During the O–H stretching vibrations, on the other hand, it is the hydrogen atoms that mainly move, and they are not arranged periodically in space. Consequently, the vibrations are not mechanically regular, and there is as yet no simple theory. They are of course also electrically irregular.

It is immediately clear that interpretations in terms of 1-, 2-, 3-, and 4-coordinated water, as suggested many years ago by Cross, Burnham, and Leighton [16] and occasionally revived [17], cannot be maintained. Attempts [12] to explain this spectrum in terms of the

ν_1 and ν_3 vibrations of the individual molecules, with ν_3 being in Fermi resonance with $2\nu_2$, also do not account for the larger number of features. There is, in fact, no evidence, as the following discussion shows, of the importance of $2\nu_2$ in the spectrum.

What can be said with reasonable assurance is as follows. The observed infrared or Raman band is the density of vibrational states modified by the absorption or scattering intensity of the fundamentals and perhaps by anharmonic coupling with other vibrations. The main Raman peak at 77°K is at 3085 cm^{-1} (2283 in D_2O), and the band starts to rise from the background some 30 cm^{-1} below this. The infrared band starts to rise from the background at about 2800 (2100) cm^{-1} and is rising steeply at 3000 cm^{-1}. There is probably, therefore, even allowing for sum and difference bands, a high density of states above 3000 (2200) cm^{-1}. This is over 230 (170) cm^{-1} below the probable position of the uncoupled ν_1 vibration at about 3230 (about 2370) cm^{-1}. Both the polarization [18] and the isotope shift [15] strongly suggest that the Raman band in this region is mainly due to the ν_1 vibration, as would be expected from Fig. 11. If the ν_1 band is symmetrical about the (assumed) position of the uncoupled ν_1 vibration, it therefore extends to at least 3460 (2540) cm^{-1}.

The observed infrared band extends to at least 3600 (2600) cm^{-1} and is strong at 3500 (2500) cm^{-1}. Presumably, the higher-frequency parts are caused largely by coupled ν_3 vibrations. The ν_3 band therefore extends about 200 or 250 (130 or 180) cm^{-1} above the frequency of the uncoupled ν_3 vibration. It seems likely that the band extends about this amount to low frequency, i.e., to about 3100 (2220) cm^{-1}.

The ν_1 and ν_3 bands probably therefore overlap about 450 (320) cm^{-1}, and, in most of the band, further coupling of the ν_1 and ν_3 vibrations probably makes a description in terms of ν_1 and ν_3 not very useful.

The density of vibrations in the band of coupled OH vibrations is surely not uniform; there are presumably regions of frequency with high, and regions with low, densities which would tend to show as features in the optical spectra. In addition, the vibrations have no doubt different intensities, and a variation in intensity across a part of the band may also cause features in the spectrum. Clearly, any number of features can be rationalized in these terms without adding greatly to the understanding. Furthermore, an absorption or scattering peak may not be caused by a group of similar vibrations; it could be due to a coincidence of two different groups. An example [14] of this kind of behavior occurs in ice III at 3300 cm^{-1}. Any detailed understanding will require help from other techniques of determining the density of vibrational states and from the theory of the vibrations of these disordered systems.

Overtone and Combination Bands

In a perfect crystal, the allowed overtones and combination bands are limited by the rule that the sum of the wave vectors shall be essentially zero. There may also be further restrictions dependent on the unit-cell symmetry. In an orientationally disordered crystal, no such selection rule holds and all combinations are allowed. Combinations undoubtedly explain the high absorption underlying the main infrared bands that is evident in Fig. 1, and, no doubt, parts of the main bands are also due to combinations. Apart from such trivial generalities, almost nothing can be said. The frequencies of the maxima in the ν_R, ν_2, and ν_{OH} bands are in roughly the ratio 1:2:4, so that any combination band above 1600 cm^{-1} has many possible assignments. Since the fundamental bands are so broad, the possible combinations from the different parts of each band have a wide frequency range and it seems safe to suggest that little is known about the detailed assignment of the overtones. The most useful measurements in this region would be on the uncoupled O–H and O–D bands, but none appears to have been reported.

THE O — H AND O — D STRETCHING BANDS OF LIQUID WATER

Uncoupled O–H and O–D Stretching Vibrations

Measurements of these bands have been reported recently in the Raman spectrum by Wall and Horning [19] and at the Eighteenth Mid-America Symposium on Spectroscopy by Walrafen and in the infrared by Falk and Ford [20] and by Hartman [21]. Earlier references are given in these papers. The band is similar by absorption and scattering, with the peak near 3400 (2500) cm^{-1}, corresponding to most probable static field shifts of 300 (225) cm^{-1}, or about 0.7 of the corresponding shifts in ice I. The band is continuous with little or no sign of fine structure and is about 260 (160) cm^{-1} in half width. The half width is much greater than that in ice I and undoubtedly reflects the much greater variability in the environment (as described by O— — —O distance, O–H— — —O angle, and other parameters) of the O–H and O–D groups in the liquid compared with the solid.

The evidence, in fact, suggests that the O— — —O distance is the most important factor determining the frequency. There is a good correlation between the distribution curve of near–neighbor O— — —O distances determined by X-ray diffraction and by Raman and IR spectroscopy with the use of a relation between frequency and O−−−O distances derived from a number of different hydrogen-bonded crystals [19, 22].

Several authors have commented that the uncoupled O−H band provides strong evidence for there being a continuous distribution of hydrogen-bond strengths (see Ref. 20 for a list of references), perhaps the earliest being Bulanin [23].

The O−H Stretching Bands in Pure Water

The infrared and Raman spectra of liquid water are well investigated, and a typical spectrum is shown in Fig. 12. Three characteristic frequencies are usually quoted, the peak near 3440 (2515 in D_2O) cm^{-1} and two shoulders, the more prominent at 3190 (2360) cm^{-1} and the less prominent at 3650 (2680) cm^{-1}.

The explanations advanced for this band have usually been based on a comparison with the vapor frequencies ν_1 and ν_3 of the molecules. Cross, Burnham, and Leighton [16] suggested that these vibrations were perturbed by hydrogen bonding either to the lone pair or to the hydrogen atoms in various ways corresponding to various degrees of coordination from 0 to 4. There is, however, no evidence of such discrete structures in the uncoupled O−H and O−D vibrations, and this explanation is untenable.

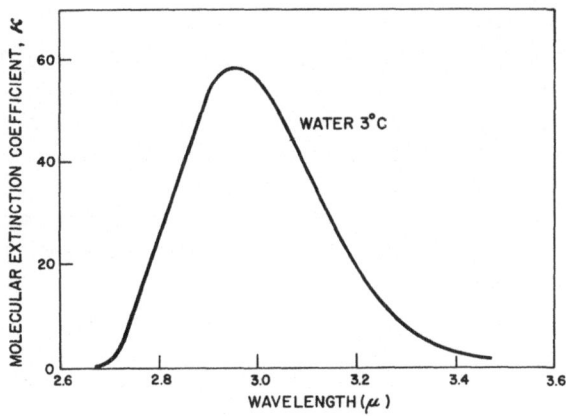

Fig. 12. O−H stretching band of liquid H_2O [after J. J. Fox and A. E. Martin, Proc. Roy. Soc. (London) A 174:234 (1940)].

A second explanation, due to Ellis and Sorge [24] and advocated by Hornig and co-workers [11, 12], is that the three components are due to ν_1, ν_3, and $2\nu_2$ in Fermi resonance with ν_1. While it is by no means impossible that $2\nu_2$ makes an appreciable contribution, the only evidence for it is that there are apparently three components to the band. In fact, three components are expected on other grounds as well, and there is no more evidence for $2\nu_2$ in the liquid than in the solid.

An interpretation of the spectrum is probably best started, not from the vapor vibrations ν_1 and ν_3, but from the uncoupled O—H (O—D) stretching vibrations, in the manner described for ice. We first allow the two O—H groups in the same molecule to couple. Because of the large half width of the uncoupled bands, many molecules will have two quite different uncoupled O—H frequencies. The coupling in these cases may not be large, and a description of the vibrations in terms of the ν_1 and ν_3 vibrations of the isolated molecule may not be useful. Intermolecular coupling of the vibrations will further complicate the spectrum. The coupling will tend to be stronger, the closer the frequencies are, and not dependent on the character (ν_1-like or ν_3-like or intermediate) of the motions causing them. This would further make a description in terms of ν_1 and ν_3 less useful.

The similarity between the liquid and solid spectra is striking; the shapes are similar, but, because the liquid has broader bands, less detail shows. Water is no doubt four-coordinated but in a less regular manner than ice, and it is not surprising that the spectra are similar. The normal vibrations are no doubt more localized than those in ice owing to the greater disorder but are still highly coupled and are otherwise rather similar. Since the O—H stretching vibrations of ice are not understood in detail, those of the liquid are even less well understood. There is therefore no useful structural information at present in the band.

A point usually overlooked is that it is *a priori* quite possible for the intermolecular coupling to be strong enough to carry some of the uncoupled vibrations to a frequency higher than the vapor frequency. The observation, then, of absorption or scattering at the vapor frequency is no evidence that there are vaporlike molecules unless the maximum coupling shift is known.

Combinations of the O — H Stretching Bands

The simplest overtones and combinations in principle are those of the uncoupled O—H stretching vibration. A study has recently been reported by Worley and Klotz [25] which shows several overlapping bands near 6500 cm^{-1}. These were interpreted as overtones of several discrete kinds of O–H having different kinds of hydrogen bonding. But it is not easy to see why discrete O–H bands should show as overtones but not as fundamentals. In any case, there are other explanations possible for the bands, such as combination of the O–H stretch with other vibrations in the vicinity, and no structural evidence can be obtained from the bands without a much deeper understanding.

All possible combinations and overtones of pure water are infrared and Raman active in the liquid, as in the solid. There are many possibilities, and it is not necessary that the strongest fundamentals should give rise to the strongest combinations. Detailed assignments, such as that of Buijs and Choppin [26], which are often quoted as evidence for particular structures in water, can have no scientific basis.

CONCLUSION

The main conclusion from this discussion is that we perhaps know a good deal less than we should like to think about the origin of the observed optical bands in ice and water. There is so much interest from so many different points of view that are all very important, in the structures of ice and particularly of water, that we must guard ourselves from reading, in our enthusiasm, more into our data than is justified. It is surely better to know nothing than to be wrong.

ACKNOWLEDGMENT

The ideas and conclusions presented in this paper owe much to discussions with my colleagues. Dr. J. E. Bertie, in particular, has played a large part.

REFERENCES

1. W.S. Benedict, N. Gailar, and E.K. Plyler, J. Chem. Phys. 24:1139 (1956).
2. R.G.W. Wyckoff, Crystal Structures, 2nd Ed., Vol. 1, Interscience Publishers, New York (1963), p. 322.
3. J.E. Bertie and E. Whalley, J. Chem. Phys. 46:1271 (1967).
4. J.E. Bertie, H. J. Labbé, and E. Whalley, unpublished work.

5. E. Whalley and J. E. Bertie, J. Chem. Phys. 46:1264 (1967).
6. E. Whalley and J. E. Bertie, J. Colloid Interface Sci. 25:161 (1967).
7. P. A. Giguère and K. B. Harvey, Can. J. Chem. 34:798 (1956); see also D. F. Hornig, H. F. White, and F. P. Reding, Spectrochim. Acta 12:338 (1958); and J. E. Bertie and E. Whalley, J. Chem. Phys. 40:1637 (1964).
8. R. W. Blue, J. Chem. Phys. 22:280 (1954).
9. P. Flubacher, A. J. Leadbetter, and J. A. Morrison, J. Chem. Phys. 33:1751 (1960); also A. J. Leadbetter, Proc. Roy. Soc. (London) A 287:403 (1965).
10. J. E. Bertie and E. Whalley, to be published.
11. F. P. Reding, Doctoral thesis, Brown University, June, 1951; also H. F. White, Doctoral thesis, Brown University, June, 1952; and D. F. Hornig, H. F. White, and F. P. Reding, Spectrochim. Acta 12:338 (1958).
12. C. Haas and D. F. Hornig, J. Chem. Phys. 32:1763 (1960).
13. J. E. Bertie and E. Whalley, J. Chem. Phys. 40:1637 (1964).
14. J. E. Bertie and E. Whalley, J. Chem. Phys. 40:1646 (1964).
15. M. J. Taylor and E. Whalley, J. Chem. Phys. 40:1660 (1964).
16. P. L. Cross, J. Burnham, and P. A. Leighton, J. Am. Chem. Soc. 59:1134 (1939).
17. Z. A. Gabrichidze, Opt. i Spektroskopiya 19:575 (1965), for example; for Eng. transl. see Opt. Spectry. (USSR) 19:319 (1965).
18. N. Ockman, Advan. Phys. 7:199 (1958).
19. T. T. Wall and D. F. Hornig, J. Chem. Phys. 43:2079 (1965).
20. M. Falk and T. A. Ford, Can. J. Chem. 44:1699 (1966).
21. K. A. Hartman, J. Phys. Chem. 70:270 (1966).
22. C. L. van P. van Eck, H. Mendel, and J. Fahrenfort, Proc. Roy. Soc. (London) A 247:472 (1958).
23. M. O. Bulanin, Opt. i Spektroskopiya 2:557 (1957); for Engl. transl. see Natl. Res. Council of Can. Tech. Transl. 773 (1958).
24. J. W. Ellis and B. W. Sorge, J. Chem. Phys. 2:559 (1934).
25. J. D. Worley and I. M. Klotz, J. Chem. Phys. 45:2868 (1966).
26. K. Buijs and G. R. Choppin, J. Chem. Phys. 40:3120 (1964).

Pollution Studies

A Comparative Study in Eutrophication

Ursula M. Cowgill

Osborn Memorial Laboratory
Yale University
New Haven, Connecticut

Linsley Pond, a small body of water, 14 m deep located in North Branford, Connecticut, has been the subject of investigation for some 30 years. Over the past years, like many lakes, it has suffered from steadily increasing pollution. During the year August, 1965, through August, 1966, the entire water column of 14 m was sampled weekly and temperature and oxygen data were simultaneously monitored, with the idea in mind of making as complete a geochemical study as possible of all detectable elements in the water and how they move in a thermally stratified lake.

The general process of heating described for 1937 to 1938 appears to have been modified, so that, late in the heating period in 1966, a vertically undisturbed layer from 5 to 7 m apparently transmitted heat downward solely by molecular conduction. Below this, heating was probably controlled by profile-bound chemical density currents.

The total amounts of calcium, magnesium, phosphorus, and sulfur, though variable throughout the season, are in general from two to eight times greater than in 1937 to 1938. During stagnation, sulfur tends to decrease in the deeper waters owing to precipitation as ferrous sulfide. The iron concentration of the deep hypolimnetic water is markedly lower than at an equivalent season in 1938, while manganese, which is not precipitated as a sulfide at ordinary pH values, is strikingly higher in the deeper water than formerly. Phosphorus is likewise greatly enriched at the present time but is still less than the stoichiometric equivalent of the ferrous iron in the deep water, so that, during the autumnal circulation, it tends to sediment out as the iron is oxidized. The copper content of the water is lower than that recorded by Riley 30 years ago. Any increase in copper in the deep hypolimnion during stratification is irregular and insignificant compared with that of iron and manganese, as is to be expected from the insolubility of copper in the presence of ferrous sulfide. The very large changes in phosphorus observed throughout the season in 1938 were also noted in 1966.

In 1937, Hutchinson [1] began a study of the movement of calcium, phosphorus, magnesium, iron, manganese, and sulfur in the thermally stratified waters of Linsley Pond, a small lake in North Branford, Connecticut. During the same period, Riley [2] studied the movement of copper in this lake. More recently, a study was initiated to make as complete a geochemical investigation as possible of the water of this pond. As the project progressed it became evident that these elements had undergone extreme changes over the past years, doubtless as a result of eutrophication. In 1938, there were six summer homes and two farms, both maintaining cattle, while in 1966 there were about 300 dwellings in the basin, all apparently occupied throughout the year. It was noted during the recent study that human sewage was draining into the lake and, toward the end of the study, the effluents of some 15 cesspools had been diverted from emptying into the pond. It is, therefore, not surprising that the lake has suffered from what is now usually politely called hypereutrophication. It is the intent of this paper to compare the two periods of study, and all data gathered about these elements in the more recent investigation are included.

FIELD METHODS

The field methods used in the earlier studies have been published elsewhere [1, 2]. The more recent investigation began in August of 1965 and continued through August of the following year. Consecutive half-meter samples were taken of the entire water column by using a Van Doren bottle and amalgamating the water in 20-liter samples of the following levels: surface to 2.5 m, 2.5 to 5.0 m, 5.0 to 8.0 m, 8.0 to 11.0 m, and 11.0 to 14.0 m. Twenty-liter samples were also collected from the inlet and the outlet. During the period that the lake was ice covered, January 16 to February 27, large samples of ice were also collected. Samples were gathered weekly for 1 year. Bi-weekly samples were taken of Cedar Lake for half the year of study. This is the lake that flows into Linsley Pond. Oxygen and temperature determinations were also made at half-meter intervals. These results have been published elsewhere [3].

LABORATORY METHODS

The laboratory procedures employed in the earlier investigations have been described elsewhere [1, 2]. Since it was the purpose of the more recent study to investigate all detectable elements and since it was desirable to obtain as many elements as possible, the 20-liter samples were evaporated to dryness in beakers that were placed in heating mantles. The sediment in the bottom of the beakers was scraped out with a gum-rubber policeman and the material was dried at 110°C for 24 hr. The sediment was then ground in an agate mortar and pestle to pass 200 mesh. The sieves that were used in this study were nylon. The material was then placed in cadmium-plated rings and pressed at 1.6 metric tons/cm². The filled rings were stored in plastic boxes.

The samples were examined by an X-ray emission spectroscope with a General Electric Company XRD-6 high-voltage power control, detector, and pulse-height selector. For copper, iron, and manganese, a scintillation counter and a proportional-flow counter filled with PR gas were used simultaneously, while the other elements were monitored only with the proportional-flow counter. A platinum-target tube was used in the study of iron and manganese, a molybdenum-target tube in the study of copper, while a chromium target was employed in the investigation of the other elements. Both chromium and platinum targets were operated at 75 kV and 40 mA, and the molybdenum was used at the same kV rating but with 50 mA. A helium path was employed in the study of all elements in this investigation. Calcium, iron, and manganese were determined with a lithium fluoride analyzing crystal, sulfur with a sodium chloride crystal, magnesium with an ammonium dihydrogen phosphate crystal, phosphorus with a germanium crystal cut to a 111 plane, and copper with a silicon crystal cut to a 111 plane. A 0.025-cm Soller slit was used throughout the study.

A basic attempt was made in all cases to accumulate at least 10,000 counts and wherever possible 100,000 counts on peak and background. Thus, all elements were determined to a precision within less than 1% except for phosphorus, which was estimated with a precision within less than 2%, and sodium, within less than 5%.

None of the elements here studied suffered interference from major lines with the exception of manganese, whose K_α line is close to the K_β of chromium, and copper, whose K_α conflicts with the K_β of nickel. Both manganese and copper results are therefore corrected for β interference [4, 5]. Methods for standardization of these elements have been published elsewhere [5].

LIMNOLOGICAL CONDITIONS, 1965 TO 1966

Though the oxygen and temperature data for the period of summer stagnation have been published elsewhere [3], the results are necessary for the chemical discussion and hence will be summarized. During the summer months of 1966 there was no dissolved oxygen from 4 to 14 m, whether it was determined by the Winkler method or by an electrode.   Though the oxygen was deficient in 1938 [1], it was not entirely absent during that period.   There has been no significant change in temperature over the years, but the clinolimnion, as orig- inally described by Hutchinson in Linsley Pond, was not formed in 1966 and there was, during the later part of the heating period, a non- turbulent layer in the thermocline through which the heating between 5 and 8 m must have been entirely due to molecular conduction.   Below 8 m, a density current is presumably operative since it appears un- likely that significant turbulence would occur below 8 m when the region above is nonturbulent.

Early in the study, during the week prior to September 19, ground was broken on the scarp side of the lake for several houses; a road was built to facilitate the transportation of building materials to these sites, and, prior to sampling, the precipitation was unusually heavy. Though not all elements exhibited a sharp increase and decrease in concentration during this period, it is suggested that those that did were probably reflecting disturbance in the littoral zone.   During these few weeks, only the hypolimnetic values are of any immediate interest.

During the week prior to February 20, there was a fair amount of melting of ice in the littoral zone of the lake, which apparently re- sulted in a density current (Hutchinson [1]).   This seems to have been responsible for a spectacular increase in iron and less striking rises in some but not all other elements.   The nature of this phe- nomenon will become clearer as additional elements are determined, and it is hoped to revert to it in a later paper.

THE SULFUR CYCLE

Pink sulfur bacteria, probably *Thiopedia*, and hydrogen sulfide, as evidenced by its odor, neither of which were observed in the earlier study, were present when the study began August 29, 1965, and remained in the lake water until the autumnal homothermal period began about November 14, the hydrogen sulfide returned after the spring homo- thermal on June 18, 1966, when the lake was already fully stratified. Its presence more or less coincided with the existence of acute oxygen deficiency.   As the lake became more and more deoxygenated, sul- fate reduction occurred further toward the surface waters, as evi- denced by the odor of hydrogen sulfide.   During summer stagnation, it was prevalent from 4 to 14 m.

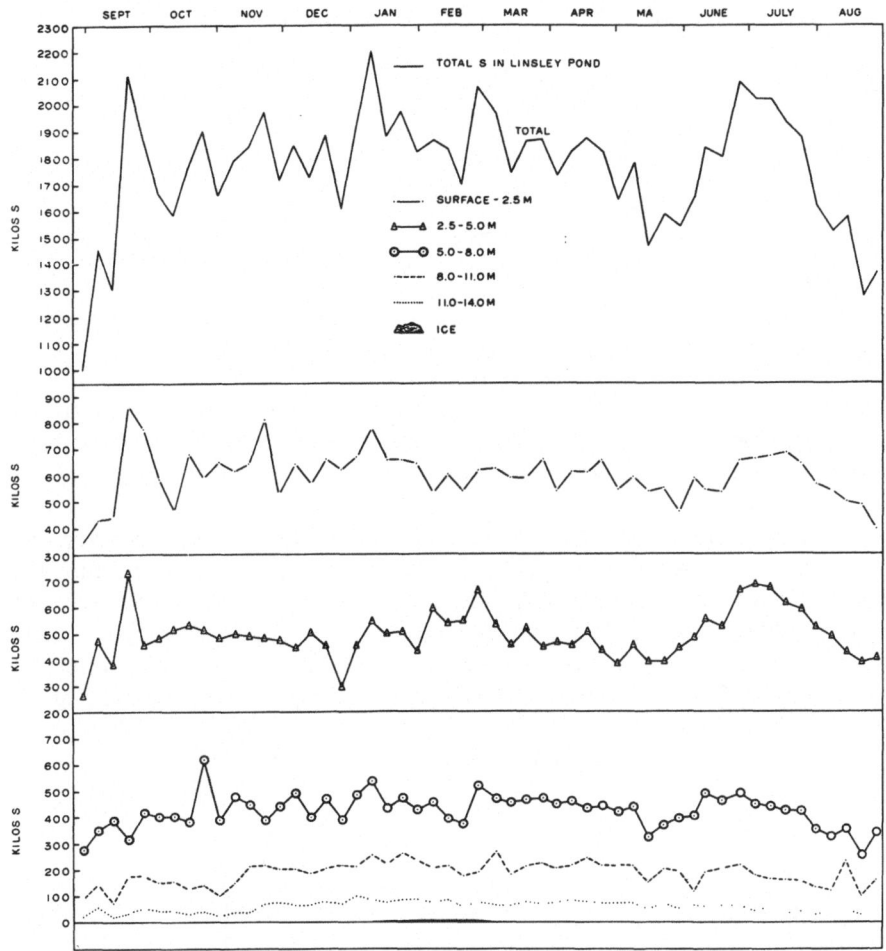

Fig. 1. The distribution of sulfur in Linsley Pond and in the various levels sampled.

Figure 1 shows the total mass of sulfur in the lake for the period of study and its distribution as well for the various levels sampled. The rise in sulfur on September 19 is probably the result of human disturbance. Though some of the other elements reflect the effect of the density current prior to February 20, sulfur shows a decline at this time. The huge rise in concentration appearing at the end of June coincides with the beginning of a summer algal bloom, and, presumably, the concurrent increase in the 5- to 11-m region is the result of dying algae carrying sulfur with them as they fell.

The total sulfur became greatly reduced in the hypolimnion owing to the precipitation of ferrous sulfide. During the latter part of the

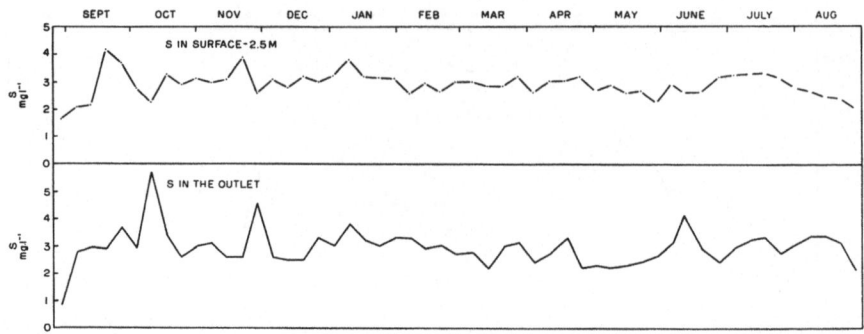

Fig. 2. The concentration of sulfur in the outlet and the surface to 2.5-m region.

study, the water was black from 5 to 14 m presumably owing to the precipitation of this mineral. Much sulfur left the upper waters during the late summer owing, no doubt to the sedimentation of the dying algae.

The rise in total sulfur that appeared at the end of February and beginning of March was probably due to runoff when some melting occurred in the basin. The sulfur content of the ice varied from 0.6 to 4.8 kg. On April 17, there was a slight rise in sulfur, probably due to an increase in concentration, initially at the south end of the lake and reflected in the outlet water during the past week, Fig. 2,

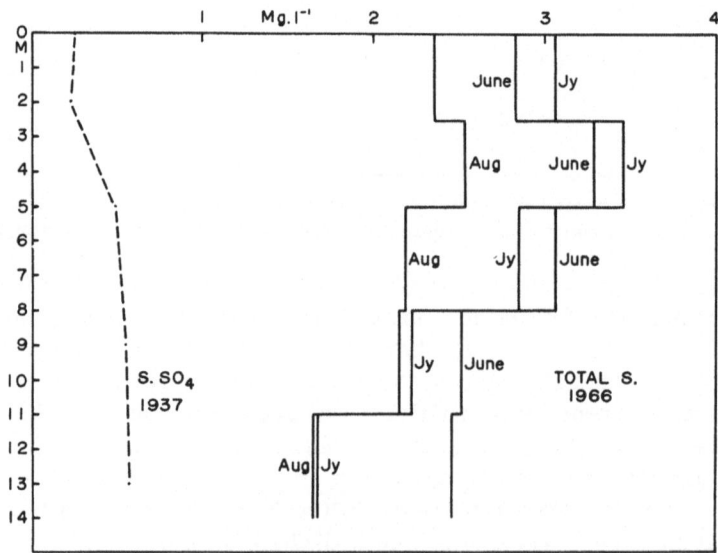

Fig. 3. The vertical distribution of sulfur during summer stagnation in 1937 and 1966.

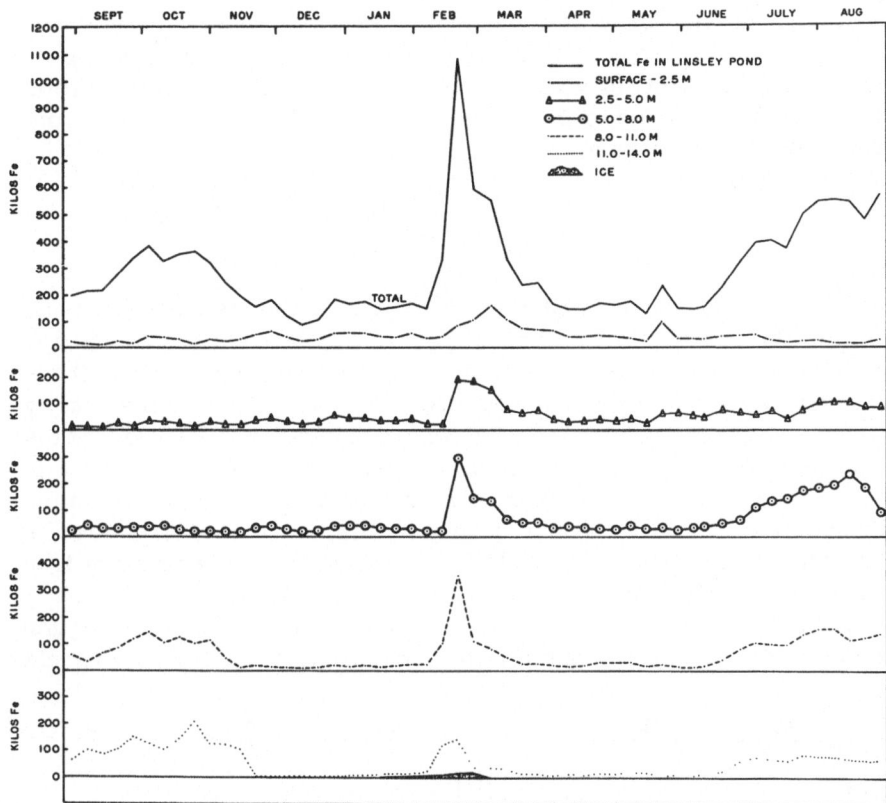

Fig. 4. The distribution of iron in Linsley Pond and in the various levels sampled.

possibly as a result of the decomposition of littoral vegetation. During the week of April 24 and May 1, there was a decline in sulfur concentration not only in the central section of the lake but in the outlet as well, probably brought about by the spring renewal of vegetation, which doubtless took up some sulfur from the surface waters. This was also observed in Cedar Lake, which drains into Linsley Pond during part of the year. It would appear that the vegetation in the two lakes was behaving uniformly with respect to sulfur during this period.

Figure 3 shows comparative data for sulfur for the two periods studied. In the earlier investigation, the oxygen was depleted, though not to the extent it is now; there was no hydrogen sulfide being produced, and sulfur as sulfate was accumulating in the lower waters during the height of summer stagnation. During the recent year of study, the region from 4 to 14 m was void of dissolved oxygen during the summer months, hydrogen sulfide was present throughout this zone, and the sulfur was depleted in the lower waters owing to the precipitation of ferrous sulfide. The total sulfur has increased eight-fold during the past 30 years. This increase in sulfur concentration has probably been due to the increased content in rain resulting from progressive urbanization.

THE IRON CYCLE

Figure 4 shows the distribution of iron in the lake as a whole as well as in the various composite water samples studied. Iron apparently does not reflect the effect of human disturbance on the shore initiated the week prior to September 19. The effect of the density current, noticed on February 20, was spectacular, indicated by the huge rise in concentration in the deeper part of the lake but not in the surface to the 2.5-m region. The increase in concentration in the surface waters that appeared in the beginning of March may be attributed to melting and runoff from the basin. The content of iron in the ice varied from 3.9 to 19.6 kg during this study.

During the summer months of 1966, there was no dissolved oxygen from 4 m downward. During this period there appeared to be no great

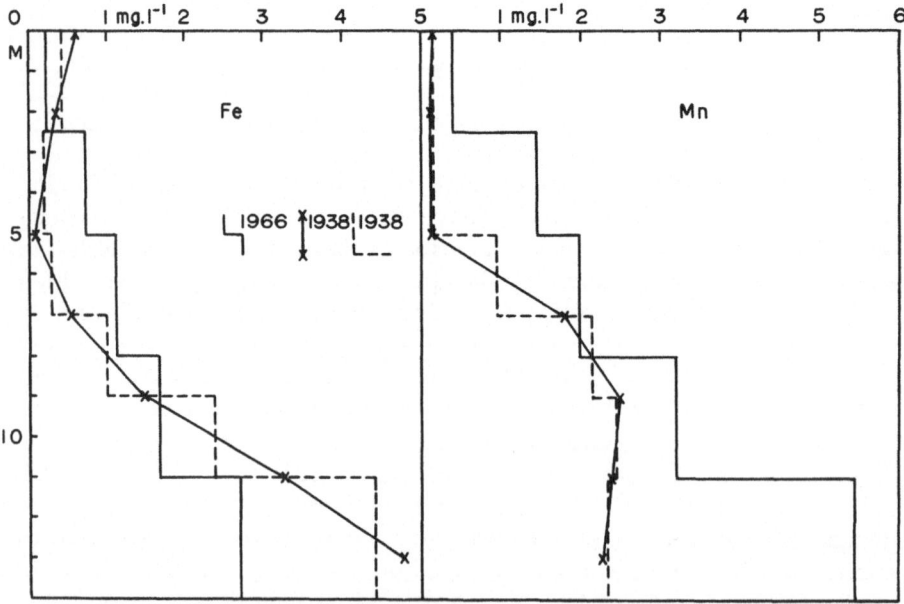

Fig. 5. The vertical distribution of iron and manganese during summer stagnation in 1938 and 1966.

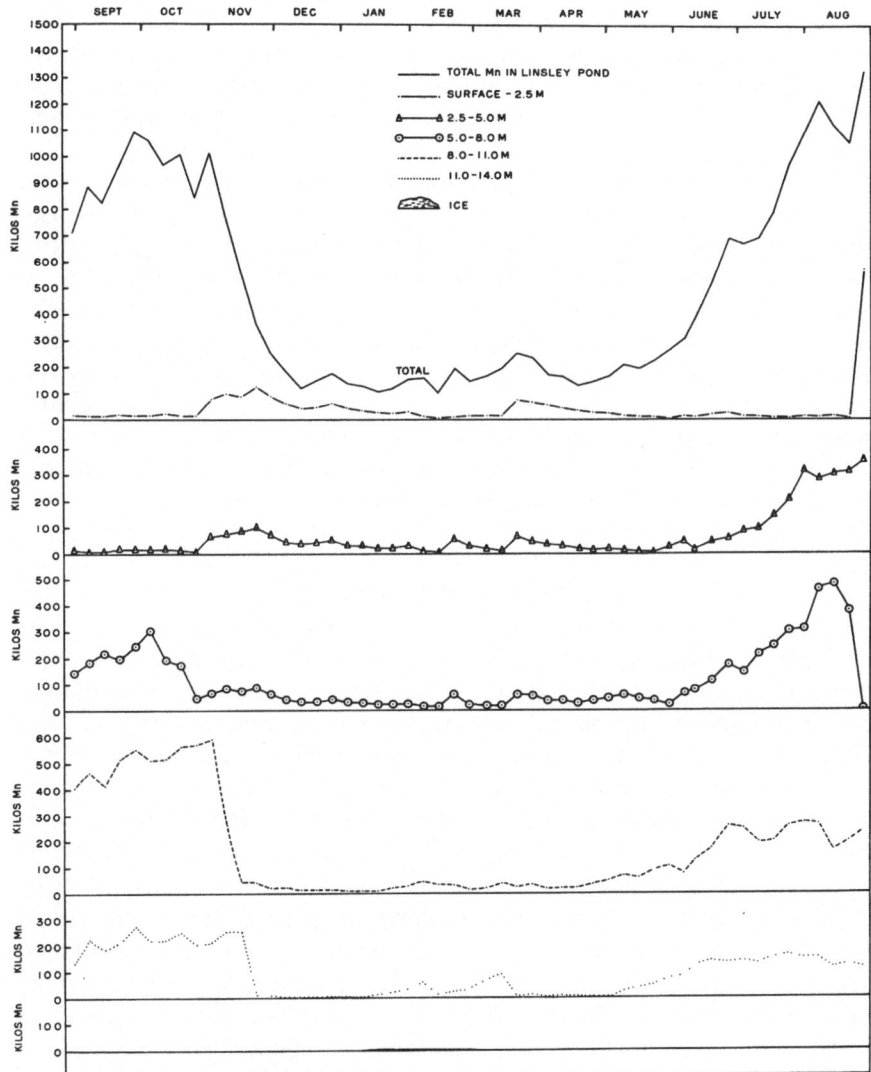

Fig. 6. The distribution of manganese in Linsley Pond and at the various levels sampled.

change in the surface waters; however, there was a small rise in concentration in the 2.5- to 5-m region, probably reflective of the reduced
conditions that developed in the lower water of this zone. From about
the beginning of June until the autumnal turnover toward the end of
November, ferrous ions were diffusing from the reduced mud surface
into the upper waters. These ions, once released, formed ferrous

sulfide which precipitated out and returned to the mud surface. There was a decrease in concentration of iron in the lower waters during the autumnal turnover and a subsequent rise in the upper waters, from which iron was then lost, doubtless precipitated out as ferric phosphate.

Figure 5 shows the stratigraphic distribution of iron for the two periods of study during summer stagnation. During the earlier investigation, the lack of oxygen and the absence of hydrogen sulfide resulted in an accumulation of ferrous ions in the hypolimnion. In the more recent study, the even greater oxygen deficit coupled with the presence of hydrogen sulfide throughout most of the column of water permitted the precipitation of ferrous ions as ferrous sulfide which resulted in a decreased concentration of iron in the hypolimnion compared with that found in the earlier study. As Linsley Pond has become more and more eutrophic over the past 30 years, the enrichment of iron in the deeper waters has become less and less.

THE MANGANESE CYCLE

Figure 6 illustrates the distribution of total manganese in Linsley Pond and in the various regions of the water column under investigation. The effect of the initial human disturbance does not appear to be reflected by a rise in manganese concentration. During the period of the density current, there was a slight rise in concentration in all but the surface waters, but this was not nearly as spectacular as in the case of iron. The melting of the ice and subsequent runoff does not appear to have greatly affected the concentration in the waters. However, during the spring homothermal period some manganese does appear to have been circulating throughout the lake. About the beginning of June, as the mud surface had undergone reduction, manganous ions began to diffuse from the mud surface into the waters up to the 2.5- to 5-m zone. Toward the end of August, 1966, the concentration in the 5- to 8-m region decreased but there was a spectacular rise in the surface to the 5-m region. Temperature data indicate a little mixing down to about 6 m, which may account for this phenomenon. In early September, 1965, the upper waters were low in manganese, but the lower waters showed a distinct accumulation. During the autumnal homothermal period, the manganese concentration dropped in the hypolimnion, and the element was clearly circulating throughout the lake.

Since manganous manganese does not precipitate as a sulfide at

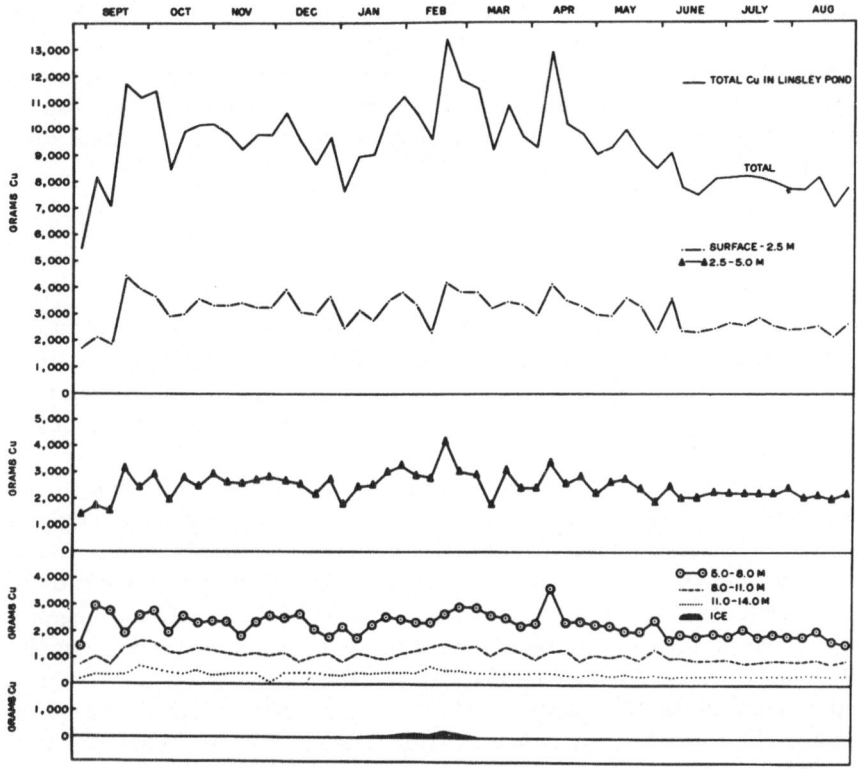

Fig. 7. The distribution of copper in Linsley Pond and at the various levels sampled.

ordinary pH values, it tends to accumulate in the lower waters during summer stagnation. Figure 5 shows the stratigraphic distribution during this time for the two studies. During the earlier investigation manganese concentrated in the lower waters. Manganese, being reduced to Mn^{++} at a redox potential higher than iron is reduced to Fe^{++}, can begin to diffuse from the mud surface before complete reduction has taken place. Clearly, the more recent study shows the mud to be more reduced and, as a result, a greater accumulation of manganese appears in the hypolimnion. Unlike iron, which precipitates as the sulfide and hence becomes reduced in concentration in the lower waters, this manganese accumulates in solution in the deeper water. The greater eutrophication of the pond has resulted in a greater enrichment of manganese in the hypolimnion. In fact, iron and manganese over the past 30 years have reversed their stratigraphic distribution. There still appears, however, to be a stoichiometric excess of iron over phosphorus, so that the phenomenon suggested by Hasler and Einsele [6], who point out that replacement of iron by manganese can promote solubility of phosphorus at the autumnal circulation period, does not yet occur.

THE COPPER CYCLE

Figure 7 shows the distribution of copper in the lake and in the various levels studied. Copper appears to reflect the effect of human disturbance evident on September 19. An effect of the density current in February is apparent throughout the lake with the exception of the 11- to 14-m zone.

Copper appears to be rather invariant during summer stagnation and exhibits its lowest concentration during July and August throughout the lake. In September, the copper content in the southern end of the lake tends to be somewhat higher than during the previous month, which suggests that copper is being released by decomposing vegetation; some of this would be distributed by wind action throughout the epilimnion. Even for April 17, Fig. 8, there is a slight increase in copper content in this region, which is also reflected by other elements. The decline in concentration during the last week of April and first week of May suggests that renewed vegetation is extracting some copper from the surface waters.

An earlier study of the copper cycle in Linsley Pond was made by Riley [7]. His data were enormous for autumn and early winter, the quantity in the lake being as high as 135 kg. No such phenomenon appeared during the recent study, nor did Riley encounter it again in the subsequent autumn. With the exception of one sampling period

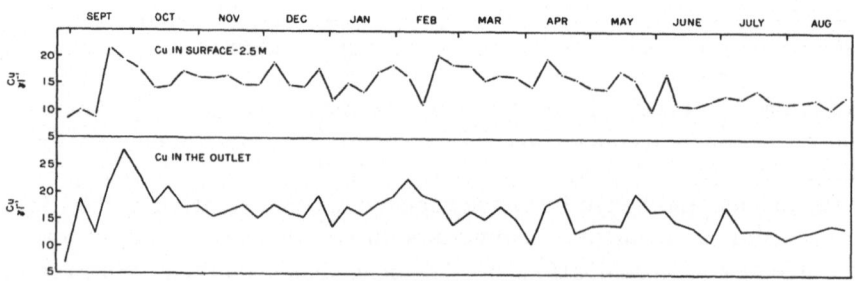

Fig. 8. The concentration of copper in the outlet and the surface to 2.5-m region.

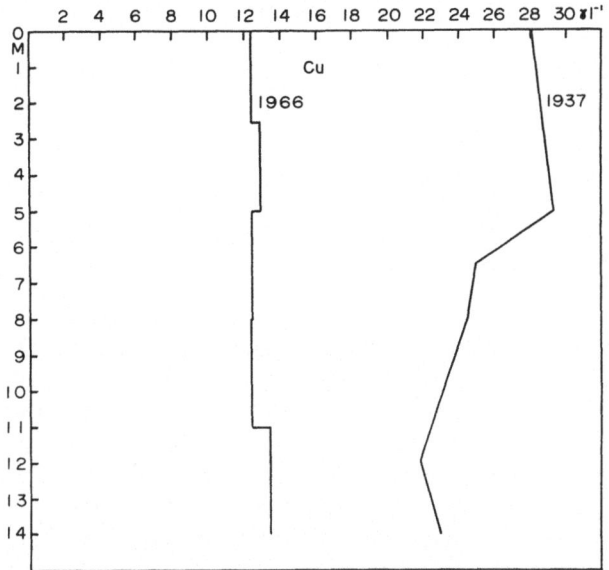

Fig. 9. The vertical distribution of copper during summer stagnation for 1937 and 1966 (earlier data from Riley [7]).

in February and another in early July, Riley's total figures for copper content of the water were always higher than were those in the more recent study. It would appear that the copper of the basin is now less mobile than it was previously. In the more recent study, both during summer stagnation and during the fall turnover period, there was a very slight increase in concentration in the lower waters. The increase of copper in the vertical distribution series, Fig. 9, was, however, insignificant when compared with that of iron or manganese. This is to be expected in view of the insolubility of copper in the presence of ferrous sulfide [8, 9]. Apart from the great autumnal rise recorded by Riley early in his studies, his data show the same sorts of irregular variations that were observed in the more recent study.

THE PHOSPHORUS CYCLE

It would be reasonable to suppose that the balance of phosphorus in the pond is largely by uptake from littoral sediments by rooted vegetation which, on decay, liberates the element into the surface waters. This process, neglected in modern limnological writing, was in fact suggested by Kofoid in 1903 [10], and Pond in 1905 [11]. The vegetation

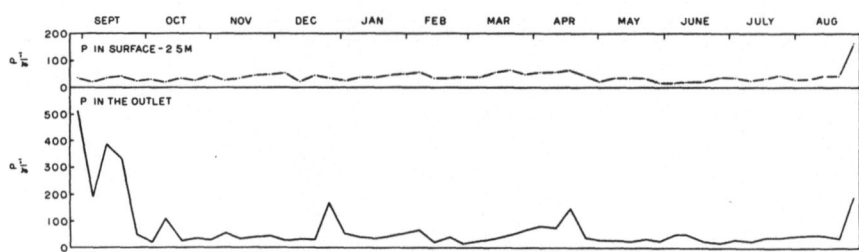

Fig. 10. The phosphorus concentration in the outlet and the surface to 2.5 m.

is most developed at the south end of the lake, from which the outlet flows, and the outlet water, which gives a measure of what happens in that part of the lake, from which water will be moved into the center of the lake whenever a south wind is blowing. Large quantities of phosphorus were found in the outlet water sample, beginning about April 3, Fig. 10. For example, on April 17, the outlet contained the enormous amount of phosphorus of 158 gamma/liter. Similarly, at the end of August, 1966, when plants were again decaying, the concentration of the outlet water reached 192 gamma/liter.

There is presumably also uptake of phosphorus from the water by littoral plants or epiphytic algae associated with them. During the week of April 24 to May 1, the phosphorus dropped 15.5 kg, presumably in this way. A similar decline occurred in Cedar Lake, where the phosphorus concentration of the open water dropped from 57 gamma/liter to 14 gamma/liter over a 2-week period, which suggests that the uptake of phosphorus by the developing vegetation was general for this time. It is unlikely the plankton could have sedimented rapidly enough to account for this change.

Figure 11 shows the seasonal distribution of phosphorus in Linsley Pond. There is no rise in concentration during the period of human disturbance, and phosphorus exhibits only a small increase as a result of the density current. During summer stagnation, the phosphorus diffuses from the mud into the lower waters. The great increase in the surface at the end of August is, as has been indicated, the probable result of plant decay in the littoral zone. During the autumnal homo-thermal period, the phosphorus is precipitated out presumably as ferric phosphate, though some phosphorus remains circulating through the lake. One of the lowest quantities of phosphorus encountered during the recent study appeared on December 12, more or less coincident with the autumnal turnover when, presumably, much of the soluble phosphate phosphorus present in the water during the previous week had precipitated out as ferric phosphate and sedimented to the bottom. Between December 5 and 12, the lake lost 12 kg of phosphorus.

There was an increase in phosphorus of the order of 7 kg between February 27 and March 6, which may be attributable to runoff as the basin thawed. By March 13, the spring circulation period, the total amount of phosphorus had only dropped 0.9 kg, which indicates very little precipitation of free phosphate phosphorus at that time.

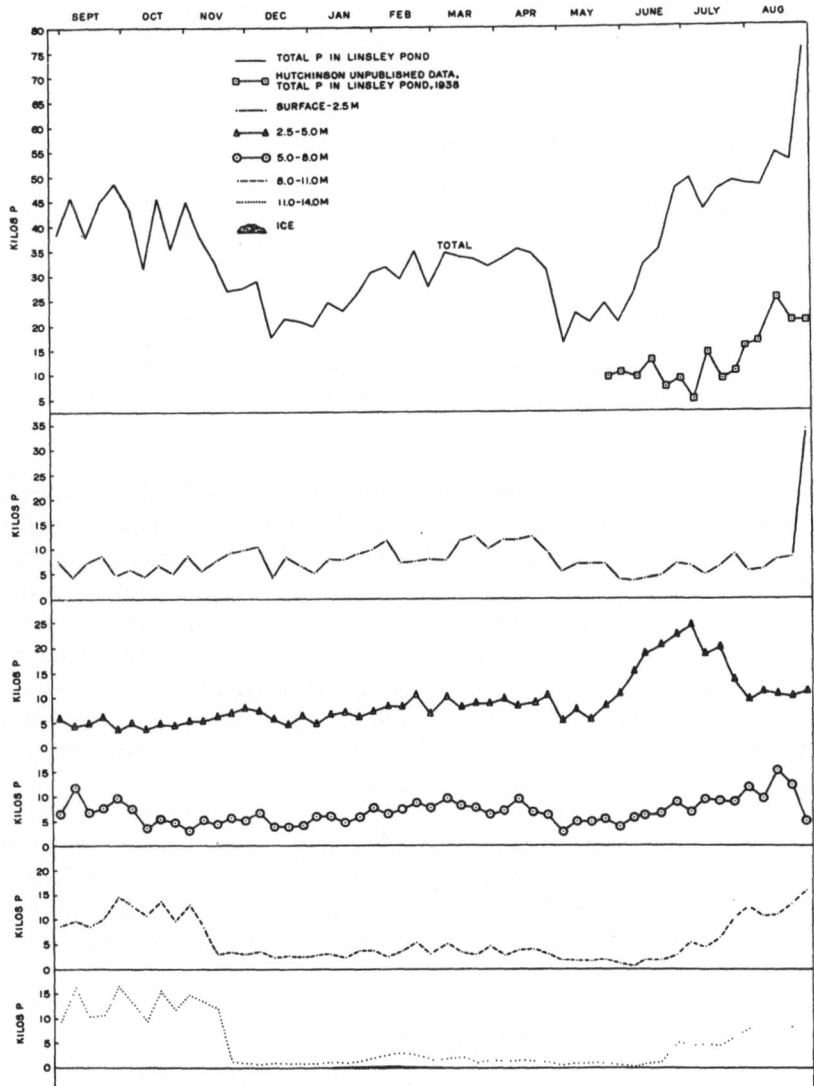

Fig. 11. The distribution of phosphorus in Linsley Pond and at the various levels sampled. The distribution of total phosphorus found by Hutchinson in 1938 during the period of summer stagnation is added for comparison.

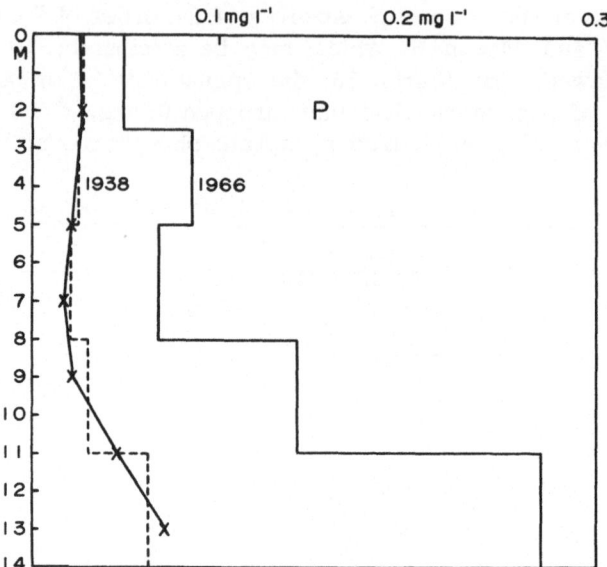

Fig. 12. The vertical distribution of phosphorus during summer stagnation for 1938 and 1966.

The variations in total phosphorus for the whole lake are great and apparently quite rapid. For example, between August 21 and 28, the concentration of phosphorus rose from 8 to 34 kg in the surface waters. The phosphorus concentration in the hypolimnion can drop as much as 11 kg in a week, as it did at the time of autumnal turnover. At least some of these variations are presumably brought about by horizontal water movements that carry phosphorus from the reduced mud into the free water and by continuous sedimentation of dying organisms and their fecal pellets. During summer stagnation, the amount of fecal pellets observed was considerable.

Figure 12 shows the vertical distribution of phosphorus for the two studies. The concentration at all times seems greater than formerly and has increased fourfold in the hypolimnion during the summer. The spectacular changes in phosphorus recorded earlier during the summer stagnation period, Fig. 11, clearly still occur.

THE CALCIUM CYCLE

The distribution of calcium in the pond is shown in Fig. 13. The greatest rise in calcium appears on September 19, presumably as a

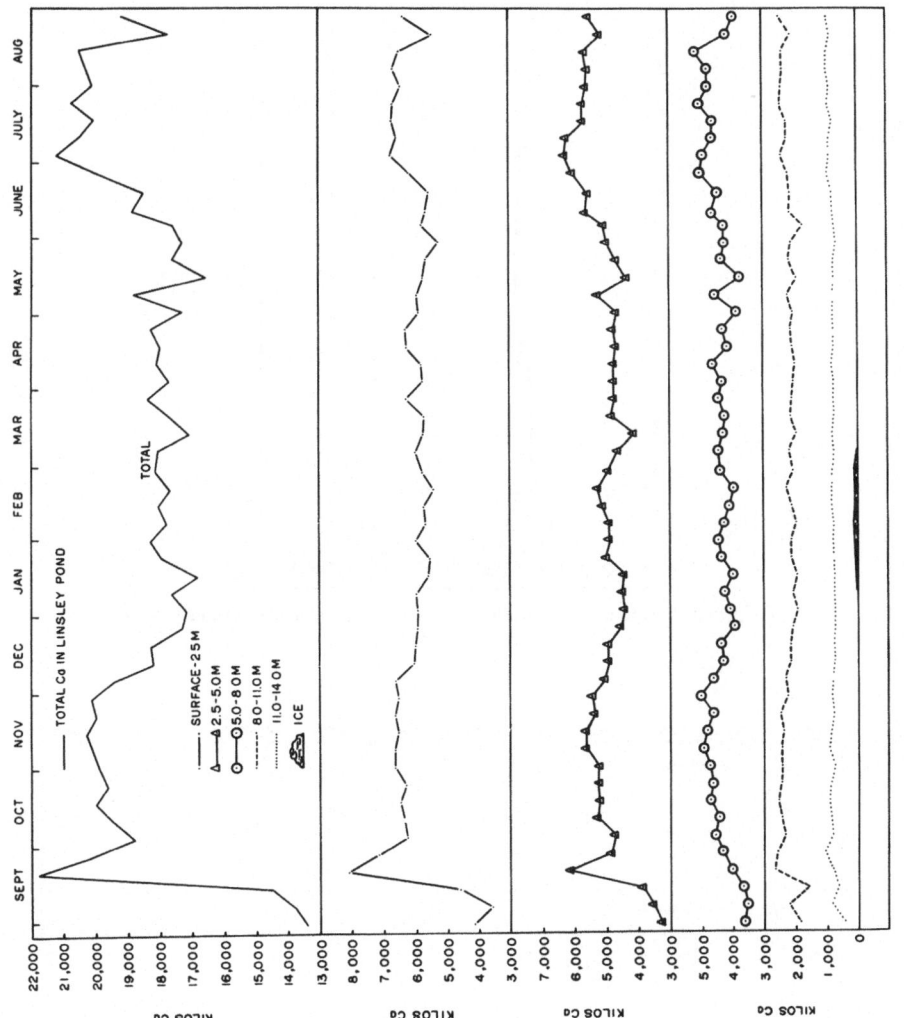

Fig. 13. The distribution of calcium in Linsley Pond and in the various levels sampled.

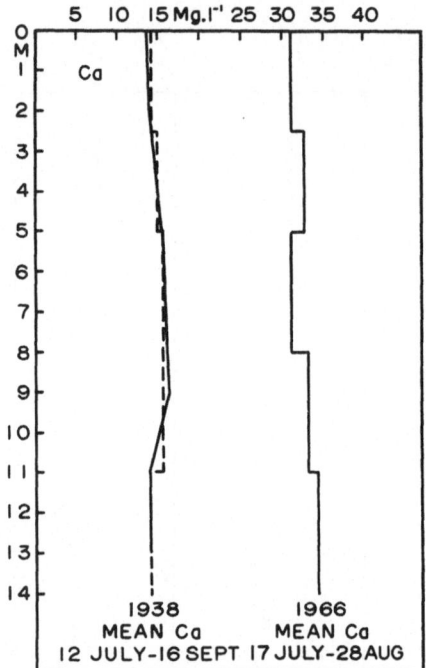

Fig. 14. The vertical distribution of calcium during summer stagnation in 1938 and 1966.

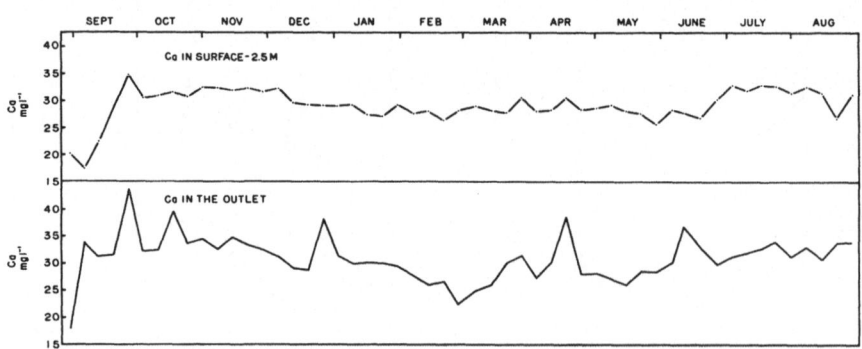

Fig. 15. The calcium concentration in the outlet and the surface to 2.5 m.

result of human disturbance. The added material appears to be notice-able in the surface to 5-m zone and is in slight evidence in the 8- to 11-m region. The density current reflected so spectacularly by some of the other elements is not indicated by calcium. The rise that ap-pears during summer stagnation apparently accompanies a most pro-ductive algal bloom in the euphotic zone. The zone beneath this shows sporadic peaks and troughs possibly due to photosynthetic precipitation and resolution; the variation possibly indicates changes in phyto-plankton populations, but the data are inadequate to elucidate this further.   It is interesting to note that the calcium content of the hypolimnion varies little throughout the recent year of investigation.

Figure 14 shows comparative data for the period of summer stagnation for the two years of study.   The most noticeable change that has occurred during the 30 years between the two investigations is that the concentration of calcium has doubled as a result of in-creased eutrophication of the lake.   There appears to be a slight tendency for calcium to stratify regularly in the lower waters during the most recent study, which is not apparent in the older work.

With the exception of the effect of human disturbance in the fall and the algal bloom in the summer, the variation in calcium does not tend to be pronounced in either year of study, though, like phosphorus but not as spectacular, similar changes appear in the outlet waters, Fig. 15.

THE MAGNESIUM CYCLE

The distribution of magnesium is shown in Fig. 16. The effect of disturbance on the shore is exhibited from the surface to 11 m on September 19.   Magnesium, like calcium, does not show the effect of the density current. The rise in magnesium that appears concurrently with the algal bloom during the summer stagnation may in part be due to magnesium being released from the mud in the littoral zone and transported into the main section of the lake by wind action. On August 14, magnesium showed a spectacular rise; at this time, the algal bloom had already disappeared, and the amount of dissolved oxygen in the euphotic zone gave no indication of the beginning of another bloom.   During this rise from July 10 to August 14, when the magnesium increased 1360 kg, the inlet was not functioning. It would therefore appear reasonable to suggest that magnesium was diffusing from the littoral mud. The magnesium content varies little in the hypolimnion.

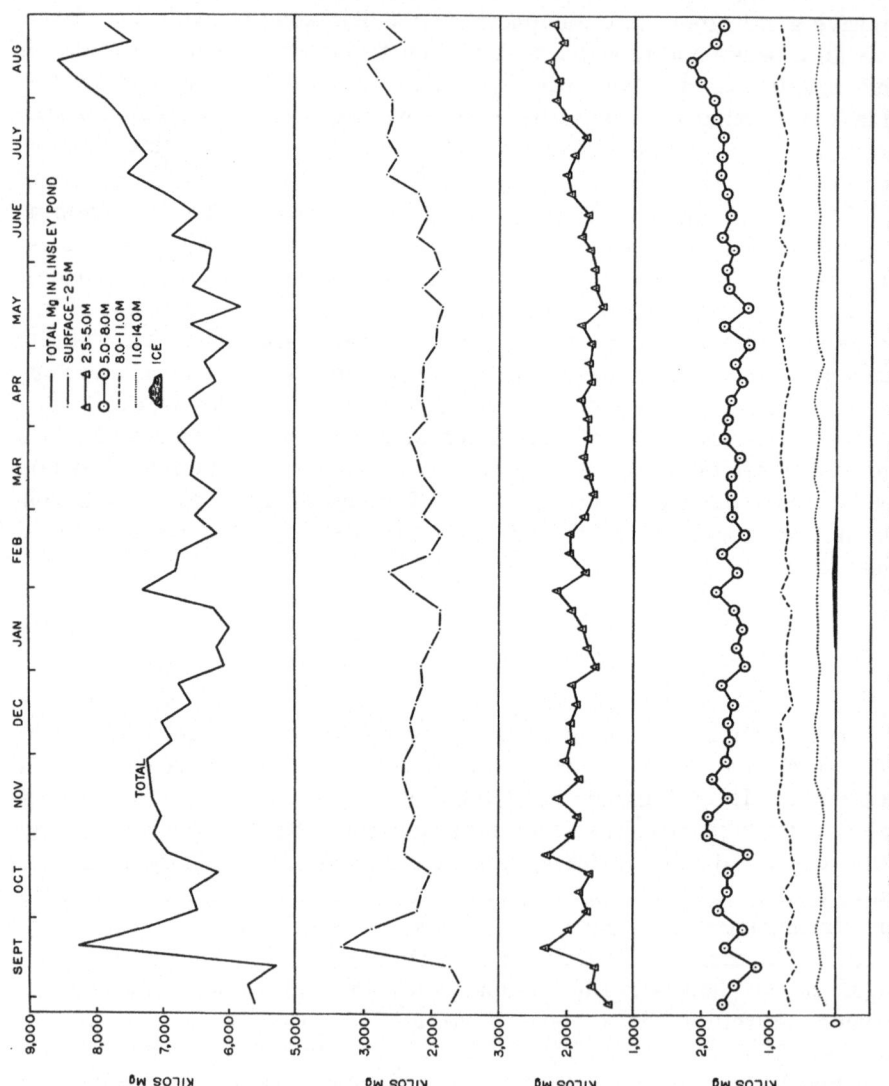

Fig. 16. The distribution of magnesium in Linsley Pond and in the various levels sampled.

Figure 17 shows comparative data for the period of summer stagnation for 1938 and 1966. The element still tends to stratify slightly though the difference between the 8- to 11-m and 11- to 14-m zone is almost negligible. It would appear that the concentration has more than doubled between 1938 and 1966.

The variation in magnesium has not been very pronounced in either year of study, with the exception of the rises on September 19 and during summer stagnation, and the outlet tends to indicate by its changes in concentration the coming and going of littoral vegetation, Fig. 18.

SUMMARY

1. It would appear that there exist several mechanisms by which elements become distributed throughout Linsley Pond. The development and death of littoral vegetation certainly play a significant role in the nutrient supply of the epilimnion, as is apparent in the case of sulfur, phosphorus, and possibly copper. Similarly, organisms living in the surface waters take up these elements and release them on death, adding to the quantities in the mud or the lower waters. The decaying populations appear to have a noticeable effect on the distribution of sulfur, phosphorus, calcium, and magnesium. The reduced mud surface largely controls the movement of iron, manganese, and phosphorus, though the presence of hydrogen sulfide now causes the iron to precipitate out in the deeper layers. The reoxygenation at the autumnal turnover evidently precipitates the phosphorus out as ferric phosphate; manganese tends to remain for a time in solution.

2. As the lake has become more eutrophic over the past 30 years, a number of noteworthy changes in elemental behavior as well as concentration have taken place. In the case of manganese and phosphorus, the hypolimnion has undergone an increase in concentration, which, at the time of autumnal turnover, has resulted in a circulation of manganese in the epilimnitic waters and a precipitation of phosphorus as the ferric compound. On the other hand, iron and sulfur concentrations in the lower waters have undergone a decline as a result of the precipitation of ferrous sulfide. Both calcium and magnesium have increased over the years, and copper to some extent has suffered a decline in concentration due to its probable precipitation as the sulfide; however, their overall distribution has undergone no significant change since the earlier studies.

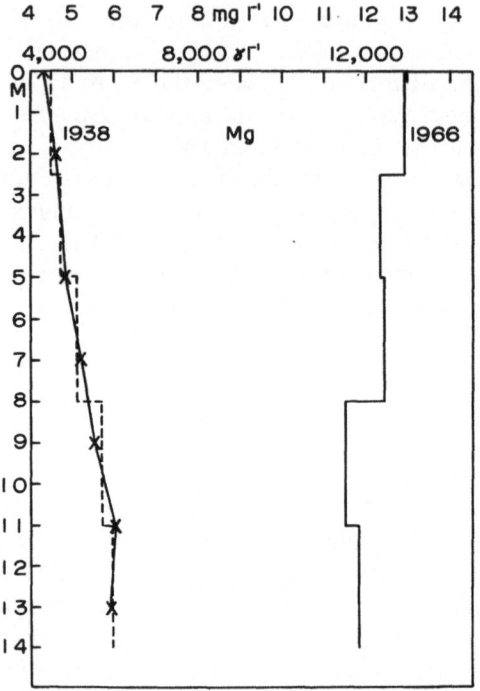

Fig. 17. The vertical distribution of magnesium during summer stagnation in 1938 and 1966.

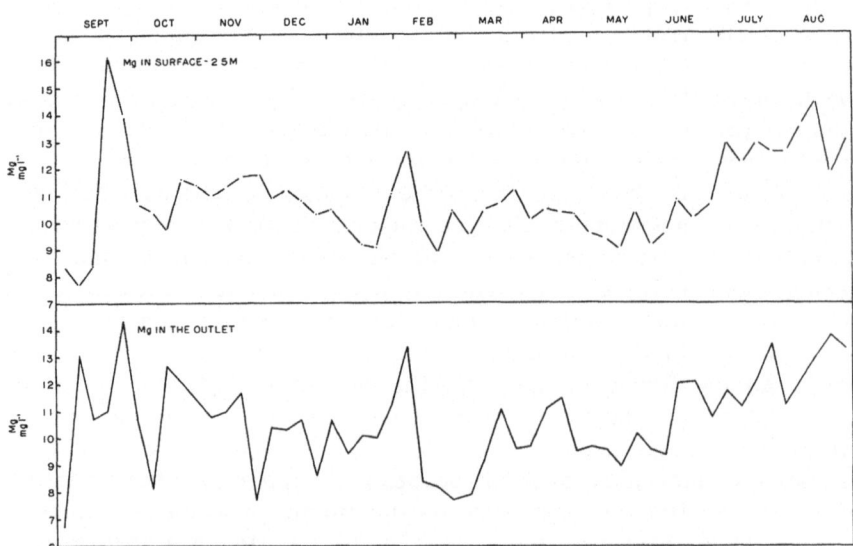

Fig. 18. The magnesium concentration in the outlet and the surface to 2.5 m.

ACKNOWLEDGMENTS

I gratefully acknowledge the support for this study from the National Science Foundation (Grant No. 4309). I wish to thank property owners Mr. Forbes Sargent and Dr. T. Evans of the Linsley Pond area for their constant interest in the study and for the use of their property for the purpose of gathering samples. The field work could not have been completed without the loyal help of Dr. Robert H. Stavn who assisted with the weekly collections during the entire recent study. I appreciate the loan of an oxygen electrode provided by Dr. John L. Brooks. The constant interest and support of Prof. G. E. Hutchinson is gratefully acknowledged.

REFERENCES

1. G.E. Hutchinson, "Limnological studies in Connecticut: IV. Mechanism of intermediary metabolism in stratified lakes," Ecol. Monographs 11:21–60 (1941).
2. G.A. Riley, "Limnological studies in Connecticut," Ecol. Monographs 9:53–94 (1939).
3. U.M. Cowgill, "Heat transfer solely by molecular conduction in the metalimnion," Proc. Nat. Acad. Sci. U.S. 57:198–200 (1967).
4. P.D. Zemany, "Line interference corrections for X-ray spectrographic determination of vanadium, chromium, and manganese in low-alloy steels," Spectrochim. Acta 16:736–741 (1960).
5. U.M. Cowgill, "Use of X-ray emission spectroscopy in the chemical analyses of lake sediments, determining 41 elements," Developments in Applied Spectroscopy, Vol. 5, Plenum Press, New York (1966), pp. 3–23.
6. A.D. Hasler and W.G. Einsele, "Fertilization for increasing productivity of natural inland waters," Trans. 13th N. Amer. Wild Life Conf., March 8–10, 1948, pp. 527–554.
7. G.A. Riley, "The copper cycle in natural waters and its biological significance," Ph.D. thesis, Yale University, New Haven (1937), (Typescript), 239 pp.
8. G.E. Hutchinson, A Treatise on Limnology 1, John Wiley and Sons, Inc., New York (1957), 1015 pp.
9. R.M. Garrels and G.L. Christ, Solutions, Minerals, and Equilibria, John Wiley and Sons, Inc., New York (1965), 450 pp.
10. C.A. Kofoid, "Plankton Studies: The plankton of the Illinois River, 1894–1899, with introductory notes on the hydrography of the Illinois River and its basin, Part I. Quantitative investigations and general results," Bull. Illinois State Lab. Nat. Hist. 6:95–629 (1903).
11. R.H. Pond, "The relation of aquatic plants to the substratum," Rept. U.S. Fish Comm. 21:483–526 (1905).

The Analysis of Trace Constituents in Water by Spectroscopic Methods

S. C. Caruso, H. C. Bramer, and R. D. Hoak

Mellon Institute
Pittsburgh, Pennsylvania

This paper will describe techniques for the detection, separation, and identification of pollutants present in surface waters in very low concentration. The procedure consists of the concentration of the organic compounds by extraction with a high-purity solvent, the gas chromatographic separation of the extracts to produce a trace or "fingerprint" of the contaminants, and the application of spectrometric methods for the identification of the separated components.

The procedure has been applied to tracing sources of organic pollution in rivers and lakes, to determining the extent of pollution in a river basin, to demonstrating the assimilation capacity of a river, to evaluating the efficiency of waste treatment systems, and to identifying taste- and odor-producing compounds in surface waters.

INTRODUCTION

In recent years, increased concern has been expressed over the role of trace contaminants in surface waters. This anxiety has brought about added emphasis in the development of techniques for the detection, determination, and identification of organic compounds present in very low concentrations in potable water. The main problem has been to increase the concentration of these substances to levels detectable by instrumental methods without alteration of their chemical composition or artifact formation. At present, the major concentration methods in use are adsorption on activated carbon [1] followed by elution with organic solvents, freeze concentration [2, 3], and solvent extraction [4]. Although each of these methods has advantages and limitations, the authors [5] have found that small-scale solvent extraction provides a suitable method by which analyses can be conducted with sufficient rapidity to be useful in river and lake surveys.

The procedure consists of extracting 3 to 6 liters of a composite water sample with a high-purity solvent. Sodium chloride is added to the sample prior to extraction to reduce the solubility in water of the organic compounds present and to minimize emulsion formation. Relatively complete extraction is ensured by using a volume of solvent equal to one-third of the sample volume in three serial extractions. The pH of the sample is adjusted to 4 and an extraction made; the pH of the sample is then adjusted to 10 for a second extraction. This accomplishes the extraction of acidic, basic, and neutral compounds and performs an initial partial separation if the two extracts are not combined. The extracts are concentrated to 1 ml by distillation before they are subjected to gas-liquid partition chromatography. Two to ten μl of the concentrate are chromatographed with the use of dual Apiezon L columns (10% on acid-washed chromosorb W, 60 to 80 mesh) and hydrogen-flame ionization detectors with temperature programming from 60 to 250°C at 4°/min. It is necessary to use highly conditioned columns and pre-bled septums to operate at the required high-sensitivity electrometer settings. Nitrogen is used as the carrier gas at a flow rate of 82 ml/min. The flow rates for hydrogen and air are 63 and 445 ml/min, respectively. The method is capable of separating mixtures of components with a boiling point range of 80 to 375°C and can detect many compounds present in the original sample in the low parts per billion range. For example, naphthalene, durene, biphenyl, hexadecane, octadecane, anthracene, triphenyl methane, and pyrene can be detected in the 2 to 5 ppb concentration range, whereas 10 to 25 ppb are necessary for some of the more volatile compounds, i.e., benzene, toluene, and ethylbenzene. These sensitivity limits are based on using the programmed temperature conditions to 250°C. If isothermal or programmed conditions below 220°C can be used for specific samples, the electrometer sensitivity settings can be increased by a factor of 4 or 8. In practice, the lower limits of sensitivity are generally determined by column bleed, septum bleed, and solvent purity rather than by the detector.

The authors have successfully applied this technique to tracing sources of pollutants; demonstrating the rate of assimilation of organic compounds in a stream; assessing the magnitude of the organic load in a river or lake; and determining the efficiency of waste treatment systems.

The utility of the chromatographic "fingerprint" alone will be illustrated by tracing the source of a polluting material which caused a taste and odor problem at a waterworks and by determining the fate of organic compounds through a secondary, activated-sludge sewage-treatment plant.

TRACING THE SOURCE OF A TASTE AND ODOR INCIDENT

Recently, a waterworks in western Pennsylvania experienced a taste and odor problem. Over the last few years, this water-treatment plant has had a working relationship with a large upstream company to receive help in solving their odor problems. Representatives of the industrial concern collected samples of the raw and finished water during the taste and odor incident to have them analyzed by the solvent extraction—gas chromatographic technique. They also prepared a sample of a suspected contaminant. The results are shown in Fig. 1. The chromatogram of the raw water indicates a grossly polluted supply. This chromatogram contains over 38 discernible peaks mostly of intermediate and high-boiling compounds. The second chromatogram represents a sample prepared by mixing 0.2% of fuel oil with water, separating the phases, extracting 3 liters of the water phase only with ethyl ether, and analyzing by the above procedure. It is readily apparent that there is a high degree of similarity between the raw water supply and the fuel oil sample, which indicates this material is the primary source of the problem. However, the "profile" of the finished water shows that the contaminant was almost completely removed by the water-plant treatment process.

During an extensive survey in the same river basin, the technique has been used in a similar way for tracing intermittent pollution loads directly to offending outfalls. Also, chromatographic profiles have been obtained of the major industrial and municipal outfalls in the river basin to aid in assessing the contribution of contaminants from the

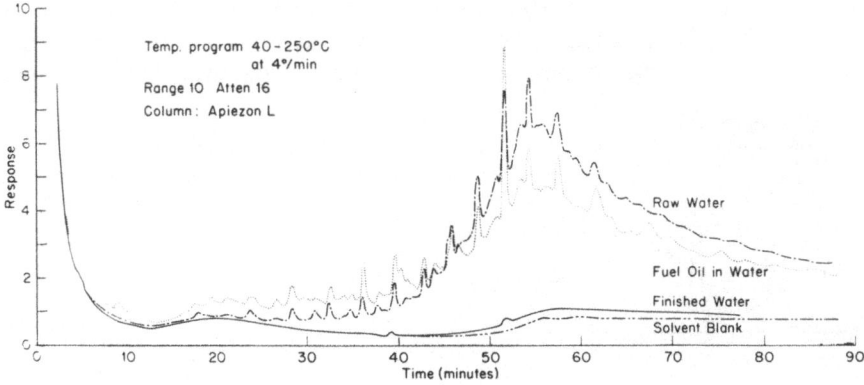

Fig. 1. Tracing the source of a T/O incident.

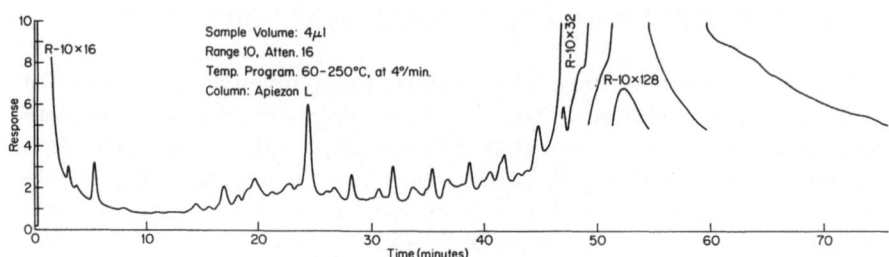

Fig. 2. Waste water influent.

various sources to the total pollution picture and to serve as a reference file for comparative purposes as tastes and odor problems
develop.

FATE OF ORGANICS IN SEWAGE-TREATMENT PROCESS

Samples were taken at four points in a secondary activated-sludge
sewage-treatment plant in eastern Ohio. The sampling points were
as follows:

1. Raw sewage influent
2. Primary settling-basin effluent
3. Digestor supernatant liquor
4. Final clarifier effluent

Figure 2 shows the chromatogram of the raw waste water entering
the treatment plant. The organic compounds consist largely of high-
boiling compounds with some intermediate- and low-boiling substances
also present. After primary treatment, Fig. 3, the effluent contains
a large number of organic compounds approximately evenly distributed over the range of the chromatographic program. Some of these

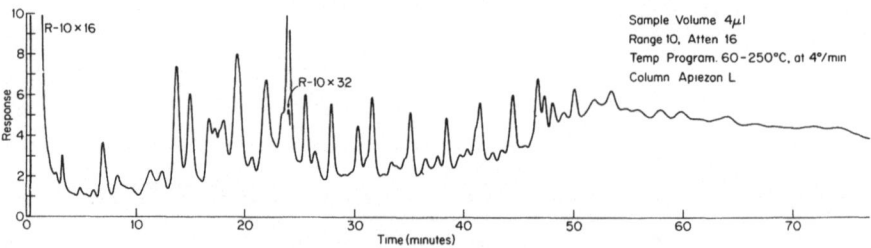

Fig. 3. Primary settling-basin effluent.

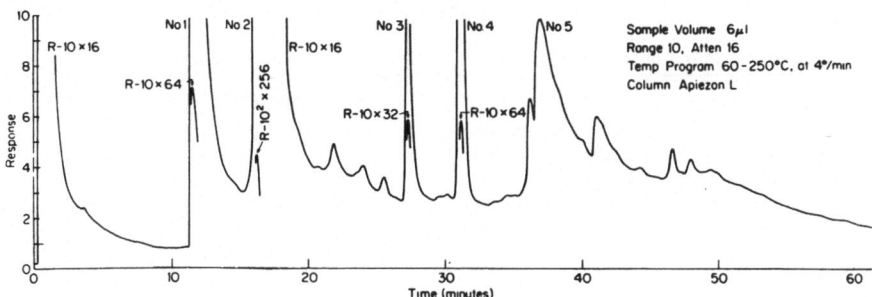

Fig. 4. Digestor sample.

compounds are presumably formed by biodegradation of solids and the high-molecular-weight material present in the raw waste.

The chromatogram of the supernatant liquid from the digestor, Fig. 4, shows the presence of 5 major components and 18 minor components. High-boiling compounds compose only a small part of this sample. Area measurements reveal that the five major components compose over 90% of the compounds represented in the chromatogram.

The final clarifier effluent produced a chromatogram, Fig. 5, which appears to be relatively free of organic compounds. Only one significant peak is present on the chromatogram, which demonstrates a highly efficient removal of the organic compounds present after primary treatment. These results are summarized in Table I.

If this sewage-treatment plant is typical, then primary treatment only can contribute a large number of organic compounds to a receiving stream, whereas secondary treatment removes substantially all

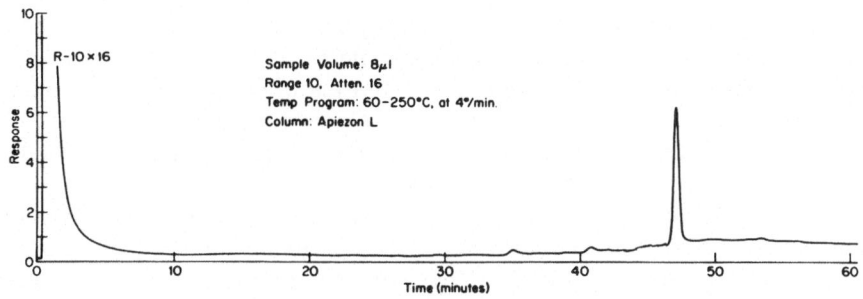

Fig. 5. Final clarifier effluent.

TABLE I

Added between stages*		Disappeared between stages*
Raw waste	6, 8, 9, 12, 25, 29, 31, 33, 37, 38, 41, 46,	16, 20, 23, 27, 34, 61
Primary	49, 50, 51, 53, 54, 55, 57, 58, 59, 60, 65	
Primary		3, 4, 5, 6, 7, 8, 9, 11, 12, 14, 15, 17,
	. . . .	18, 19, 21, 22, 24, 25, 26, 28, 29, 30, 31, 32,
Secondary		33, 35, 36, 37, 38, 39, 40, 41, 42, 43, 44, 45, 46, 48, 49, 50, 51, 52, 54, 55, 57, 58, 59, 60, 63, 65

*Retention time of peaks in minutes.

of the organic substances detectable by the gas chromatographic technique.

IDENTIFICATION OF COMPOUNDS

Although a great deal of information can be obtained by the solvent extraction–GLC procedure alone, unequivocal identification of specific components in a sample is often required. Gas chromatography is extremely valuable as a separation technique, but it furnishes little, if any, structural information concerning the separated components. Sometimes, with additional information about the sample, one can arrive at a tentative identification by comparison of retention-time data with standard compounds on more than one column. However, in the study of trace contaminants in surface water, the organic mixtures are generally very complex, present in low concentrations, originate from multiple sources, and subject to dynamic change in their aqueous environment owing to microbiological and chemical action. For these reasons, unless extensive preliminary class separations are carried out, characterization of the constituents present in a sample usually requires auxiliary analytical methods.

Modern developments in analytical instrumentation and techniques have made it possible to obtain structural information on quantities of a few micrograms or less. Thus, it is possible to use gas–liquid chromatography to separate small amounts of a complex mixture into

its components and to identify these by their response to various spectrometric techniques. Illustrations of the use of these techniques will
be given with a synthetic sample, a sample of digestor liquor from a
sewage-treatment plant, a sample from a stream carrying industrial
wastes, and a sample from Lake Michigan near the southern shore.

INFRARED SPECTROMETRY

At present, infrared spectrophotometry is perhaps the most widely
used spectrometric technique since instrumentation is relatively inexpensive and generally available to most laboratories. An IR spectrum
is highly specific and yields useful structural data. In addition, extensive spectral libraries and structure-correlation data are available
for comparative purposes; it is a nondestructive technique, and the
sample can be used for additional tests. This is important when a
limited amount of sample is available for analysis.

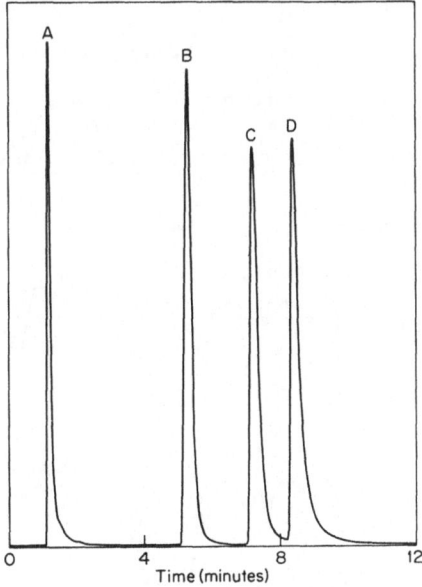

Conditions: sample (A) propyl acetate, (B) anisole,
(C) o-chlorophenol, (D) m-cresol
electro. set   range 10, atten 16
program   90-200°C at 8°/min

Fig. 6. Chromatogram of the mixture.

S. C. CARUSO, H. C. BRAMER, AND R. D. HOAK

Recently a rapid-scan IR spectrophotometer has become commercially available. This instrument promises to be of considerable use in the identification of compounds eluted from a gas chromatograph. It is a single-beam spectrophotometer designed to accept the effluent of a gas chromatograph through a heated tube and record the spectrum from 2.5 to 14.5 μ in scans of 5 or 12.5 sec.

Figure 6 shows a chromatogram of a mixture of propyl acetate, anisole, o-chlorophenol, and m-cresol. By means of a sample splitter nine-tenths of the chromatographed fractions were put through a Beckman Model 102 rapid-scan IR spectrophotometer. The IR spectra of these fractions are shown in Fig. 7.

Although the sensitivity of the rapid-scan IR spectrophotometer is not much greater than that obtainable with other common IR techniques,

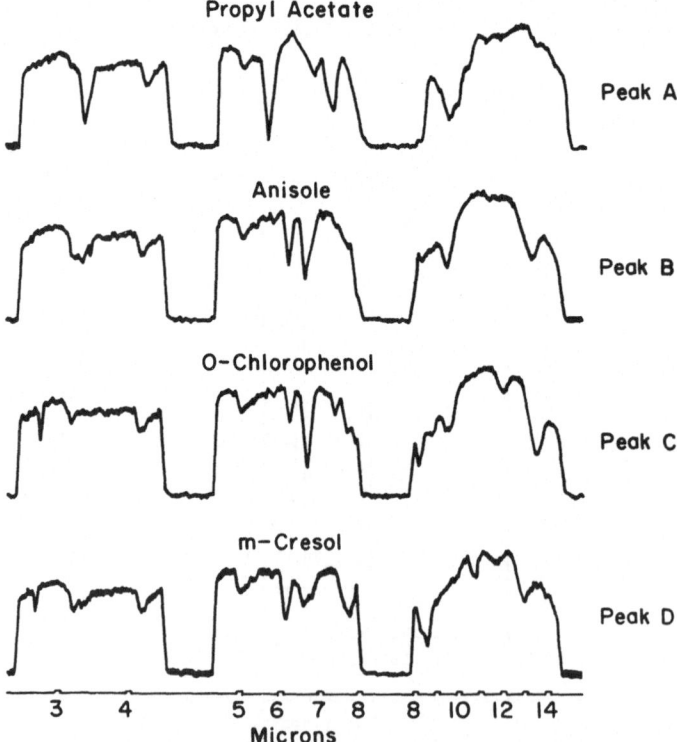

Fig. 7.   Infrared spectra of separated components obtained with a Beckman IR 102 rapid-scan spectrophotometer.

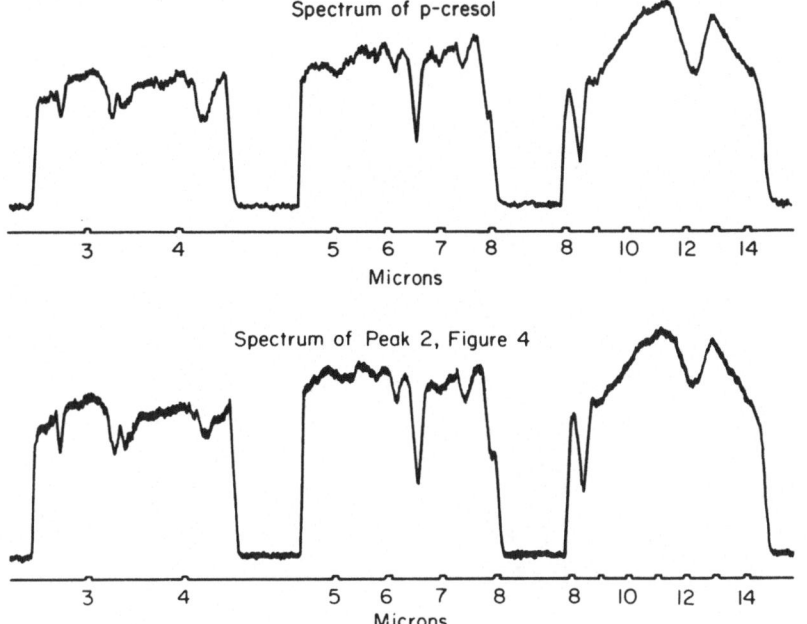

Fig. 8. Identification of a component separated by gas chromatography with the aid of rapid-scan IR spectrophotometry.

manipulative problems with the latter give the rapid-scan instrument certain practical advantages. These are:

1. Rapid analyses; in the example given, the components were separated and the spectra recorded within 10 min.
2. Avoids losses usually associated with trapping, especially of volatile components.
3. Trapping for other tests can be carried out after the component has traversed this instrument.

The substance represented by peak 2 in Fig. 4 was present in sufficient concentration that identification by the rapid-scan IR spectrophotometer was possible with an aliquot of the sample. The elution time of this material and deductions concerning functional groups present from the IR spectrum suggested that it might be a cresol. The spectra of the three cresols were obtained under the same conditions and compared with that of the unknown. Figure 8 shows that the spectrum of p-cresol and that of peak 2 are identical. Retention-

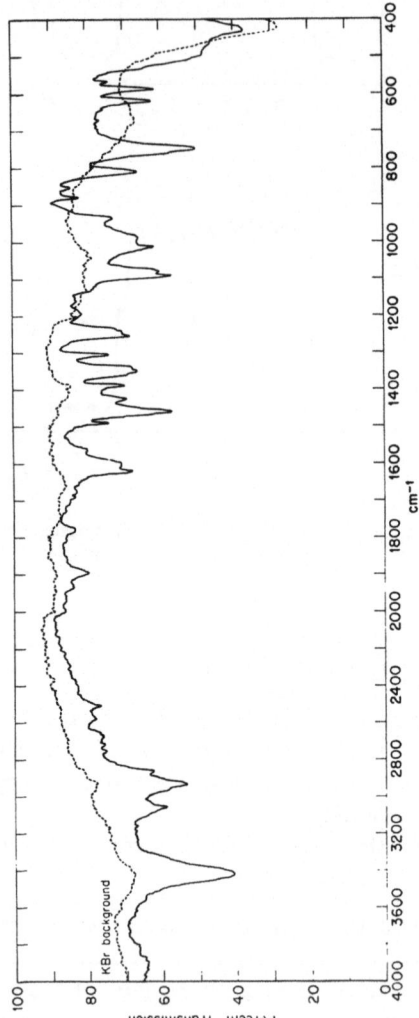

Fig. 9. Infrared spectrum of peak 3 by the KBr pellet technique.

time data and comparison of the ultraviolet spectra corroborate this finding.

Another very useful IR technique for obtaining spectra of microgram quantities has been widely used by workers in the biomedical field. This procedure involves the trapping of gas chromatographic fractions in KBr powder and pressing the powder into a micropellet by using about 80,000 psi pressure. A spectrum of the transparent pellet can then be obtained with a microbeam condensing lens system with a conventional spectrophotometer. This technique was used to aid in the analysis of the components represented in peaks 3, 4, and 5 of the digestor-liquor chromatogram. Figure 9 shows an example of the type of spectra obtained with this procedure. The component in peak 3 has been definitely identified as 3-methyl indole, component 4 appears to be a hydroxyindole, and component 5 is a 3-methyl hydroxyindole. Good spectra have been obtained with sample sizes of 10 to 15 μg.

Although beam-condensing systems have proved highly satisfactory for IR analysis of microsamples, the increasing need to identify even smaller samples in microanalytical investigations in the biomedical, electronics, and environmental fields has required investigators to press for even greater sensitivity. The combination of an IR microscope with a microspectrophotometer [6] allows the analysis of samples of 0.1-μg size. However, this procedure can be used only for samples that can tolerate an increase in temperature of approximately 70°C. Computer addition, statistical-averaging, and curve-smoothing techniques have also been employed with the IR microscope and beam-condensing attachments. Samples of the order of 10 nanograms have been analyzed by this procedure [7]. These advanced ultramicrotechniques should prove highly rewarding for the analysis of trace components present in water and air.

ULTRAVIOLET SPECTROMETRY

Ultraviolet spectra can also be used to aid in the identification of trace materials. Although the spectrum obtained is characteristic of the chromophoric moiety rather than of the complete molecule, it can be useful to corroborate other data or to decide among several alternative structures. It, too, is a very sensitive procedure since most compounds which absorb in the quartz ultraviolet region produce a strong spectrum in concentrations of a few parts per million. Furthermore, it is a nondestructive technique, and the sample is available for additional testing.

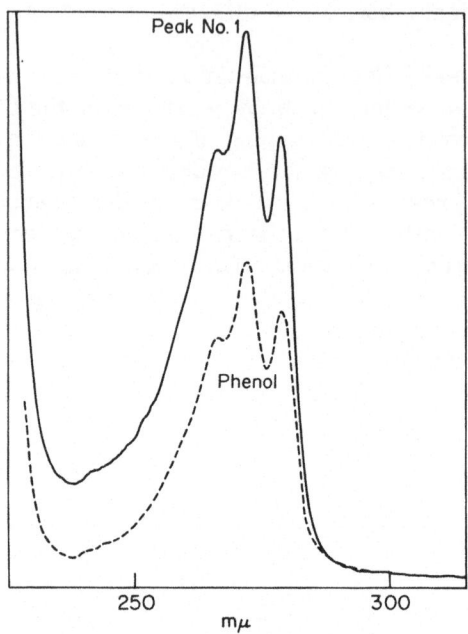

Fig. 10. Ultraviolet spectra of peak 1, Fig. 4, and phenol.

This technique was found helpful in the identification of peak 1 of Fig. 4. The component was trapped in about 1 ml of solvent, and the ultraviolet spectrum was obtained in a 1-cm microcell. This spectrum indicated the presence of a simple aromatic system. The phenol spectrum was selected from a compilation of spectra as matching that of peak 1, Fig. 10. Retention-time data on two columns also agree favorably.

MASS SPECTROMETRY

Mass spectrometry has become one of the most powerful tools available to the organic chemist for structural elucidation of organic molecules. In recent years, a great deal of progress has been made in the correlation of spectral fragmentation behavior with chemical structure. Many investigators [8] have recognized the great potential of the gas chromatography–mass spectroscopy combination to the solution of analytical problems. However, very little published data of the use of mass spectrometry has appeared in the field of trace contaminants in water. This is especially surprising since the analysis of volatile hydrocarbons from waste water was reported as long ago as 1953 [9]. The authors have applied this technique to the solution of analytical problems for several years [10]. Perhaps the most significant feature of mass spectroscopy is the extreme sensitivity available to yield such a rich harvest of molecular information. Special techniques make it possible to obtain useful spectra with microgram or submicrogram sample sizes.

Recent developments in instrumentation and techniques.have increased the utility of mass spectrometry for trace analysis. The direct solid-sample introduction probe [11], which can introduce a solid directly into the ion source chamber, provides a convenient means for obtaining spectra of small samples and allows the analysis of compounds of molecular weights up to 2000. McLafferty et al. [12] have described the use of capillary tubes containing gas-chromatography column packing for trapping components separated by gas chromatography and analyzing the sample with the direct-sampling probe.

A sample taken from a stream in eastern Ohio carrying industrial waste effluents was extracted by the above procedure. The chromatogram of the concentrated extract is shown in Fig. 11. Several peaks were trapped (A, B, C, and D) by the capillary-tube method and analyzed by using a heated probe with an AEI (Associated Electrical Industries, Ltd.) MS-9 mass spectrometer. This is a double-focusing high-resolution (> 33,000) instrument capable of mass measurements with an accuracy of 2 ppm.

Figure 12 shows the spectrum for the component in peak B. Area measurements indicate that less than 15 μg of material was trapped for analysis. This component was identified as quinaldine by interpretation of the mass spectrum only and verified by comparison with the spectrum of a known sample. Similarly peaks A, C, and D have

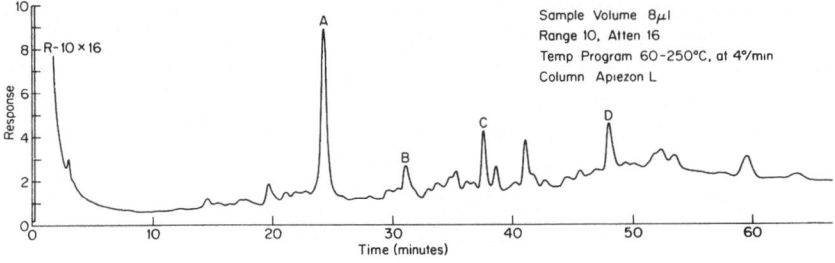

Fig. 11. Stream sample.

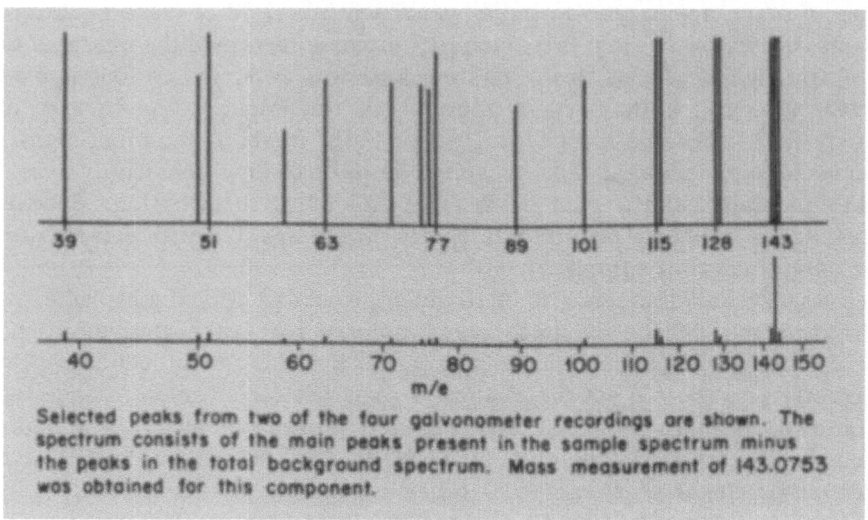

Selected peaks from two of the four galvonometer recordings are shown. The spectrum consists of the main peaks present in the sample spectrum minus the peaks in the total background spectrum. Mass measurement of 143.0753 was obtained for this component.

Fig. 12. Mass spectrometric analysis of peak B, Fig. 11.

been identified as Δ^9-octalin (tentative), biphenyl, and anthracene. Biphenyl and anthracene were also verified by obtaining spectra of authentic samples.

Another exciting development has been the commercial introduction of a new compound instrument in which a mass spectrometer is connected directly to a gas chromatograph to monitor continuously the separated components of a sample. Ryhage [13] has accomplished this successful instrument combination by using two molecule separators [14] in series to increase effectively the sample-to-carrier gas ratio at least a hundredfold. He reports that good mass spectra can be obtained with less than 1 μg of material over the mass range of m/e 12 to 500 with a 1- or 2-sec scan. The high sensitivity and rapidity with which the separation and spectra of complex mixtures can be obtained suggest that this type of instrumentation can be of great value in the trace analysis of water contaminants. As a demonstration of its utility 10 μl of an extract of 6 liters of a Lake Michigan sample were injected into an LKB Model 9000 gas chromatograph-mass spectrometer. Nineteen spectra of components were obtained in approximately 1 h. Although these spectra have not been completely analyzed yet, preliminary interpretation indicates that the main constituents are hydrocarbons.

Biemann [15] and his associates have developed an important new approach to the interpretation of high-resolution spectra, which they call "element mapping." By this procedure, very accurate measurements of mass are obtained for most of the ions formed, the possible elemental formulas are calculated by use of a computer, and the data are presented in a comprehensive table. In this format, the first column presents the nominal mass of the ion species; the second column contains the elemental formulas for the ions which have C and H only; and the remaining columns are arranged according to their heteroatom content. Each column is arranged in ascending order of numbers of carbon and hydrogen atoms present in the ion species. In this scheme, the molecular ion is found in the lower right-hand corner of the table. This arrangement makes assimilation and utilization of the mass spectral data more direct, rapid, and efficient. One author [16] reports that 100 spectra per day can be reduced to element-map form.

Several investigators [17] have recently shown that fast scanning and electrical recording of high-resolution spectra are feasible. This allows the use of a double-focusing, high-resolution mass spectrometer for direct monitoring of gas-chromatographic effluents. Another important advantage of an electrical recording system is that the output signal is in a form most suitable for digitizing either on line or after recording on magnetic tape for computer processing. In addition, reference high-resolution spectra can be compiled for matching spectra. Thus, the instrument combination of gas chromatography with high-resolution mass spectrometry and the use of the computer for spectral interpretation and search promises to be the most powerful tool available for the rapid analysis of trace contaminants in the environment. At least for substances that are separable by gas chromatography, such a system is the nearest approach to a "universal analyzer" that modern technology has yet developed. The most prohibitive feature is the expense of this instrumental combination. Although it is beyond the budget of most laboratories, certainly large research centers can afford to assemble the necessary hardware and perhaps make it available to the scientific community on a sample fee basis.

This study has shown that spectrometric methods are sufficiently sensitive in many cases to provide identification of trace contaminants extracted from relatively small volumes of water with a high-purity solvent and separated by gas chromatography. In addition, recent developments in high-resolution mass spectrometry and the use of computers to aid in structural elucidation of components eluted from a gas chromatograph promise to furnish rapid information concerning trace contaminants in the environment.

ACKNOWLEDGMENTS

This paper is a contribution from the Water Resources Research Project that has been sustained at Mellon Institute since 1938 by the American Iron and Steel Institute. The authors wish to thank Dr. Charles C. Sweeley of the Graduate School of Public Health, University of Pittsburgh, and Mr. R. E. Rhodes of Mellon Institute for their help in obtaining the mass spectral data.

REFERENCES

1. H. Braus, F. M. Middleton, and G. Walton, Anal. Chem. 23:1160 (1951).
2. J. Shapiro, Science 133:2063 (1961).
3. R. A. Baker, J. Water Pollution Control Federation 37:1164 (1965).
4. R. D. Hoak, Intern. J. Air Water Pollution 6:521 (1962).
5. S. C. Caruso, H. C. Bramer, and R. D. Hoak, Intern. J. Air Water Pollution 10:41 (1966).
6. H. L. Dinsmore and P. R. Edmondson, Spectrochim. Acta 15:1032 (1959).
7. R. W. Hannah and R. C. Gore, Paper presented at the Fifth National Meeting of the Society for Applied Spectroscopy, Chicago, Ill. (1966).
8. F. A. J. M. Leemans and J. A. McCloskey, J. Am. Oil Chemists' Soc. 44:11 (1967).
9. F. W. Melpolder, C. W. Warfield, and C. E. Headington, Anal. Chem. 25:1453 (1953).
10. R. D. Hoak, Science and Human Progress, Mellon Institute, Pittsburgh, Pa. (1963), p. 203.
11. R. S. Gohlke, Chem. Ind. (London) 946 (1963).
12. J. W. Amy, E. M. Chart, W. E. Bartinger, and F. W. McLafferty, Anal. Chem. 37:1265 (1965).
13. R. Ryhage, Anal. Chem. 36:759 (1964).
14. E. Stenhagen, Z. Anal. Chem. 205:109 (1964).
15. K. Biemann, Pure Appl. Chem. 9:95 (1964).
16. F. W. McLafferty, Science 151:641 (1966).
17. C. Merritt, Jr., P. Issenberg, M. L. Bazinet, B. N. Green, T. O. Merron, and J. G. Murray, Anal. Chem. 37:1037 (1965).

A Comparison of Trace Elements in Natural Waters, Dissolved versus Suspended

John F. Kopp and Robert C. Kroner

U.S. Department of the Interior
Federal Water Pollution Control Administration
Water Quality Activities
Cincinnati, Ohio

Through a continuing program of the Water Pollution Surveillance System of the Federal Water Pollution Control Administration, raw river waters of the United States are routinely monitored for 19 trace elements with the use of a direct-reading emission spectrograph.

Recent investigations have attempted to characterize differences be tween elements in solution and elements associated with suspended matter. Of the programmed elements, strontium, boron, barium, zinc, copper, aluminum, manganese, and iron were found in the majority of samples in either solution or suspension or in both; lead, nickel, beryllium, and chromium were detected less frequently, while cobalt, silver, cadmium, and vanadium rarely occurred at measurable levels.

INTRODUCTION

When the Federal Water Pollution Control Administration's sur-veillance system initiated the routine measurement of trace elements in surface waters, it was concerned primarily with elements in solu-tion because any suspended material would be removed by the water-treatment plant before it reached the consumer. While the philosophy of this approach is valid, it does not measure the total trace-element load in the stream. This study was performed, therefore, as a pre-liminary attempt to classify the distribution of the trace elements between the dissolved and suspended fractions of the streams selected for investigation. All analyses were performed on a direct-reading emission spectrograph with the rotating disc and high-voltage spark excitation.

SAMPLING STATIONS

Table I lists the percent of frequency of detection for dissolved trace elements in waters of the United States as determined from the analysis of over 1400 surveillance system samples. Barium, boron, and strontium are found in measurable concentrations in nearly all samples, while iron, zinc, copper, and manganese occur 55 to 80% of the time. Aluminum, molybdenum, chromium, and lead are detected in less than 35% of the samples; silver, beryllium, vanadium, cadmium, and cobalt in less than 10%. From a study of these data, it is possible in many instances to predict beforehand what elements will be found in a particular stream. Trace-element levels in some streams remain remarkably consistent; others fluctuate considerably.

A group of seven sampling points was selected for preliminary study. They included three sampling points on the Delaware, one each on the Allegheny and Monongahela Rivers at Pittsburgh, Pennsylvania; one on the Ohio River below Addison, Ohio; and one on the Kanawha River at Winfield Dam, West Virginia. Their choice was based both on the variety of trace elements as well as on the concentration ranges previously observed. This paper is a report on the trace elements in solution versus those in suspension at these seven selected stations.

The Delaware River rises in the western slope of the Catskill Mountains of east-central New York and flows southerly into Delaware

TABLE I

Per Cent of Frequency of Detection for Trace Elements
in Solution*

Element	Percent	Element	Percent
Aluminum†	33.0	Lead	21.8
Arsenic	6.0	Manganese	55.1
Barium	99.2	Molybdenum	33.3
Beryllium	5.7	Nickel	16.7
Boron	94.0	Phosphorus	46.7
Cadmium	2.7	Silver	7.4
Chromium	26.7	Strontium	99.9
Cobalt	3.1	Vanadium	3.5
Copper	74.1	Zinc	76.8
Iron	76.5		

*1410 determinations (Oct. 1, 1962–Dec. 31, 1966).
†1290 determinations (Oct. 1, 1962–Dec. 31, 1966).

Bay. The lower reach of the Delaware is polluted by industrial and also municipal wastes. The three sampling points on the Delaware are located at Martins Creek, Pennsylvania; Trenton, New Jersey; and Philadelphia, Pennsylvania.

The Allegheny River begins in north-central Pennsylvania, flows into New York, re-enters Pennsylvania, and flows southward to Pittsburgh. Oil field brines, acid mine drainage, and steel mill wastes are discharged into the headwaters of the Allegheny and tributary Kiskiminetas Rivers.

The headwaters of the Monongahela River are in northern West Virginia, from whence the river flows north to Pittsburgh. Both the Monongahela and its principal tributary, the Youghiogheny, are polluted by acid mine drainage resulting in unusually low pH values.

The Ohio River is formed by the junction of the Monongahela and Allegheny Rivers at Pittsburgh and starts out as an acid stream with low pH values. Municipal and industrial wastes from Pittsburgh and the basic steel industries and additional acid mine drainage are discharged to the upper reach of the river. By the time the Ohio has reached Addison, however, it has established itself as a more typical alkaline stream.

The Kanawha River drains the central-western half of lower West Virginia and enters the Ohio River at Point Pleasant. In the Charleston area, waste discharges from a large chemical industry complex result in organic, inorganic, and thermal pollution.

PROCEDURE

The data presented in this paper cover the period of October, 1965, through December, 1966, and include 101 monthly composites. In all, 202 analyses were performed. The sample processing procedure as described in an earlier publication [1] is employed, except that the plastic membrane used to filter the sample is not discarded. After a measured volume of sample has been filtered, the plastic membrane containing the suspended material is placed in a 250-ml beaker and 2 ml of redistilled nitric acid is added. The beaker and contents are warmed slightly until the membrane has been destroyed. The beaker is covered with a watch glass and additional acid added as needed until the material has been completely digested. The wet-ashed material is then diluted with distilled water and nitric acid to a final acid concentration of approximately 10%. The samples are then ready for spectrographic analysis. The spectrographic procedure, which has also been described earlier [1], employs the rotating disc method of analysis with high-voltage spark excitation. All analytical results are expressed on a weight per volume basis as micrograms per liter.

While 19 elements are routinely measured in the soluble fraction, only 16 were included for measurement in the suspended fraction. High levels of iron, which are frequently found in the suspended material, result in interference to the arsenic line at 2349.84 Å and also the phosphorus 2535.65 Å line. In addition, the 2816.18 Å line of aluminum interferes with the molybdenum line 0.03 Å away when high levels of aluminum are present. Therefore, arsenic, phosphorus, and molybdenum are not reported in this study. The elements that were routinely measured in the suspended fraction and their respective wavelengths are listed in Table II.

Of the 16 programmed elements, strontium, boron, barium, zinc, copper, aluminum, manganese, and iron were observed in the majority of samples in either solution or suspension or in both. The range of positive values, the frequency of occurrence, and the mean concentration for these elements at the seven selected stations are shown in Table III.

RESULTS AND DISCUSSION

Strontium

One of the most abundant minor constituents in igneous rock is strontium. As the carbonate and sulfate are only slightly soluble, it would not be expected to occur in high concentrations in surface water. Strontium was, however, detected in solution in all samples at a mean concentration of 0.1 mg/liter. The mean observed suspended value was only 2 μg/liter with a frequency of only 17%.

TABLE II

Element	Wavelength, Å	Element	Wavelength, Å
Ag	3382.8	Cu	3273.9
Al	3961.5	Fe	2599.4
B	2497.7	Mn	2949.2
Ba	4554.0	Ni	3414.7
Be	3130.4	Pb	4057.8
Cd	2288.0	Sr	4607.3
Co	3453.5	V	4379.2
Cr	4254.3	Zn	2138.5

TABLE III

Mean Concentration and Per Cent of Occurrence of Trace Elements*

Element	Dissolved			Suspended		
	Range of positive values, μg/liter	Percent of occurrence	Mean, μg/liter	Range of positive values, μg/liter	Percent of occurrence	Mean, μg/liter
Strontium	4–520	100	104	2–3	17	2
Boron	2–750	100	72	3–108	86	16
Barium	5–195	100	43	1–65	97	11
Zinc	6–97	91	63	5–151	90	30
Copper	3–280	80	24	3–66	85	10
Aluminum	1–1875	25	85	3–1440	87	316
Manganese	1–3230	56	330	2–442	98	100
Iron	2–144	70	19	7–>2500	100	>395

*At seven selected stations.

Boron

Soluble boron was found in all samples, generally at levels below
200 ppb. The average boron concentration in the Delaware River increased between Martins Creek and Philadelphia, both in the dissolved
and in the suspended portions. The highest soluble concentrations
in this study were observed in the Ohio and Monongahela Rivers. In
general, the ratio of soluble boron to that in suspension was 4:1.

Barium

Barium occurred in all soluble fractions for a mean frequency of
100%. Concentrations were generally below 75 μg/liter except for
isolated instances. The mean concentration for barium in the soluble
fraction was 43 μg/liter. Barium was also observed in all but four
suspended fractions for a frequency of 97%. Generally, the average
barium concentration in solution was two to three times that found in
suspension; in the Kanawha River, this ratio was 10:1 and in the
Allegheny, 7:1.

Zinc

Soluble zinc was observed in all but 9 samples (91%) and in all but
10 suspended samples (90%). The average soluble zinc concentration
remained constant in the Delaware River, but the suspended portion
doubled between Martins Creek and Trenton and tripled between
Martins Creek and Philadelphia. The Kanawha and, especially, the
Monongahela were much higher in soluble zinc. The mean zinc concentration in solution was 63 μg/liter and in suspension, 30 μg/liter
at the seven selected stations.

Copper

Levels of copper routinely observed in natural surface waters
were generally below 0.05 mg/liter. Thus, higher levels are usually
the result of pollution attributable to the corrosive action of the water
on copper and brass tubing, to industrial effluents, or to the use of
copper compounds for the control of undesirable aquatic organisms.

Copper was observed in both the dissolved and suspended fractions
with about the same degree of regularity (80%). The soluble copper
concentration in the Delaware River decreased at Trenton from 24 to

10 μg/liter and to 7 μg/liter at Philadelphia, while the suspended portion remained about the same at all three Delaware stations, 7 to 13 μg/liter.

The soluble copper concentrations in the Monongahela were some of the highest observed (5 to 280 μg/1), probably owing to the presence of acid mine drainage. The insoluble levels, however, were more nearly equal to the other stations.

Aluminum

Except for the Monongahela River, aluminum was found predominately in the suspended material. Soluble aluminum was detected in only 21% of the Delaware River samples in concentrations ranging from 1 to 40 μg/liter. The Allegheny, Kanawha, and Ohio showed only a 13% frequency of detection for soluble aluminum; however, the concentrations observed were somewhat higher (24 to 120 μg/liter). The Monongahela contained a mean concentration of 170 μg/liter soluble aluminum ranging from 30 to 1875 μg/liter, and a frequency of detection of 67%. Aluminum in the suspended material at the sampling points in this study ranged from 3 to 1440 μg/liter with a mean of 316 μg/liter and a frequency of detection of 87%. Again, the presence of high aluminum in the Monongahela River is undoubtedly due to acid drainage.

Manganese and Iron

Manganese resembles iron both in its chemical behavior and in its occurrence in natural waters. It is, however, much less abundant in rocks than iron is and, as a result, is found in water at lower concentrations. In most natural waters, the concentration of manganese was less than 20 μg/liter but can be higher when mining or industrial wastes are involved, as is the case in both the Allegheny and the Monongahela Rivers. Manganese concentrations above 1 mg/liter may also result where manganese-bearing minerals are attacked by water under reducing conditions or where certain types of bacteria are active.

Manganese frequently accompanies iron in ground waters, and both are commonly reported together. Manganese was almost always found in the suspended fraction (98%) and frequently in solution (56%). It was found in solution in the Delaware 32% of the time ranging from 1.1 to 6.2 μg/liter for an average of 3.4 μg/liter. Soluble concentrations were much higher in the Allegheny, where it occurred 60%

of the time at concentrations ranging from 70 μg/liter to over 3 mg/liter. In the Monongahela River, soluble manganese was observed in 93% of all samples ranging from under 10 μg/liter to as much as 2 mg/liter (Table IV).

The concentration of soluble iron in well-aerated waters is generally low, rarely exceeding 75 μg/liter on the rivers studied. However, the pH may be such that high concentrations of iron do occasionally remain in solution. Table V lists a mean concentration for soluble iron of only 19 μg/liter and a frequency of detection of 70%. Suspended iron was observed in all samples, with a mean of over 400 μg/liter. Concentrations of suspended iron in excess of 2.5 mg/liter were observed in the Monongahela River at Pittsburgh.

Rarely Occurring Elements

Table VI lists the range of positive values and the frequency of detection for some of the less seen elements. This group includes vanadium, cadmium, silver, cobalt, chromium, beryllium, nickel, and lead.

Vanadium, cadmium, and silver were detected in only 5% or less of the samples, including both soluble and suspended portions. Cobalt, chromium, beryllium, nickel, and lead were observed more frequently, 7 to 39%. The positive occurrences of cadmium, silver, chromium, beryllium, and nickel shown in this table were detected mainly in the Delaware River. Two elements shown in this table are of special interest, namely, lead and beryllium.

Lead

Measurable concentrations of soluble lead were detected in only 25% of the samples studied, ranging from traces to 0.2 mg/liter (Table VI). While certain lead salts such as the acetate and chloride are readily soluble, the carbonate and hydroxide are insoluble, and the sulfide is only slightly soluble. Thus, lead would not be expected to remain long in solution in natural waters.

Approximately 50% of the positive occurrences of lead, in solution as well as in suspension, were observed in the Kanawha River at Winfield Dam. Table VII lists the concentrations of lead, dissolved and suspended, measured at this station over a 15-month period. The mean concentration in solution was 0.10 mg/liter; the mean concentration in suspension, 0.34 mg/liter. (See Addendum, page 350).

TABLE IV

Occurrence of Manganese

Sampling point	Dissolved			Suspended		
	Average of positive values, µg/liter	Range of positive values, µg/liter	Percent of occurrence	Average of positive values, µg/liter	Range of positive values, µg/liter	Percent of occurrence
Delaware River						
At Martins Creek, Pa.	2.9	1.3–4.0	40	18	3–51	93
At Trenton, N.J.	3.2	1.1–6.2	27	57	10–107	100
At Philadelphia, Pa.	4.2	3.2–5.8	31	68	29–120	100
Allegheny River						
At Pittsburgh, Pa.	>1000	74.0–3230	60	205	18–1500	100
Monongahela River						
At Pittsburgh, Pa.	607	6.6–2150	93	73	2–442	87
Ohio River						
Below Addison, Ohio	57	7.8–180	51	238	140–400	100
Kanawha River						
At Winfield Dam, W. Va.	44	4.6–115	87	35	2.5–84	100
Mean	>330	1.0–3230	56	100	2–1500	98

TABLE V

Occurrence of Iron

Sampling point	Dissolved			Suspended		
	Average of positive values, µg/liter	Range of positive values, µg/liter	Percent of occurrence	Average of positive values, µg/liter	Range of positive values, µg/liter	Percent of occurrence
Delaware River						
At Martins Creek, Pa.	9.0	2–55	93	>105	7–>500	100
At Trenton, N.J.	7.7	3–13	67	>336	78–>750	100
At Philadelphia, Pa.	27.0	5–144	77	>685	130–>1080	100
Allegheny River						
At Pittsburgh, Pa.	24.0	5–60	73	>293	70–>500	100
Monongahela River						
At Pittsburgh, Pa.	30.0	9–68	73	>425	18–>2500	100
Ohio River						
Below¹ Addison, Ohio	25.0	8–70	54	>652	110–>1500	100
Kanawha River						
At Winfield Dam, W. Va.	11.0	6–26	47	>268	100–>625	100
Mean	19.0	2–144	70	>395		100

TABLE VI

Concentration Ranges and Per Cent Occurrence for Some Lesser Elements*

Element	Dissolved		Suspended	
	Range of positive values, µg/liter	Percent of occurrence	Range of positive values, µg/liter	Percent of occurrence
Vanadium	15—54	3	12, 13	2
Cadmium	3—11	4	6, 16, 35	3
Silver	0.7—3.0	4	0.5—2.5	5
Cobalt	13—48	12	8—13	8
Chromium	2—25	23	3—13	38
Beryllium	0.03—3.2	30	0.04—2.35	39
Nickel	3—86	35	5—900	20
Lead	6—205	25	10—625	35

*At seven selected stations.

TABLE VII

Occurrence of Lead in the Kanawha River
(See Addendum)

Period	Dissolved, mg/liter	Suspended, mg/liter
October, 1965	0.09	0.55
November	0.04	0.34
December	< 0.04	0.60
January, 1966	0.17	0.46
February	0.09	0.20
March	0.07	0.12
April	0.09	0.55
May	0.21	0.63
June	0.14	0.38
July	0.07	0.33
August	0.14	0.44
September	< 0.04	0.10
October	< 0.01	0.10
November	0.08	0.19
December	0.02	0.08
Mean	0.10	0.34

Soluble lead was observed in the Delaware River with a frequency of 21%, concentrations ranging from 6 to 32 μg/liter. Suspended lead in the Delaware occurred with a frequency of 33%, ranging from 10 to 34 μg/liter. Occurrences in the Allegheny, Monongahela, and Ohio Rivers were less frequent. Excluding the Kanawha River, the highest recorded value in this study was 48 μg/liter of soluble lead in the Ohio River at Addison.

Addendum. Subsequent review of these data indicates possible error due to improper sampling resulting in higher values.

Beryllium

Another infrequently found element is beryllium. It is a comparatively rare element not likely to be found in natural waters in greater than trace amounts because of the relative insolubility of the oxide and hydroxide at the normal pH range. Table VIII lists the mean concentrations, the range, and the percent of occurrence at the seven selected stations.

The highest single observed value for soluble beryllium, 3.2 μg/liter, as well as the highest mean concentration in solution, 0.62 μg/liter, were recorded in the Monongahela River. While it was de-

TABLE VIII

Occurrence of Beryllium

Sampling point	Dissolved			Suspended		
	Average of positive values, µg/liter	Range of positive values, µg/liter	Percent of occurrence	Average of positive values, µg/liter	Range of positive values, µg/liter	Percent of occurrence
Delaware River						
At Martins Creek, Pa.	n.d.*	n.d.*	0	0.06	0.06	7
At Trenton, N.J.	0.27	0.03−1.47	87	0.55	0.09−2.35	100
At Philadelphia, Pa.	0.05	0.03−0.09	31	0.22	0.05−0.55	85
Allegheny River						
At Pittsburgh, Pa.	0.25	0.08−0.42	27	0.08	0.05−0.10	33
Monongahela River						
At Pittsburgh, Pa.	0.62	0.07−3.2	60	0.08	0.06−0.09	13
Ohio River						
Below Addison, Ohio	n.d.*	n.d.*	0	0.09	0.04−0.12	38
Kanawha River						
At Winfield Dam, W. Va.	n.d.*	n.d.*	0	n.d.*	n.d.*	0

*n.d., not detected.

tected more frequently in the Delaware River at Trenton, the mean concentration in solution was only 0.27 μg/liter.

For suspended beryllium, the Delaware River at Trenton and Philadelphia contained both the highest frequency of detection and the highest mean concentration. The source of this beryllium has been shown previously [2] to be the Lehigh River which enters the Delaware at Easton, Pennsylvania.

SUMMARY

A nationwide surveillance program to characterize chemical and biological trends in water quality has been in progress since 1957. A part of this interdisciplinary approach to water-quality assessment has been the use of a direct-reading emission spectrograph for measuring the concentrations of 19 trace elements in solution in the major rivers of the United States.

Recent investigations have attempted to classify the distribution of 16 trace elements between the dissolved and suspended fractions at seven selected sampling stations. Results have shown that elements such as boron, barium, zinc, and copper are found in both the suspended and dissolved fractions of river water. Other elements which are found predominately in the suspended matter are aluminum, manganese, iron, and lead. Strontium is rarely observed in suspension but is found in solution in all samples.

REFERENCES

1. John F. Kopp and Robert C. Kroner, A Direct-Reading Spectrochemical Procedure for the Measurement of Nineteen Minor Elements in Natural Water, J. Soc. Appl. Spectry. 19:155 (1965).
2. John F. Kopp and Robert C. Kroner, Tracing Water Pollution with an Emission Spectrograph, J. Water Pollution Control Fed. 39:1659 (1967).

Application of Spectroscopy and Chromatography in Water Quality Analysis*

W. DeWitt Johnson and Preston W. Kelly

Organic Chemistry and General Chemistry Laboratories
U. S. Department of the Interior
Federal Water Pollution Control Administration
Great Lakes Region – Chicago Program Office
Chicago, Illinois

This paper discusses spectroscopy and chromatography as they are being applied to obtain factual data and develop information for a comprehensive water-pollution control program. Spectroscopy and chromatography are employed in water-quality determinations to ascertain the levels of constituents such as nitrogens, phosphorus, gross minerals, organic content, and pesticides with the use of existing methods and modified or new methods developed in the Federal Water Pollution Control Administration's Great Lakes Region–Chicago Program Office (GLR-CPO) laboratories. This paper also discusses the adaptation of automation in using an Auto-Analyzer for determining several water-quality parameters and touches upon future applications of spectroscopy and chromatography.

INTRODUCTION

The Federal Water Pollution Control Administration, Chicago Program Office laboratories were initially established in September 1960. The mission of the project was to formulate a comprehensive long-range plan for water quality management in the Great Lakes – Illinois River Basins. The Chicago Program Office laboratories are proceeding with studies of existing water quality in these basins.

The laboratories are equipped with the several spectroscopic and chromatographic instruments, including visible and ultraviolet spectrophotometers, gas chromatographs, an AutoAnalyzer, and an infrared spectrophotometer.

---

* Disclaimer: Mention of products and manufacturers is for identification only and does not imply indorsement by the Federal Water Pollution Control Administration or U.S. Dept. of the Interior.

This paper will discuss the application of these spectroscopic and chromatographic instruments as employed by the project.

LABORATORY MEASUREMENTS

Several of the pollutants now routinely determined must be detected in extremely low concentrations. Measuring extremely low concentrations of trace components in milligrams per liter to micrograms per liter is accomplished by the use of sensitive spectrophotometric methods and is extended to the nanograms per liter range via gas chromatographic techniques. This new instrumentation provides increased sensitivity, precision, and accuracy. Greater reliability than with older methods is thus obtained in determining whether test waters meet water-quality criteria and standards.

ORGANIC CONTAMINANTS

A large number of synthetic organic pollutants present problems by their persistence and are damaging to water quality. Organics enter water in four important ways: industrial wastes, sewage, farmland runoff, and direct application as in fish control.

Organics in trace quantities produce taste and odor problems, toxicity problems, and oxygen- and chlorine-demand problems. The total amount of many organics in a liter of water is usually insufficient for "wet" or classical methods of testing. The powerful team of spectroscopy and chromatography have made possible the analysis and identification of smaller quantities.

Gross organics may be estimated by various "wet" and "dry" oxidation methods, such as the gravimetric—burning off of organic matter—via the "difference" method, the chemical oxygen-demand test, and the biochemical oxygen-demand method. While these tests are still very useful, the more definitive or specific spectroscopic and chromatographic tests are desirable for assigning exact measurements of water-quality constituents. For example, even though the sensitive spectroscopic aminoantipyrine method is commonly employed for phenol in the micrograms per liter range, only gross phenols are given; yet in solving such problems as taste and odor in water, specific compounds must be identified.

To facilitate water-quality analysis further, the carbon adsorption method is employed to obtain large amounts of organics by filtering water through carbon. This is done by passing from 300 to 5000 gal of water through a carbon column of approximately 3- by 18-in. size to adsorb the organic contaminants, according to Briedenbach et al. [1]. The carbon is dried at 40°C, extracted with chloroform, then with an alcohol, and reported as carbon chloroform and alcohol extracts.

Approved or treated drinking water should contain less than 200 μg/liter of carbon chloroform extract, according to the U.S. Public Health Service Drinking Water Standards of 1962 [2].

The carbon chloroform extract is separated into fractions: acid, neutral, and basic. The neutrals present the major persistent pollution problems. Through the use of column chromatography, the neutral fraction is separated into aliphatic (petroleum), aromatic (odorous coal-tar hydrocarbons), and oxygenated fractions. The aromatic is generally the most interesting from a pollutional standpoint since pesticides, phenyl ethers, and o-nitro compounds can be isolated from this group. The oxygenated portion includes aldehydes, ketones, and esters.

The alcohol extractives include water-soluble materials—proteins, carbohydrates, surface-active materials, and natural substances. Gas chromatography and infrared spectroscopy has made it possible to identify and quantify rapidly macro and even micro quantities of organic substances.

Carbon adsorption is simply a tool for concentrating organics from water; however, different organic compounds can be expected to vary in adsorption and extraction rates [1]. Following carbon adsorption, infrared spectroscopy and gas chromatography analyses permit the identification and quantification of the damaging organic compounds.

PESTICIDES

Pesticides are receiving increasing attention because of the extreme toxicity and persistence of the chlorinated hydrocarbon pesticides in the aquatic environment.

The GLR-CPO laboratory is able to detect chlorinated pesticides such as lindane, heptachlor, aldrin, dieldrin, endrin, and DDT in concentrations ranging from micrograms per liter to nanograms per liter. The pesticides are extracted by liquid—liquid extraction and concentrated to a small volume. The extract is then "cleaned up" by means of column chromatography with a Florisil column and then analyzed by injecting a few microliters of it into a gas chromatograph with an electron-capture detection system. Further verification of results are made by gas chromatography with a microcoulometric detection system, thin-layer chromatography, and, where possible, infrared identification [1, 3].

In the analysis for pesticides in soil, mud, aquatic plants, and water, some of the problems of removing varying amounts of interfering substances before the application of chromatography and spectroscopy were solved as follows [1, 3, 4, 5].

1. Water samples were extracted with hexane, which tends to extract pesticides more than some of the interfering substances. The use of a $^3/_8$ × 12 in. macro-clean-up-column with a mixture of 1:2 carbon–Florisil, eluted with 15% petroleum ether in ethyl ether, facilitates interference removal.

2. Soil, mud, and aquatic plants were dried at room temperature to immobilize some of the soluble aqueous interfering substances. These samples were extracted in a Soxhlet extractor with 10% acetone in hexane; the extract was cleaned up as previously described. However, further clean up was done by thin-layer chromatography, especially for infrared identification.

3. The optimization of gas chromatography, especially with the microcoulometric titration system, was done by making the following critical adjustments: The flow rate was set at 130 ml/min with nitrogen carrier gas, the temperature of the column was set at 200°C, and resistance was set at 500 ohms.

INORGANIC CONTAMINANTS

Some of the inorganic pollutants determined spectroscopically are: ammonia nitrogen, phosphorus (total and soluble), iron, magnesium, nitrate nitrogen, potassium, sodium, silica, and chloride.

Nitrogen is important because it is a nutrient and contributes to the fertilization of lakes and streams. It is a product of the biochemical decomposition of nitrogenous matter.

The method used at GLR-CPO for the determination of ammonia nitrogen with the AutoAnalyzer is a modification of the Ternberg and Hershey [6] method involving sodium nitroprusside as a catalyst. O'Brien and Fiore [7] automated this method for water analysis. The minimum detectable concentration is 0.01 mg/liter [6, 7, 8].

The determination of phosphorus is another important factor to consider. As a nutrient, along with the nitrogen, phosphorus plays a leading role in eutrophication of lakes and streams [9]. Algal blooms, which may occur as a consequence of eutrophication, cause bad tastes and odors in waters. The associated prolific growth of aquatic and shoreline plants fills in lakes and interferes with recreational uses.

Total soluble phosphorus and total phosphorus are determined by the potassium antimonyl tartrate method [10] with a minimum detectable concentration of 0.01 mg/liter. Interference in samples from iron and chromium in the phosphorus test was minimized by proper dilution of the sample, by employing the potassium antimonyl tartrate method.

The chloride analysis is performed with the AutoAnalyzer by using the mercuric thiocyanate method; the minimum detectable concentration is 1.0 mg/liter [7, 11].

Iron occurs in water as ferrous and ferric iron, depending on factors such as dissolved oxygen and pH. Excessive iron concentrations give water a bad taste and render it unfit for cooking and laundering.

Iron is spectrophotometrically determined by employing the *o*-phenanthroline method with a minimum detectable concentration of 0.02 mg/liter [11].

Magnesium occurs in natural water as a common salt and, along

TABLE I

Use of Chromatography at Chicago Program Office

Chromatography method	GLR-CPO	Reference
Column Florisil, alumina, silica gel, etc.	Clean up and separation of organic mixtures for gas, and thin-layer chromatography analysis also for IR identification and pesticide analysis	(3)
Thin layer	Organic chemical identification and clean up, especially pesticides	(3)
Paper	Preliminary organic screening, infrequently used	(5)
Gas Electron capture	a. Pesticides b. Fatty acids and misc. organics c. Industrial wastes	(3)
Microcoulometric	a. Chlorinated pesticides b. Phosphated pesticides	

TABLE II

Enumeration of Parameters

Parameter	Instrument	Wavelength, mμ	Method	Reference number
Arsenic	Spectrophotometer	535	Silver diethyldithiocarbamate	(11)
Chloride	AutoAnalyzer	480	Mercuric thiocyanate	(11, 7)
Color (true)	Eye–Hellige Aqua Testor	(White light with filters)	Visual comparison	(11)
Cyanide	Spectrophotometer	625	Pyridine-pyrazolone	(11)
Fluoride	Spectrophotometer	570	SPADNS	(11)
Iron	Spectrophotometer	510	o-phenathroline	(11)
Magnesium	Spectrophotometer	525	Brilliant yellow	(11)
Magnesium	AutoAnalyzer	655	Magnesium, blue	(12)
Manganese	Spectrophotometer	525	Persulfate	(11)
Methylene blue active substance	Spectrophotometer	655	Extraction, methylene blue	(11)
Methylene blue active substance	AutoAnalyzer*	655	Extraction, methylene blue	(11)
Nitrogen–ammonia	Spectrophotometer	425	Nesslerization	(11)
Nitrogen–ammonia	AutoAnalyzer	630	Sodium phenate-hypochlorite	(6, 7)
Nitrates	Spectrophotometer	410	Phenol-disulfonic acid	(11)
Nitrates	AutoAnalyzer	520	Zinc-sulfanilic acid	(7, 11)
			Naphthylamine-hydrochloride	
Nitrite	Spectrophotometer	520	Sulfanilic acid	(11)
			Naphthylamine-hydrochloride	(8, 11)
Nitrite	AutoAnalyzer	520	Naphthylamine-hydrochloride	
Total Kjeldahl nitrogen	Spectrophotometer	425	Nesslerization	(11)
Organic nitrogen	Spectrophotometer	425	Nesslerization	(11)
Phenols	Spectrophotometer	460	Chloroform extraction method (aminoantpyrine)	(11)

Parameter	Instrument	Wavelength/Value	Method	Reference
Total and soluble phosphorus	Spectrophotometer	880	Molybdenum blue (ascorbic acid) (potassium antimonyl tartrate)	(10, 11)
Silica	Spectrophotometer	650	Heteropoly blue	(11)
Silica	AutoAnalyzer	660	Heteropoly blue	(11)
Sodium potassium	Flame photometer	589, 768, respectively	Flame emission	(11)
Sulfate	Spectrophotometer	420	Turbidimetric (Sulfa Ver)	(11)
Sulfate	AutoAnalyzer	465	Turbidimetric (Sulfa Ver)	(11)
Total sulfide	Spectrophotometer	650	Methylene blue	(16)
BIOLOGY				
Phytoplankton productivity	Spectrophotometer	630, 645, 665	Chlorophyll a, b, c; estimation	(13)
FIELD MEASUREMENT				
Current flow	Fluorometer G. K. Turner III	Absorb, 550 Reemit, 590	Rhodamine B	(14)
Light penetration	G. M. Mfg. Co. Submarine photometer	Visible range	Photoelectric cell	(13)
Light penetration	Eye—Secchi disc	Visible range	Light extinction	
ORGANICS				
Pesticides	Gas Chromatograph with microcoulometric titration and electron capture		Gas—liquid	(3)
Pesticides	IR spectroscopy		IR scan absorption	(3)
Pesticides	Thin-layer chromatography		Liquid—solid	(3)
Carbon chloroform, alcohol extracts, and special analyses	IR spectroscopy and chromatography		Absorption, thin-layer, and gas chromatography	(1)

*Technicon AutoAnalyzer methodology.

with calcium, is the major cause of hardness in water. Magnesium is determined on the AutoAnalyzer by the magnesium-blue method with a minimum detectable concentration of 1.0 mg/liter. This method was developed by GLR-CPO and has replaced the brilliant yellow method [12].

Sodium along with potassium is determined by flame photometry, each having a minimum detectable concentration of 0.1 mg/liter [11].

Sulfates are determined by a turbidimetric method employing the AutoAnalyzer. A 2% solution of Sulfa/Ver* in 0.05% polyvinyl alcohol corrects a "coating" problem, which occurred in the spectroscopic flow-through cell, and results in a stable nondrifting base line. The minimum detectable concentration is 1.0 mg/liter. Results in the range of 1 to 50 mg/liter follow Beer's law; higher values cause base-line drift [11].

Silica is determined by the AutoAnalyzer with the use of the ammonium molybdate aminonaphthol-sulfonic acid method. The manual spectroscopic method is slow and tedious. The minimum detectable concentration for the automated method is 0.02 mg/liter [11].

BIOLOGY

Phytoplankton densities are useful in accessing the state of eutrophication of lakes and streams. This water-quality criterion can be measured by extracting chlorophyll from the plankton, reading the intensity of the color in a spectrophotometer at the proper wavelengths, and calculating the standing crop of algae in terms of chlorophyll a, b, and c [13]. High chlorophyll content is generally characteristic of eutrophic lakes or streams; low values usually indicate oligotrophic conditions [9].

FIELD MEASUREMENTS

The GLR-CPO employs dye-tracer studies to determine hydrologic, flow-direction, and mixing conditions in streams and lakes with the use of a fluorometer [14]. The studies assist in determining the rate, quantity, and direction of movement of a contaminant.

Light-penetration measurements made with the Submarine photometer indicates the transparency or the depth of the phototropic zone in a lake, which is inversely proportional to suspended matter.

*The 2% Sulfa-Ver is a commercial reagent.

FUTURE APPLICATIONS

Plans for atomic absorption spectrophotometry are under way at the GLR-CPO laboratory. Atomic absorption is a method by which metallic and semimetallic elements in solution or in a very fine suspension can be determined [4, 6, 15]. The use of atomic absorption spectroscopy is planned to measure copper, cadmium, nickel, zinc, lead, arsenic, iron, and manganese.

Most of these metals have been analyzed by employing the polarographic technique with a high degree of reliability. However, this method is considerably slower than the atomic absorption technique.

SUMMARY

In general, Table I summarizes the use of chromatography at GLR-CPO, with references. Table II summarizes the laboratory application of spectroscopy at GLR-CPO, with references given.

CONCLUSIONS

1. Spectroscopy and chromatography have made possible the rapid measurement of low-level pollutants in complying with water-quality criteria and standards. These methods ensure a higher degree of sensitivity, accuracy, and precision than do the classical manual methods.

2. Atomic absorption spectroscopy promises increased detection capability and time savings.

3. The AutoAnalyzer is producing precise and accurate values at GLR-CPO, with significant economies in manpower.

REFERENCES

1. A.W. Breidenbach et al., The Identification and Measurement of Chlorinated Hydrocarbon Pesticides in Surface Waters, USDI, FWPCA, WP-22, Washington, D.C. (1966).
2. U.S. Dept. of Health, Education, and Welfare, Public Health Service Drinking Water Standard, PHS Pub. No. 956, Supt. of Documents, Washington, D.C. (1962).
3. H.P. Buchfield, D.E. Johnson, and E.E. Storres, Guide to the Analysis of Pesticide Residues, Vols. I and II, U.S. Dept. of Health, Education, and Welfare, Supt. of Documents, I (1)−VIII, E.3.4 (1), Washington, D.C. (1965).
4. Beckman, An Introduction to Spectrophotometry, Bull. 295B, Beckman Instruments, Inc., Fullerton, Calif. (1966).

5. R. Stock and C.B.F. Rice, Chromatographic Methods, Reinhold Publishing Corp., New York (1963), pp. 1–206.
6. J.L. Ternberg, M.D. and F.B. Hershey, M.D., Colorimetric Determination of Blood Ammonia, J. Lab. Clin. Med. 56:766–776 (1960).
7. J.E. O'Brien and J. Fiore, Ammonia, Nitrate, Nitrite, Chloride Determination by Automatic Analysis, Wastes Eng. (July 1962).
8. N.S. Zaleiko, Simultaneous Application of Chemical Analysis and Physical Parameter Instrumentation to Wastes, Instrument Society of America, 21st Annual ISA Conf. and Exhibit, New York, October 24–27, 1966 (preprint).
9. W.B. Sarles, "Algae and Metropolitan Wastes," in Madison's Lakes and Urbanization, U.S. Dept. Health, Education, and Welfare, Public Health Service, Robert A. Taft SEC, Cincinnati, Ohio (1960), pp. 10–22.
10. E.P. Gail, A.H. Molaf, and R.W. Shuman, Determination of Orthophosphate in Fresh and Saline Water, J. Am. Water Works Assoc. 57:917 (1965).
11. APHA, AWWA, and WPCF, Standard Methods for the Examination of Water and Wastewater, 12th ed., New York (1965).
12. P.W. Kelley and F.D. Fuller, "An Automated Method for Determining Magnesium in Water and Wastewater," in Automation in Analytical Chemistry, Technicon Controls, Inc., Ardsley (Chauncey), New York (1966), pp. 266–269.
13. Water Pollution, Biology Field and Laboratory Manual, Great Lakes Region–Chicago Program Office, FWPCA, U.S. Dept. Health, Education, and Welfare, Chicago, Ill. (1966).
14. V.H. Farmer, A Preliminary Study of Surface Currents along the Southern Shore of Lake Ontario, Unpublished internal report, Dept. Health, Education, and Welfare, U.S. Public Health Service, Great Lakes–Illinois River Basins Project, Chicago, Ill. (1965).
15. M.J. Fishman and S.C. Downs, Methods for Analysis of Selected Metals in Water by Atomic Absorption; Spectrographic Analysis of Natural Water, U.S. Geological Survey Water Supply Paper 1540-C, U.S. Govt. Printing Office, Washington, D.C., pp. 23–45.
16. Chemistry Laboratory Manual, Great Lakes Region–Chicago Program Office, U.S. Dept. I, FWPCA, Chicago, Ill. (1967).

Application of Atomic Absorption Spectroscopy in a Water–Pollution Control Program

Alfred M. Tenny

Department of Research and Control
The Metropolitan Sanitary District of Greater Chicago, Illinois

The use of atomic absorption spectroscopy has allowed the Metropolitan Sanitary District of Greater Chicago to expand its stream surveillance and industrial-waste monitoring in the number of parameters measured per sample without any increase in professional personnel.

Presently, copper, nickel, iron, manganese, sodium, potassium, calcium, magnesium, chromium, zinc, cadmium, and lead are checked routinely. Both the analytical methodology and pollution control significance of many of these described elements will be discussed. Methodology will include sample preparation, effect of interferences, ease of analyses, and use of automatic data processing.

INTRODUCTION

The Metropolitan Sanitary District of Greater Chicago (the District), initiated a comprehensive water-quality investigation of the waterways under its jurisdiction in 1965. This study, which included the measurement of various physical, chemical, and bacteriological parameters, is still continuing. Atomic absorption spectroscopy was used to measure sodium, potassium, calcium, magnesium, iron, manganese, zinc, cadmium, copper, nickel, chromium, and lead. The data collected has a multitude of uses. First, the obvious is to assess current water quality. Second, variations in parameters or ratio of parameters can indicate and help detect pollution. The data can also be used to evaluate progress of the water-pollution control program, and lastly it can help devise means of upgrading the system to allow a higher quality of water usage.

The waterway system consists of 85 miles of navigable water, including canals, channels, and rivers. In addition, hundreds of miles of secondary tributaries and storm sewers also drain into the main watercourse. Three major sewage-treatment plants and nine small treatment plants add effluent to the drainage basin. The District covers 858 square miles, including nearly all of Cook County, and serves a population of $5\frac{1}{2}$ million and an industrial-waste load equivalent to another 3 million.

The purpose of the work herein is to demonstrate how atomic absorption spectroscopy was applied in a water-quality program.

EXPERIMENTAL PROCEDURE

The most important step in any analysis is sample collection. This is especially true in pollution-control enforcements, where laboratory results must be available for legal action. The present group of Industrial Waste Ordinance (1946, 1962) limit, among other parameters, the discharge of toxic materials by industry to sewers, streams, and drainage areas [1, 2].

Samples for metal analyses are collected either by use of an automatic sampler or manually and sent to the laboratory in polyethylene bottles which contain 1% by volume of concentrated nitric acid.

Nitric acid is used to prevent decreases in metal concentration by plating out on the walls of the container, which occurs at prolonged storage. Tests, in the laboratory, on river and wastewater samples stored several months after acid treatment and rechecked show no significant changes. Standards stored and rechecked after 3 weeks gave identical readings.

Upon being received in the laboratory, most samples can be analyzed directly on the atomic absorption spectrophotometer without any further sample treatment. Occasional industrial wastewater samples containing a very high concentration of suspended solids must be digested. The usual digestion procedure is to heat with nitric acid for 30 min or until the sample is clear. Samples having visible floating oil require removal of the oil before analysis. Since the oil cannot be easily digested, it is simpler to use a separatory funnel to remove the oil and only analyze the water phase.

Approximately one hundred comparisons between digested and nondigested samples showed little difference. Consequently, as the number of samples continued to increase, it was finally decided to analyze nondigested samples routinely.

Even with digestion, atomic absorption has the advantage of being able to use one digested sample for twelve or more different metal determinations.

Other reasons for not digesting samples were:

1. Toxic metals must be in a soluble or easily solubilized form to have an effect on the environment.

2. Though results between digested and undigested samples may vary, the undigested value would be on the low side. Any industry found in violation cannot reasonably challenge results over lack of digestion because results favor them.

3. Time saved in the direct reading of samples without digestions exceeds by a factor of 2 to 3 the total analytic time needed when digestion is used. The accomplished work load is doubled or tripled without additional manpower.

The District uses two Perkin–Elmer Model 303 atomic absorption spectrophotometers for measurements. Accessories include a digital concentration readout and a recorder readout with range expansion. Instrument settings recommended in the manufacturer's manual [3] are used. Variations in the acetylene–air ratio and auxiliary pressure are used for optimizing sensitivity for each metal. The detection limit (as given in the literature and verified in our laboratory) for the 12 metals cited earlier is between 0.05 and 0.001 ppm.

Unless the digital concentration readout accessory is used, data must be converted from per cent of absorption to absorbancy and determined graphically.

CALIBRATION

Calibration curves for eight selected metals are shown in Fig. 1. A six-point, or more, calibration curve is run with every set of samples, and standards are rechecked during each run. With unfiltered samples, the sample aspirator has been known to become partially clogged and so give low readings. Consequently, the standard should be checked occasionally to verify proper operation. Frequent rechecks ensure accurate results.

Another point worth mentioning is the use of multimetal standards. Since one metal has little or no effect on the determination of another metal, one stock solution can contain several metals.

INTERFERENCES

There are three types of interferences mentioned in atomic absorption studies. This classification is based on the nature of the

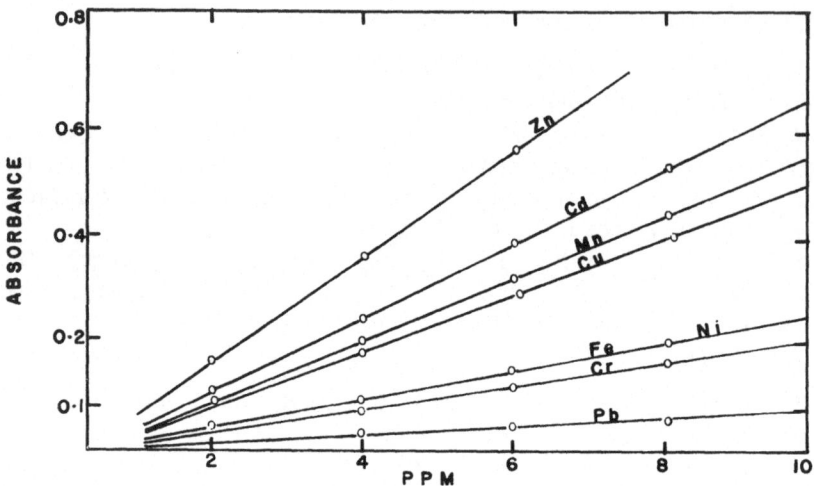

Fig. 1. Calibration curves for eight metals at IX scale expansion.

effect, whether spectral, chemical, or physical. The laboratory usually used single-element lamps, and no spectral overlaps or spectral effects have been noted. In aqueous solution, the physical effects usually due to viscosity differences between samples have not caused problems. The salt concentrations of most waterways samples are within the same order of magnitude, and the concentration of acid in all samples and standards is the same. In nonaqueous solvents, the viscosity is more critical. Even in Lake Michigan water, which is concentrated twentyfold before trace-metal analyses, the salt concentration is still in the same range as many waterway and industrial wastewater samples.

A number of chemical interferences have been listed in the literature for various metals in various matrices. In waterways and wastewater analyses, only the effect of phosphate on the calcium determination caused problems and required the use of a releasing agent. Lanthanum nitrate is added and the sample diluted to give a final lanthanum concentration of 1% by weight per volume. Both samples and standards must contain the same concentration of lanthanum.

Silica is reported to interfere with the manganese and iron determinations. Their presence gives low results. This can be overcome by adding calcium ion to the sample [4]. In most waters in this area, sufficient natural calcium is present to allow the determination of iron and manganese.

Aluminum is reported [4] to interfere with the magnesium determination and give low results. Investigations show as little as 1 ppm can give results 10% lower, and 10 ppm, almost 20% lower. Lanthanum is recommended as a releasing agent. Studies show as little as 50 ppm

of calcium added to the samples eliminates interferences in the presence of up to 10 ppm of aluminum. Except for rare industrial wastewater, all other samples, including those of the lake and river waterways, have less than 1 ppm of aluminum and over 35 ppm of calcium.
Because of these conditions in this area, magnesium is determined
directly on all samples.

DETECTION LIMIT—PRACTICAL

One can define the detection limit in several ways, but, for this
presentation, we shall limit the discussion to workable and practical
levels of concentration. The practical approach can be defined as that
which does not require sample concentration, allows a technician to
perform a large number of analyses, and, lastly, gives the information
needed. The information needed is the real key. For such elements as
sodium, potassium, calcium, and magnesium, atomic absorption easily
measures levels of interest which are in the parts per million. For
heavy metals, especially cadmium, copper, lead, chromium, and nickel,
the usual concentration present is below the practical detection limit
of atomic absorption. The question is: At what level is the concentration of these metals significant? This can only be answered by asking:
What is the future use of the water being analyzed? Since water-use
criteria were not set at the time these laboratory data were being
collected, the highest possible use was assumed. This is a municipal
water supply. If metal pollutants could be measured to at least below
the limits set in the U.S. Public Health Service 1962 Drinking Water
Standards, it was felt this would be sufficient for the study. Copper,
chromium, cadmium, zinc, and manganese could all be measured
easily at or below Drinking Water Standards directly with atomic absorption. No standards are given for nickel, and, until recently, lead
could not be measured below 0.1 ppm without special treatment. By
using a new atomic absorption spectrophotometer, lead can now be
measured directly at 0.05 ppm. There is very little nickel discharged
in industrial wastewater, and it is felt the detection limit of 0.03 ppm
is satisfactory for this area.

RESULTS

The ease of operation of atomic absorption allowed the laboratory
to determine metals on numerous samples. This includes the analysis
of natural waters in the area, such as rivers, lakes, and canals. Nu-

merous industrial wastewater discharges were sampled as part of technical studies involving a water and sewer protection program. Besides water and wastewater samples, a large number of samples related to sewage treatment are analyzed. These include preliminary influent, preliminary effluent, final effluent, and various sludge streams.

The mean concentration of the various elements measured by atomic absorption as part of the waterway surveillance program is presented in Table I. While samples were collected from 67 sampling points, only 3 are presented in this table.

The Lake Michigan data are based on 24 samples from 7 sampling points in the Chicago vicinity. Lake Michigan is the major water source for this area—supplying water for domestic pumpage, industrial use, and diversion.

The Stickney Sewage Treatment Plant (S.T.P) effluent is the largest single source of "used" water in the area, and data are based on approximately 125 samples from 3 sampling points. The low concentration of most toxic metals in the plant effluent should be noted. The Stickney S.T.P. provides approximately 50% of the flow of the Sanitary and Ship Canal at the point of effluent discharge. Any significant concentration of metal in the final effluent would significantly affect stream quality.

Lockport was selected as the third sampling point to present since it is the last downstream point within the District's jurisdiction. It represents the quality of water leaving the canal system. Its evaluation is based on over 30 samples collected weekly for over 7 months.

TABLE I

Mean Concentration of Metals (all results in ppm)

Metals	Lake Michigan at Chicago	Effluent at Stickney S. T. P.	S & S Canal at Lockport
Sodium	4.90	83.2	51.3
Potassium	1.27	10.34	6.74
Calcium	35.6	46.5	58.5
Magnesium	12.0	15.2	18.4
Iron	0.15	0.49	0.806
Manganese	0.010	0.085	0.072
Zinc	0.015	0.28	0.09
Cadmium	< 0.001	< 0.01	< 0.01
Copper	0.009	0.055	< 0.03
Nickel	< 0.003	0.091	< 0.03
Chromium	< 0.002	< 0.02	< 0.02
Lead	< 0.005	< 0.1	< 0.1

Samples from the North and South Branches of the Chicago River, Calumet and Little Calumet Rivers, Des Plaines River, and the three canals showed little evidence of "heavy metal" buildup, including the waterway samples collected in heavy industrialized areas; only manganese, iron, and zinc were among heavy metals measured frequently above detection limits.   Other heavy metals were below detection limits in the majority of samples analyzed.   Out of 568 samples, chromium gave 1 value above 0.02 ppm, nickel gave 5 values above 0.03 ppm, cadmium gave 4 values above 0.01 ppm, and copper gave 20 values above 0.03 ppm.

CONCLUSIONS

Atomic absorption spectroscopy provides a rapid and accurate means of measuring many metals found in waterway wastewater and discharges.   The ease of measurement allows one to both expand and accelerate a pollution-control program, without additional professional staff.   Qualified laboratory technicians can analyze several hundred samples per day with automatic data processing and efficient laboratory management.

By detailed evaluation of both waterways data and related industrial wastewater analyses, considerable additional information can be obtained to expedite pollution control.   This information can include concentration of various metals to be expected in waterways, approximate quantity of each metal disposed to drainage basin, changes in nature of industrial loadings, effectiveness of pollution-control program, and detection of illegal wastewater discharges.   This is in addition to the primary purpose of this study, the measurement of water quality.

ACKNOWLEDGMENTS

The help of George Stanley, who processed much of the original data, and of William Martucci, who performed most of the actual analyses, is acknowledged.

REFERENCES

1. Ordinance for the Control and Abatement of Pollution of Water within the Metropolitan Sanitary District of Greater Chicago, MSDGC, Chicago, 1946.
2. Industrial Waste Ordinance of the Metropolitan Sanitary District of Greater Chicago, MSDGC, Chicago, 1962.

3. Analytical Methods for Atomic Absorption Spectrophotometry, Perkin-Elmer Corporation, Norwalk, Conn., Nov., 1966.

4. J. A. Platte and V. M. Macy, "Atomic Absorption Spectrophotometry as a Tool for the Water Chemist," Industrial Water Eng., May, 1965.

Gas Chromatography

Application of Gas Chromatography in the Petroleum Industry

Donald C. Ford

Sinclair Research, Inc.
Harvey, Illinois

Gas chromatography has become an important analytical tool in virtually every phase of the petroleum industry, from exploration of crude oil and refining of finished products to research on new petrochemicals. Because of the variety and complexity of sample types, petroleum chemists use a broad spectrum of gas chromatographic methods. Both laboratory and process instrumentation are employed extensively. The evolution of gas chromatographic analysis in the petroleum industry will be traced in conjunction with the development of new gas chromatographic instrumentation and specialized techniques.

It has been 15 years since Martin and James published their famous original work [1] on gas chromatography. Petroleum chemists, such as Desty, Keulemans, and Scott quickly adopted this new method of analysis and made very significant contributions. Gas chromatography provides a fast, simple, accurate method of analysis for the complex hydrocarbon mixtures found in petroleum. Many feel that gas chromatography has become the single most important analytical method in this industry. In the latest Analytical Chemistry reviews of analytical applications for petroleum [2], numerous references concerning use of gas chromatography can be noted. At the present, there are over a dozen ASTM methods employing gas chromatography for the petroleum chemist. Since there probably have been more applications to petroleum than any other industry, a comprehensive review can not be presented in this short paper. Instead, a few applications will be discussed to illustrate the development of new gas chromatographic instrumentation and specialized techniques for petroleum analysis.

REFORMATE ANALYSIS

The evolution of reformate analysis in conjunction with the development of new instrumentation can be traced from the use of constant-temperature thermal conductivity chromatographs with long packed columns to the use of temperature-programmed flame-ionization-detector chromatographs with capillary columns. Petroleum mixtures associated with reforming are complex because they contain a large number of individual compounds of different hydrocarbon types. Gas chromatographic procedures have been developed to separate and analyze for all C_7 and lighter paraffins and naphthenes and all C_9 and lighter aromatics.

The first reformate analyses were made approximately 9 years ago, by using constant-temperature thermal conductivity detector instruments with one long packed column. Because it was not possible to resolve all components with any one packed column, a number of individual columns having different aromatic-naphthene-paraffin selectivity were needed. One such procedure published by Winters, Jones, and Martin [3] used three columns: a 26-ft column containing 13.4% isoquinoline, a 40-ft column containing 3.1% 1-chloronaphthalene, and a 12-ft column containing 5% B,B'-oxydipropionitrile. Such methods required a relatively long analysis time, 8 to 12 hr. The calculations were very complex with numerous chances for error. In developing the isoquinoline column, they made use of an important factor affecting selectivity, liquid phase loading.

The development of multiple-stage chromatographs, designed for sequential analysis on different columns with one sample injection, eliminated the need for performing a number of analyses. During the analysis of complex samples, it is possible to separate out unresolved components in one stage and transfer them to another stage in which further separations are obtained concurrently. A four-stage gas chromatograph was designed and constructed at Sinclair Research specifically for the analysis of reformates. The columns employed and the instrumental conditions are given in Table I. A column having a liquid phase highly selective for individual aromatics is used in stage I. All paraffins and naphthenes are transferred to stage II, where a short boiling-point column is used to strip out heavier than C_7 components. The C_7 and lighter fraction is transferred to the main analyzer column in stage III. A fairly high resolution boiling-point column is used to make as complete separation of the paraffin and naphthene components as possible. The nonresolved component peaks are transferred to a secondary analyzing column in stage IV. Here, a moderately selective liquid phase is used to make the remaining sepa-

TABLE I
Multiple Stage Chromatograph for Reformate Analysis

Stage	Column	Column temperature	He carrier gas flow rate, ml/min
I Aromatic stripper	2-ft by $\frac{1}{4}$-in. 40% B,B'-oxypropionitrile on Chromosorb W 35−80 mesh	80°C	150
II Heavy end stripper	6-ft by $\frac{1}{4}$-in. 10% squalene on Chromosorb W 35−80 mesh	80°C	150
III Main analyzing column	20-ft by $\frac{1}{4}$-in. 2.5% cetane on Chromosorb W 35−80 mesh	30°C	150
IV Secondary analyzing column	20-ft by $\frac{1}{4}$-in. 10% di-n-decylphthalate on Chromosorb W 35−80 mesh	30°C	150

rations. Since stage analyses are performed simultaneously, the total analysis time is only approximately 1 hr. Operation of the instrument is complicated, however, and the calculations are somewhat involved.

The introduction of high resolution capillary columns and flame-ionization detector chromatographs provided a new technique eminently suited for the analysis of complex mixtures containing many close boiling components such as reformates. Initially the resolving power of very long capillary columns coated with relatively nonselective liquid phases were used. Desty, Goldup, and Swanton [4] reported using a 900-ft glass column coated with squalane. Martin and Winters [5] employed a 500-ft capillary column coated with 1-octadecene. The analyses times were relatively long, over 3 hr. Schwartz and Brasseau [6] were able to reduce analysis time and improve analysis by adjusting the column selectivity with the use of a mixed liquid phase. The utilization of the different paraffin-naphthene-aromatic selectivity properties of a fluorocarbon, Kel-F10157, in combination with a hydrocarbon, hexadecane, made the necessary separations possible. Schwartz also reported a very interesting phenomenon. The stainless steel tubing used seemed to affect the aromatic-paraffin selectivity. An adsorption capillary column, coated with a colloidal hydrophobic silica, used by Schwartz, Brasseau, and Mathews [7] appears to be promising for reformate analysis.

In developing a capillary column procedure for reformate analysis at Sinclair Research, the moderately polar liquid phase, di-n-decylphthalate, was chosen. The paraffin-naphthene selectivity of di-n-decylphthalate is a function of temperature. The column selectivity toward naphthenes increases as the temperature is raised above 30°C. In order to make the required separations, column selectivity may be adjusted by varying the temperature during the analysis. An important advantage of this liquid phase is that it may be employed at a relatively high temperature. Temperature programming may therefore be used to provide a faster and improved analysis of the C_8 and heavier components of reformate. The instrumental operating conditions for this procedure and the analysis of typical reformer feed and reformate are given in Table II.

HYDROGEN-RICH REFINERY GASES

The analysis of refinery gas streams containing hydrogen in varying amounts from trace levels to over 90% has become an important application of gas chromatography. Such samples also contain nitrogen, carbon monoxide, carbon dioxide, and C_1 through C_5 hydrocarbons. Numerous procedures have been proposed for making this relatively difficult analysis. In general, a chromatograph with thermal conductivity detector is used in conjunction with two different types of columns: an adsorption column, such as molecular sieve, for fixed gases and a partition column for hydrocarbons. The columns may be employed separately or in series.

The selection of the carrier gas is an important factor in this analysis. Helium as the carrier gas gives a small and often anomalous response for hydrocarbons. Nitrogen gives good sensitivity for hydrogen but little or no sensitivity for hydrocarbons. One approach to this detection problem was reported by Purcell and Ettre [8]. They eliminated the anomalous hydrogen-sample peak by using a mixed carrier gas of 8.5% hydrogen in helium. The hydrogen response, however, was relatively poor.

At Sinclair Research, a modified thermal conductivity detector developed by Boys [9] is used to give good sensitivity for all components present. Equal portions of a sample are injected simultaneously into identical column systems, one using helium carrier gas and connected to the sensing side of the detector and the other using nitrogen and connected to the reference side. By summing the simultaneous responses from the two sides of the detector, good sensitivity is obtained for each component.

LIGHT HYDROCARBONS

One of the first gas chromatographic applications in the petroleum industry was the analysis of the C_5 and lighter saturate and olefin fractions of gasoline to monitor gasoline volatility. Long packed columns with liquid phase combinations of different selectivities [10] can resolve all components. Since only analysis of the light fraction is desired, the backflushing technique may be used to remove heavier components from the column. The analysis of this fraction can be related to the total sample by calibration with the use of a constant sample size, an internal standard, or a measurement of the backflush peak.

The analysis of olefin distribution in polymer gasoline has become very important. Most C_7 and lighter olefins can be resolved using a long capillary column such as the 300-ft silicone-coated column of Polgar, Holst, and Groennings [11]. The C_8 and heavier fraction is too complex for isomer analysis; e.g., there are 92 possible C_8 isomers with only a few standards available for identification. A very useful technique to provide more information concerning this material is the *in situ* hydrogenation of the olefins for carbon skeleton determination, as outlined by Beroza [12].

HEAVIER HYDROCARBONS

Crude oil and petroleum fractions, such as gasoline, diesel fuel, and wax, have wide boiling ranges. Temperature programming is required for optimum analysis of such mixtures. The development of pressure programming with flame detectors should also be useful. Simulated distillation of petroleum distillates having end points above 1000°F have been made by Worman and Green [13]. True-boiling-point curves were obtained. Gaylor, Jones, Landerl, and Hughes [14] have developed a method for determining boiling range and hydrocarbon type of distribution of crude oil. Ludwig [15] has separated *n*-alkanes up to carbon number 67 in urea-adductable fractions of waxes.

Selective detectors have been used to analyze for certain types of compounds in complex petroleum mixtures. Such detectors offer high sensitivity to specific classes of compounds and are relatively insensitive to other classes. An electron-capture detector has been used for analysis of lead alkyls in gasoline by Barrall and Ballinger [16], and polycyclic aromatics in wax by Lijinsky and co-workers [17]. Coulometric titration has been applied for analysis of sulfur compounds by Martin and Grant [18] and nitrogen compounds by Martin [19].

PETROCHEMICALS

The rapid growth in the production of petrochemicals is providing numerous applications for gas chromatography. The ability of this

TABLE II

Reformate Analysis by Capillary Column

Chromatographic conditions:
    Column: 300-ft by 0.01- in. stainless steel capillary column
                            coated with di-n-decylphthalate
    Column temperature: 40°C until elution of n C_7 followed by programming
                            from 40°−115°C at 15°.
    Sample size: 0.6 μl split approximately 300:1
    Calculation: Peak area measured by Infotronics CRS−11HS

Component	% Wt			
	Reformer feed		Reformate	
Propane	0.19	0.17		
Isobutane	0.37	0.34	0.04	0.06
n-Butane	0.66	0.63	0.40	0.43
Isopentane	0.91	0.90	10.43	10.41
n-Pentane	0.85	0.81	8.78	8.88
C_5 Olefins	....	....	0.11	0.15
2,2-Dimethylbutane	0.05	0.05	1.76	1.82
2,3-Dimethylbutane	0.16	0.16	1.37	1.35
2-Methylpentane	1.03	1.03	6.38	6.54
Cyclopentane	0.11	0.08	0.38	0.36
3-Methylpentane	0.93	0.89	4.23	4.20
n-Hexane	2.43	2.28	4.14	4.20
C_6 Olefins	....	....	0.02	0.03
2,2-Dimethylpentane	0.10	0.10	0.70	0.71
2,4-Dimethylpentane	0.22	0.26	0.65	0.66
C_6 Olefins	....	....	0.02	0.03
Methylcyclopentane	1.85	1.85	0.51	0.53
2,2,3-Trimethylbutane	....	....	0.07	0.08
3,3-Dimethylpentane	0.09	0.09	0.74	0.72
2-Methylhexane	3.64	3.62	5.65	5.53
Cyclohexane	3.00	3.00	0.03	0.02
2,3-Dimethylpentane	1.60	1.59	2.04	2.07
3-Methylhexane	5.54	5.51	7.14	7.08
1,1-Dimethylcyclopentane	0.79	0.78	0.06	0.07
1-cis-3-Dimethylcyclopentane	2.71	2.72	0.12	0.11
3-Ethylpentane ⎫ 2,2,4-Trimethylpentane ⎭	0.53	0.55	0.73	0.73
1-trans-3-Dimethylcyclopentane	2.50	2.48	0.12	0.13
1-trans-2-Dimethylcyclopentane	4.79	4.76	0.16	0.16
n-Heptane	19.34	19.22	6.15	6.09
Unknown	....	....	0.06	0.08
Unknown	....	....	0.01	0.02
2,2-Dimethylhexane	0.13	0.13	0.72	0.75
Benzene	1.91	1.95	7.17	7.35

TABLE II (concluded)

Component	% Wt			
	Reformer feed		Reformate	
2,5-Dimethylhexane				
1,1,3-Trimethylcyclopentane	2.84	2.88	0.96	0.98
2,4-Dimethylhexane ⎫	1.05	1.05	1.47	1.47
1-*cis*-2-Dimethylcyclopentane ⎭	0.85	0.88		
Methylcyclohexane	19.00	19.19	0.11	0.12
2,2,3-Trimethylpentane	. . . .	. . . .	0.03	0.04
Ethylcyclopentane	1.87	1.91	0.10	0.10
3,3-Dimethylhexane ⎫				
1-*trans*-2-*cis*-4-Trimethylcyclopentane ⎭	2.45	2.51	0.88	0.68
1-*trans*-2-*cis*-3-Trimethylcyclopentane	2.40	2.42	0.06	0.06
2,3,4-Trimethylpentane	. . . .	. . . .	0.01	0.02
2,3-Dimethylhexane ⎫				
2,3,3-Trimethylpentane ⎭	0.25	0.24	0.59	0.57
2-Methylheptane	1.26	1.24	1.14	1.12
4-Methylheptane ⎫				
2-Methyl-3-ethylpentane ⎭	0.23	0.23	0.42	0.41
3-Methylheptane	0.32	0.33	0.75	0.64
3,4-Dimethylhexane ⎫				
3-Ethylhexane ⎭	0.35	0.35	0.24	0.24
n-Octane	0.07	0.08		
Unknown	0.55	0.67	0.04	0.06
Toluene	10.08	10.06	22.29	22.05

technique to analyze for trace impurities in pure organic chemicals is exceedingly valuable. A number of gas chromatographic methods for determination of purity have been published by ASTM. A typical example is ASTM-D2268, "Analysis of High-Purity *n*-Heptane and Isooctane by Capillary Gas Chromatography." Impurities of less than 0.01% can be detected with a hydrogen-flame ionization detector. Other ASTM methods are available for analysis of ethylene, propylene, butadiene, and isoprene.

There is a growing demand for the aromatic petrochemicals: benzene, toluene, xylenes, etc. All C_6 to C_{10} aromatics can be separated by using temperature-programmed capillary columns. A number of different liquid phases have been used. Miyake, Mitooka, and Matsumoto [20] achieved good aromatic separation by coating a capillary column with a mixed liquid phase consisting of 90% dinonylphthalate and 10% 2,4-dinitrochlorobenzene.

The xylene isomers are readily resolved on a capillary column; however, resolution of these isomers is also possible with packed columns. Adsorption columns of Bentone-34 have been used. The mòst successful columns modify the adsorption effects with a liquid phase, e.g., the silicone oil, Bentone-34 column used by Gent and Mortimer [21].

Polymers can also be analyzed by gas chromatography through the use of pyrolysis techniques. The chromatograms of the pyrolysis products can be used for identification and characterization of many polymers. A study of different pyrolytic methods and applications has been published by Perry [22].

Gas chromatography has become one of the most important analytical methods in the petroleum industry. Continuous improvement in instrumentation and techniques will be needed to provide analysis of the many new petroleum products of the future.

REFERENCES

1. A.T. James and A. J.P. Martin, The Analyst 77:915 (1952).
2. D.T. Tuemmler, Anal. Chem. 39:157R (1967).
3. J.C. Winters, F.S. Jones, and R.L. Martin, "Gas Chromatography Guides Development of a Paraffin-Isomerization Process," Sect. V, Fifth World Petroleum Congress, New York, June 1, 1959.
4. D.H. Desty, A. Goldup, and W.T. Swanton, "Performance of Coated Capillary Columns," ISA Third International Gas Chromatography Symposium, June 13, 1961.
5. R.L. Martin and J.C. Winters, Anal. Chem. 35:1930 (1963).
6. R.D. Schwartz and D.J. Brasseau, Anal. Chem. 35:1374 (1963).
7. R.D. Schwartz, D.J. Brasseau, and R.G. Mathews, Anal. Chem. 38:303 (1966).
8. J.E. Purcell and L.S. Ettre, J. Gas Chromatog. 3:69 (1965).
9. F.L. Boys, J. Gas Chromatog. 4:20 (1966).
10. D.C. Ford, J. Gas Chromatog. 1:36 (1963).
11. A.G. Polgar, J.J. Holst, and S. Groennings, Anal. Chem. 34:1226 (1962).
12. M. Beroza, Anal. Chem. 34:1801 (1962).
13. J.O. Worman and L.E. Green, Jr., Anal. Chem. 37:1620 (1965).
14. V.F. Gaylor, C.N. Jones, J.H. Landerl, and E.C. Hughes, Anal. Chem. 36:1606 (1964).
15. F. J. Ludwig, Anal. Chem. 37:1732 (1965).
16. E.M. Barrall and P. R. Ballinger, J. Gas Chromatog. 1:7 (1963).
17. W. Lijinsky, I.I. Domsky, and J. Ward, J. Gas Chromatog. 3:152 (1965).
18. R.L. Martin and J.A. Grant, Anal. Chem. 37:644 (1965).
19. R.L. Martin, Anal. Chem. 38:1209 (1966).
20. H. Miyake, M. Mitooka, and T. Matsumoto, Bull. Chem. Soc., Japan 38:1062 (1965).
21. P.L. Gent and J.V. Mortimer, Nature 197:789 (1963).
22. S.G. Perry, J. Gas Chromatog. 5:77 (1967).

Gas Chromatography in the Study of Pollution

Irving I. Domsky*

Abbott Laboratories
North Chicago, Illinois

Gas chromatography is used in air- and water-pollution studies for determining aliphatic and polynuclear hydrocarbons, for studying automotive
exhaust gases, and for measuring pesticides. Contaminants in the closed
atmospheres of space vehicles can also be monitored. The major problems
encountered are in sampling, where the concentration of a pollutant is
frequently in the parts per million range or less.

This review covers only data published between June, 1965, and
May, 1967. Many worthwhile contributions to pollution studies have
been presented at various meetings, but they are not included herein.
Hopefully, they will be published soon in the various journals and can
then be used by other reviewers.

Gas chromatography has been used in pollution studies to determine a wide variety of compounds, including lower aliphatic hydrocarbons in air and water, polynuclear hydrocarbons in air, carbon
dioxide in air, pesticides in air and water, and contaminants which
can develop inside of spacecraft.

Generally, the usual commercial gas chromatography equipment
can be used, but the special detectors, such as the electron capture,
microcoulometric, or other selective detectors are preferred in
studies on pollution. Burchfield and Wheeler [1] have discussed the
advantages of using a microcoulometric detector in pesticide studies.
The specificity of the detector for halogen, sulfur, or phosphorus in
pesticides makes possible the analysis for pesticides by gas chromatography, where, previously, specific methods (usually colorimetric) were used for each pesticide. The flame ionization detector
would be sensitive enough for pesticide work, but the cleanup required makes the method impractical. The electron capture detector
possesses a very high sensitivity to halogenated pesticides, but care

*Present address: Armour Grocery Products Co., 3115 S. Benson St., Chicago, Ill., 60609.

must be taken in the interpretation of the results since the response for individual compounds varies and a nonhalogenated compound can give a response which could interfere with the pesticide assay. The high sensitivity of the electron-capture detector makes it a very useful detector, however, in pollution studies on the chlorinated hydrocarbons. For water-pollution studies, where the environment is relatively simple, the electron-capture detector is used to best advantage to detect the chlorinated pesticides. Burchfield, Johnson, Rhoades, and Wheeler [2] have published a review comparing the various selective detection systems.

Frostling and Lindgren [3] have described a system for determining organic aerosols and vapors in air. In their system, samples are continuously drawn by a slight vacuum into a flame-ionization detector, and the atmosphere is always monitored for changes in aerosol or organic gas concentration.

One of the applications of gas chromatography in pollution studies where specialized instrumentation is needed is in monitoring closed environmental systems such as spacecraft. The contaminants can occur from human effluents, outgassing of instruments, and refrigerant leakage. These normally insignificant contaminants may result in significant levels of contamination during the atmospheric recycling over extended periods of time. To develop a gas chromatograph to monitor the atmosphere inside the Apollo space vehicle, Huebner, Eaton, and Chaudet [4] started with readily available equipment and modified and miniaturized the components to develop a system capable of withstanding the rigors of space flight. Three columns are used, and the carrier-gas flow rate was 12 ml/min in each column. One column was packed with molecular sieve to separate permanent gasses. The second was packed with Teflon solid support coated with 10% amine 220 and Carbowax 600 (60:40), which is capable of separating carbon dioxide, water, and ammonia. The third column, designed to separate high and medium volatility contaminants, was Chromosorb G coated with 10% Carbowax 20M. The detectors needed for space flight must be very reliable, have a wide dynamic range, and be extremely rugged. The cross-section ionization detector was chosen for this application, even though sensitivity is not as great as the other detectors. Other factors compensated for any lack of sensitivity, which was believed to be adequate. All three detector chambers can be enclosed in the same detector case, which saves space and weight. The volumes of each chamber varied to accommodate the type of sample each would handle. One recorder was used for all three detectors, with samples being injected into each column at different time intervals. By using this system, the concentration of 49 compounds which could be present in a space vehicle can be determined to levels below 10 ppm.

The collection and concentration of samples is the most difficult problem encountered in pollution studies, rather than the instrumentation problem just described. For example, a relatively simple system for measuring carbon dioxide concentrations in air samples has been described by Murray and Doe [5]. Air samples are first passed through a "precut" column, and most of the oxygen and nitrogen is vented into the atmosphere. Then the remainder of the sample is passed through the analytical column. Both columns are 0.25 in. O.D. copper columns, filled with 30 to 60 mesh silica gel. The precut column, 3.5 ft long, is maintained at 60°C, and the analytical column, 2.5 ft long, is maintained at 40°C. The carrier gas is helium. Using a 26 cm^3 gas-sampling loop, the minimum detectable amount of CO_2 that can be detected is 20 ppm. With a larger sample volume, a lower concentration of CO_2 can be detected. Perhaps this simple method can be adapted for analyzing other small molecules.

Ian H. Williams [6], at the British Columbia Research Council Laboratory in Vancouver, has devised methods for studying the concentration of a large number of organic pollutants in the atmosphere. The samples are collected in short (8- by 0.25-in.) stainless steel columns, packed with Chromosorb P partially deactivated with di-n-butyl phthalate and cooled to −80°C with dry ice. The deactivation is necessary to prevent irreversible adsorption of polar compounds. Before being collected, the air samples are dried by passing them through a drying tube containing either K_2CO_3 or $Mg(ClO_4)_2$. When K_2CO_3 is used, most classes of compounds can be collected. Carbonyl, ether, nitrile, and nitro compounds are held back by the $Mg(ClO_4)_2$. The $Mg(ClO_4)_2$-filled tube serves as a subtractive as well as a drying tube if these compounds are present.   A short column filled with molecular sieve or else with concentrated sulfuric acid on an inert substrate between the collection column and the partition column can also serve as subtractive columns. The analytical columns are 2 m long and 6.5 mm in diameter. The two packings which were used were didecylphthalate on Chromosorb P (25:75) and tri-m-tolyl phosphate on Chromosorb W (20:80). An all-glass electron-capture detector was connected in series to this electron-capture detector.

Samples are collected by pulling air samples through the system with either a water aspirator or mechanical vacuum pump. The collection rate is 0.5 liter/min, and samples are generally collected for 5 min. The collecting column is cooled for at least 10 min in dry ice before sampling. After sampling is completed, the pressure is reduced by maintaining a vacuum for 10 sec after the inlet valve is closed. The collecting column is then heated to 120°C for 6 min before the sample is flushed into the analytical column. When a subtractive column is used, the sample is flushed into it directly from the collecting column, after the 6 min heating period.

The air samples were collected on the University of British Columbia campus, which is west of the city and in a location where very few pollutants would be expected. However, 33 different pollutants were identified, including lower hydrocarbons (both aliphatic and cyclic), toluene, benzene, ethylnitrate, and CCl_4. Identification was by comparison with known compounds. The samples were collected with the use of both drying agents, both subtractive columns, two analytical columns, and both detectors. The electron-capture detector was useful only for the CCl_4, ethylnitrate, and tetrachloroethylene. Most of these compounds have been found in automobile exhaust fumes, and it is believed that this is the major source of air pollution in this area.

The greatest use of gas chromatography in air-pollution studies has been in studies on automotive exhaust gas. E. S. Jacobs [7] has recently described a method, where 85 different C_1 to C_{10} hydrocarbons can be assayed by using a single, open, tubular column and programming the analysis from −55 up to 140°C. A flame-ionization detector is used, and sensitivity is 1 ppm. The liquid phase used is DC-200 silicone oil. Retention times compared with standards were used for identifying the components in an exhaust gas. In a typical automotive exhaust gas, there can be about 35 to 40 components of the gas, ranging from methane to n-decane, and this system can complete an analysis in less than 15 min.

Swinnerton and Linnenbom [8] have found hydrocarbons in sea water and have identified a number of C_1 to C_4 hydrocarbons. The water samples, collected in the open ocean, are purged with helium to strip the dissolved hydrocarbons from the solution, and the hydrocarbons are concentrated in traps cooled to dry-ice temperature and filled with appropriate adsorbents. Alumina is used to trap all hydrocarbons except methane, which is trapped on activated charcoal. The hydrocarbons are flushed from the traps with helium into the chromatograph after warming them to 90°C. The concentration of methane was found to be higher than the other hydrocarbons. Many C_2 to C_4 hydrocarbons were positively identified, and some small peaks on the chromatograms were attributed to C_5 hydrocarbons.

A system for characterization of n-alkanes from C_{15} to C_{28} in particulate matter in air has been described by McPherson, Sawicki, and Fox [9]. The particulate matter is collected on glass-fiber filters and separated into aliphatic, aromatic, and oxygenated compounds by the method of Tabor, Hauser, Lodge, and Burttschell [10]. The unsaturated compounds are removed by bromination followed by percolation through an alumina column with pentane as the solvent. Aliquots of this eluate were introduced into the chromatograph. The column was a stainless steel column, 200 by 0.49 cm, packed with 60 to 80 mesh Gas Chrom Z coated with 10% silicone gum rubber. The sample was programmed from 100 to 325°C at 2.5°/min. After the chromatogram was obtained, an aliquot of the sample was passed through a 5A molecular sieve column to remove n-alkanes, and the eluate chromatographed in the GC unit. The difference in area between the two chromatograms corresponds to the areas due to the n-alkanes.

Using the same system for sampling air, DeMaio and Corn [11] have analyzed particulate matter for polynuclear aromatic hydrocarbons. Both SE-30 and Apiezon L were used as liquid phases, and paired columns with dual flame-ionization detectors were used to compensate for the base-line drift during the temperature programming. The relative retention time of standards was used for identifying the compounds, and they were run on both pairs of columns. Once the extract is obtained from the glass filter, an analysis can be completed within a couple of hours. The older method of separating the polynuclear hydrocarbons by column chromatography followed by identification by ultraviolet spectroscopy takes at least 2 days.

Another type of pollutant which has been analyzed from the particulate matter collected on the glass-fiber filters is the insecticides. In a study by E. C. Tabor [12] of insecticide pollution of urban air in several southern communities, large amounts of DDT and other chlorinated pesticides were found in the particulate matter which was collected during the season when crops in the particular area were being sprayed. Generally, the thiophosphates were not found, since they decompose much faster than chlorinated insecticides. All samples were analyzed first with an electron-capture detector and then with a sodium thermionic detector, which is only sensitive to the thiophosphates. For these studies, a microcoulometric detector for verifying the chlorinated pesticides was not available. All of the samples were collected on the glass-fiber filters, and the collections were made for 24-hr periods. The amount of pesticide found on each filter is, in reality, a minimum value since some of the pesticides collected at the beginning of the sampling period may volatilize or decompose before sampling is completed.

The exposure of workers who apply and handle pesticides has been studied by Wolfe, Durham, and Armstrong [13] to evaluate health hazards.   Samples were collected by attaching absorbent cellulose pads or layered gauze pads to various parts of the body or clothing of workers for specified time periods. Respiratory exposure was measured from filter pads in respirators. Air samplers were also used. The various exposure pads were extracted in Soxhlet extractors, and the pesticides were assayed by gas chromatography by using an electron-capture detector and also by other methods. The exposure of workers is generally only a fraction of the toxic dose. There are wide variations depending upon the type of clothing worn by the worker, the winds, and the techniques of the operator.

To study the amount of chlorinated pesticides present in water, either one can sample the water by taking a bottle of water (a grab sample) and extracting it with organic solvents, or one can pump water through a cylindrical cartridge filled with carbon. A settling tank is usually necessary to prevent suspended material from clogging the cartridge. One type of cartridge contains coarse and fine carbon, and between 1000 and 5000 gal of water is passed through at 0.5 gal/min.   A more recent type of cartridge employs only fine carbon, and water is pumped through at the rate of 100 ml/min. Only 1000 liters of water is pumped through, and no filter is needed to keep out suspended material.   Using the larger cartridge, one will lose the organic material adsorbed on suspended matter.   The organic material is extracted from the carbon cartridge by extraction with chloroform followed by ethanol. After a preliminary cleanup on silica gel columns, the aromatic fractions can be analyzed by TLC or GC. Using these techniques, Breidenbach, Gunnerson, Kawahara, Lichten-berg, and Green [14] have monitored pesticide levels in major river basins in the Great Lakes.   Both electron-capture and microcou-lometric detectors are used. The liquid phase used is DC-200 silicone grease on Chromosorb P. Their method gave a lower limit of sensi-tivity of 1 or 2 nanograms of pesticide per liter, depending on the pesticide.

The concentration of the more stable pesticides in areas where they are used regularly remains fairly constant, but the level of degradable pesticides varies considerably, depending on the time available for degradation before sampling the water. The pesticide found most often is dieldrin, with levels of 5 nanograms/liter occurring frequently. It appears that there has been a moderate increase in the amount of DDT over the years.

A comprehensive review of methods used for collecting and measuring the chlorinated hydrocarbon pesticides in water has been published by Breidenbach, Lichtenberg, Henke, Smith, Eichelberger, and Stierli [15].

The major problems encountered in pollution studies involve the collection and preparation of the sample. Filtering systems for both air and water sampling are not quantitative, because usually a long period of sampling is necessary to obtain a sufficient amount of sample. In this time period, a sample collected at the beginning can wash out of the filter or decompose. The instrumentation available for analyzing the samples, once they are collected, is generally adequate. What is needed is better sampling equipment.

REFERENCES

1. H. P. Burchfield and R. J. Wheeler, J. AOAC 49:651 (1966).
2. H. P. Burchfield, D. E. Johnson, J. W. Rhoades, and R. J. Wheeler, "Selective Detection Systems in the Analysis of Drugs and Pesticides," in L. R. Mattick and A. Szymanski, eds., Lectures on Gas Chromatography, 1964 Agricultural and Biological Applications, Plenum Press, New York (1965), p. 129.
3. H. Frostling and P. H. Lindgren, J. Gas Chromatog. 4:243 (1966).
4. V. R. Huebner, H. G. Eaton, and J. H. Chaudet, J. Gas Chromatog. 4:121 (1966).
5. J. N. Murray and J. B. Doe, Anal. Chem. 37:941 (1965).
6. I. H. Williams, Anal. Chem. 37:1723 (1965).
7. E. S. Jacobs, Anal. Chem. 38:43 (1966).
8. J. W. Swinnerton and V. J. Linnenbom, Science 156:1119 (1967).
9. S. P. McPherson, E. Sawicki, and F. T. Fox, J. Gas Chromatog. 4:156 (1966).
10. E. C. Tabor, T. E. Hauser, J. P. Lodge, and R. H. Burttschell, A.M.A. Arch. Ind. Health 17:58 (1958).
11. L. DeMaio and M. Corn, Anal. Chem. 38:131 (1966).
12. E. C. Tabor, Trans. N.Y. Acad. Sci. 28:569 (1966).
13. H. R. Wolfe, W. F. Durham, and J. F. Armstrong, Arch. Environ. Health 14:622 (1967).
14. A. W. Breidenbach, C. G. Gunnerson, F. K. Kawahara, J. J. Lichtenberg, and R. S. Green, Public Health Rept. 82:139 (1967).
15. A. W. Breidenbach, J. J. Lichtenberg, C. F. Henke, D. J. Smith, J. W. Eichelberger, Jr., and H. Stierli, The Identification and Measurement of Chlorinated Hydrocarbon Pesticides in Surface Waters, U.S. Dept. Interior, Washington, D.C. (1966).

The Application of Gas–Liquid Chromatography to the Analysis of Lipids

E.G. Perkins, B.L. Walker,[*]
and C.J. Argoudelis

The Burnsides Research Laboratory
The Food Science Department
University of Illinois
Urbana, Illinois

Gas–liquid chromatography has become an eminently successful technique for the analysis of lipids when used in conjunction with other methods of separation. Further information may be obtained by chemical modification of compounds prior to the analysis of gas–liquid chromatography. Examples of these applications are discussed.

The most important advance in the field of the analysis of lipids in the last two decades was the development of gas–liquid chromatography. Today, by using this technique, it is possible to achieve separations of closely related chemical compounds which were impossible a few years ago. There is no doubt that the relationship of lipids with certain diseases (e.g., atherosclerosis) was the main reason for the progress that has been made in understanding the chemistry and biochemistry of lipids.

In any consideration of the application of gas–liquid chromatography toward the analysis of lipids, several operations must be carried out before the actual analysis can be made. In this discussion, the preparation of sample for analysis, which involves the extraction procedures utilized for lipids from animal and other sources, followed by various methods of separation of lipids into lipid classes, then the conversion of, or chemical modification of compounds for gas–liquid chromatography will be illustrated. The analysis of several representative compounds by gas–liquid chromatography (GLC) will be reviewed.

*Presently with the Department of Nutrition, Guelph University, Guelph, Ontario, Canada.

The lipids in tissues exist largely bound to proteins as lipo-proteins. Bonding by water molecules plays an important part in this union. In order to rupture the lipid—protein "linkage" to obtain the lipid, a mixture of polar and nonpolar solvents is used. Various solvent mixtures, such as ethanol ether, chloroform methanol, etc., have been used for the extraction of lipids from tissues and body fluids. Although optimum conditions for extraction of lipids will vary from tissue to tissue, there are a number of conditions that must be met for good results. As an example, the use of elevated temperature along with solvents may change the lipid composition of the mixture through activation of certain lipolytic enzymes (e.g. phospholipase A, C, and D) [1, 2, 3]. Exposure to atmospheric oxygen and light (which induces peroxide formation) during extraction should be minimal to improve the quality of the final product. The important considerations to be kept in mind during the isolation of lipids from biological sources are [4]:

1. Use of nitrogen atmosphere during the isolation procedures
2. Use of purified and proper solvents
3. Rapid removal of the tissues after sacrificing the animals, to reduce enzymatic changes
4. Quick maceration of the tissues
5. Use of proper solvent to tissue ratio
6. Use of heat only when necessary
7. Removal of nonlipid impurities without loss of lipids
8. Storing lipids under conditions that minimize lipid alterations

The lipids are usually extracted as follows [5]: The tissue or tissue fraction is homogenized with 2:1 chloroform-methanol (v/v) mixture to a final dilution twentyfold the weight of the tissue sample. Brain or tissues of similar consistency are homogenized for about 3 min. Tougher tissues will require homogenization for a longer time or special handling such as grinding with a mortar and pestle at the temperature of dry ice before homogenization with the solvent mixture. Sometimes to ensure complete extraction of the lipids, the tissue which was homogenized with the solvent is subjected to acid or alkaline hydrolysis at reflux temperature. If no fatty acids are recovered from the hydrolysates, it is reasonable to assume that there are no lipids left in the tissue. As lipids tend to solubilize nonlipid materials such as sugars, free amino acids, urea, etc., the following methods have been used to remove the nonlipid components from the lipid extracts. The most reproducible route to the removal of many of the impurities cited above is that of Folch et al. [5]. In this method the lipid extract is partitioned with 0.2 volume of water or an adequate salt solution containing Na, K, Ca, or Mg. The use of salt solutions avoids the interfacial fluffy layer and also decreases, probably by a mass action effect, the loss of more acidic types of phospholipids at the interface or in the water phase. The chloroform layer contains the pure lipids. Cellulose column chromatography has been used by Lea and Rhodes [6] to eliminate satisfactorily the free amino acid found in chicken-egg-mixed phospholipids. Paper (heavy) chromatography has been used by Westley et al. [7] to free phospholipids from amino acids and peptides. Biezenski [8] has used chromatography on silicic acid-impregnated paper. The lipids were eluted from the paper with 20% methanol in chloroform with the impurities left on the paper. Wells and Dittmer [9] passed the lipid extracts through columns prepared of Sephadex. They have reported that the removal of amino acids, carbohydrates, nucleotides, HCl, and inorganic phosphate from lipid extracts by this procedure is of the order of 99%, and lipid recoveries are of 98 to 100%. Dialysis [10] has also been used for the removal of nonlipid contaminants from lipid extracts. Unless the lipid is to be used immediately, it must be stored under conditions which minimize lipid alteration, usually autoxidation. This can be hindered somewhat when extractions are being carried out, by using 0.01% α-tocopherol in the extraction solvent and by temporary storage of the material in solvents containing α-tocopherol.

The next step in the analytical sequence is the separation of the extracted lipid, usually by column chromatography with a monitoring system [11, 12], Fig. 1, or thin-layer chromatography. It is becoming extremely common to use either analytical or preparative thin-layer chromatography [13] to separate most lipid classes. For instance, in Fig. 2, spotting of thin-layer plates with a lipid sample in the form of small spots is illustrated. A more efficient way, shown in Fig. 3, to apply large quantities of sample for use and further analyses is to streak a solution of the sample across a thin-layer plate by using some type of automated apparatus. In this way, a large concentration of sample can be spotted evenly in a short time, whereas using the spotting method on the same amount of sample may take a considerably longer period of time. The thin-layer plate is then developed in a suitable solvent system (hexane, diethyl ether, and acetic acid, 90:10:1, for lipid class separation) [14] in a tank or other containers; and, after the solvent front has reached the desired distance, the plate is removed from the developing chamber, dried, and the bands visualized. The bands can be visualized by spraying with an ultraviolet fluorescing dye such as 2,7-dichlorofluorescein or visualized directly by using an ad-

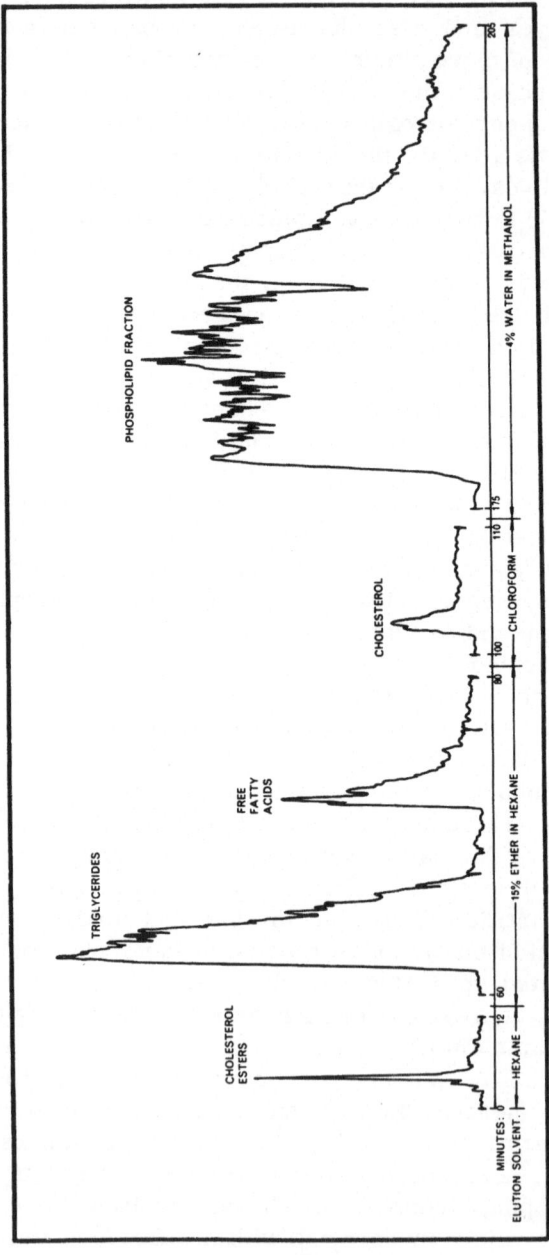

Fig. 1. Column chromatography of rat liver lipid using hydrogen–flame–ionization detection with effluent splitting.

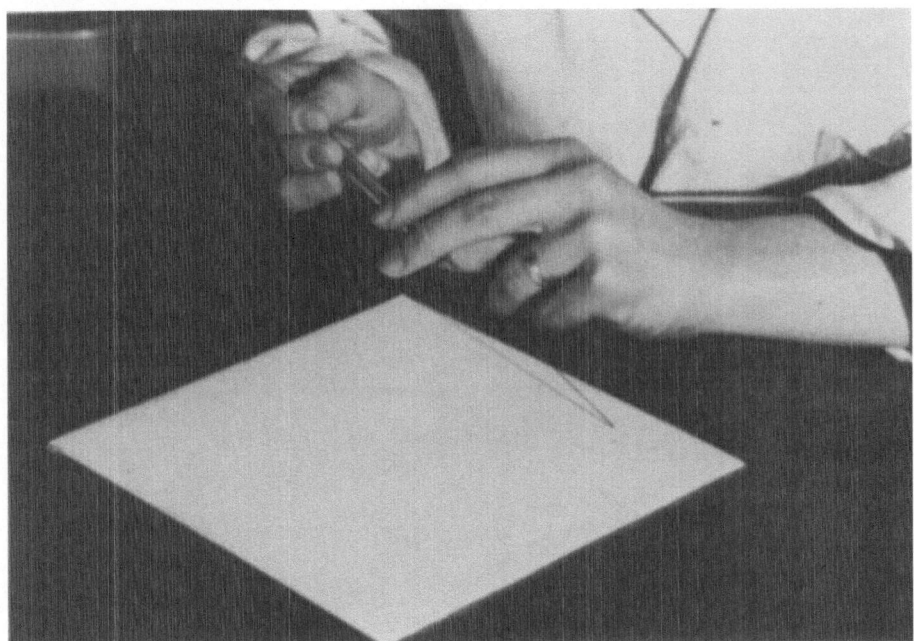

Fig. 2. Manual application of sample to thin-layer plates.

sorbent which has been impregnated with an inorganic fluorescing material, as shown in Fig. 4. The separated material then shows up as dark bands across a light background. For further analysis, these bands can be scraped off, as shown in Fig. 5, into a small vacuum device constructed from a disposable pipette. The device may be inverted to form a small funnel, and the compounds of interest eluted with solvent; a small volume of filtrate is then obtained and concentrated, and the material is ready for further use. Each lipid class is generally converted into methyl esters, usually by interesterification, and analyzed by gas-liquid chromatography.

When working with lipids, methyl esters, as well as other types of organic compounds, it is sometimes necessary to modify the compound chemically prior to gas–liquid chromatography. For instance, again in the field of lipids, unknown peaks on the gas chromatogram may result from unsaturated esters. This can be easily verified by brominating the sample of methyl esters and rechromatographing the sample. Bromination leads to the addition of bromine to double bonds, which renders the compounds less volatile and they are not usually eluted from the column; hence, any peaks appearing on the

Fig. 3. Semiautomatic application of sample to thin-layer plates.

chromatogram of the original ester mixture which do not appear on that of the brominated samples are probably unsaturated, as, e.g., a mixture of stearate and oleate. After bromination, only the methyl stearate is eluted. This method has disadvantages in that brominated compounds are not eluted from the column and will probably cause deterioration of the column packing. The method would only indicate that a given peak may represent an unsaturated compound; it will not give any indication of chain length.

Hydrogenation [15] can be more useful in this respect, as illustrated below. Unknown peaks on the gas chromatogram may result from unsaturated compounds; hydrogenation of the mixture will result in the disappearance of the unsaturated compounds and an increase in the saturated esters of the corresponding chain length. If the increase in the amount of a given saturated ester corresponds to the concentration of the given unsaturated ester which has disappeared on hydrogenation, then the chain length of the unsaturated compound is probably the same as that of the saturated compound, as is shown below. Peak C disappears on hydrogenation, and B increases by the amount of peak C.

Therefore, peaks C and B have the same chain length.

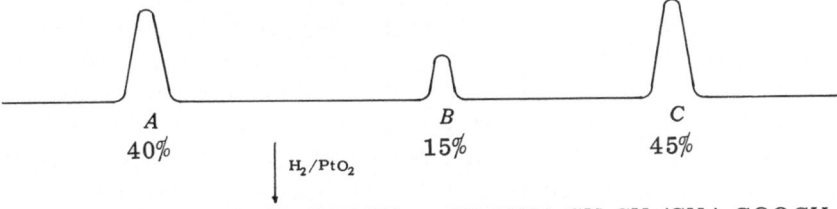

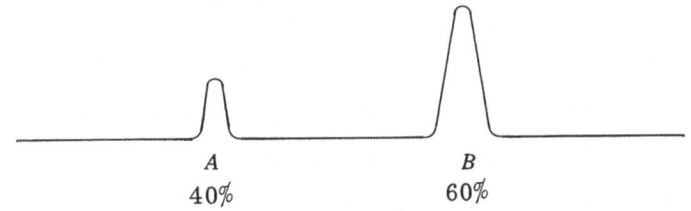

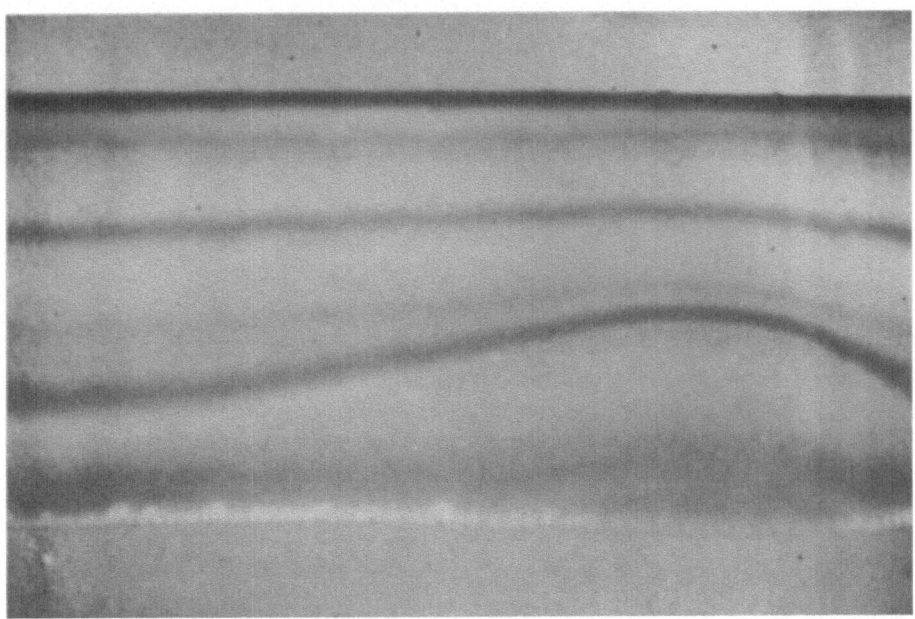

Fig. 4. Visualization of material on thin-layer plates.

Compounds having unsaturation may occur as several different isomers; therefore, it is useful to be able to locate the position of double bonds in unsaturated compounds, such as unsaturated methyl esters. These compounds react with ozone, as shown below, to give ozonides, which then may be cleaved into various products depending upon conditions [16, 17, 18]. Under oxidizing conditions, a fatty acid and monomethyl ester of a dicarboxylic acid will result. For instance, oleoyl ozonide will be cleaved to nonanoic acid, and monomethyl azaleic acid, which indicates that the double bond was originally at the 9 position. If the double bond has been in a different position, a longer-chain monocarboxylic acid or shorter one would have been isolated. These products are then collected and subjected to GLC for identification purposes and the results used in determining the structure of a particular fatty acid.

Ozonolysis of Unsaturated Methyl Esters

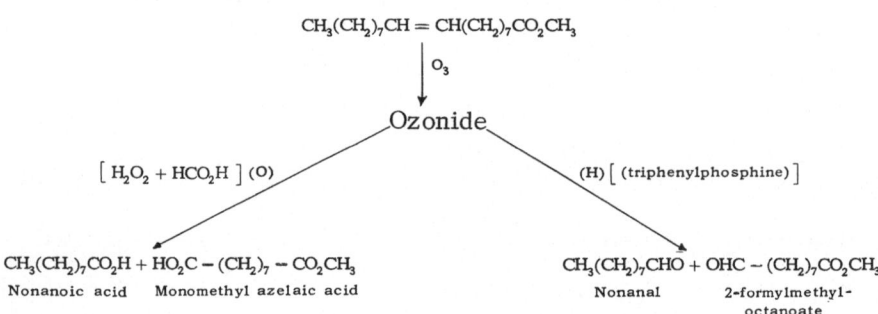

The product from reductive degradation in the presence of triphenylphosphine yields aldehydes and aldehyde esters which can be chromatographed directly. As an example, methyl palmitoleate with a double bond in the 9 position gives a C_7 aldehyde and a C_9 aldehyde ester, whereas methyl oleate has a double bond in the 9 position but it is 2 carbons longer, so the double bond position, as proven by this method, can be in the 9 position. Methyl linoleate has two double bonds, and ozonolysis can illustrate where they are located since three compounds are obtained that represent the cleavage points during ozonolysis. In a similar fashion, the double bonds in methyl linolenate containing three double bonds and methyl arachidonate which contains four double bonds can be located.

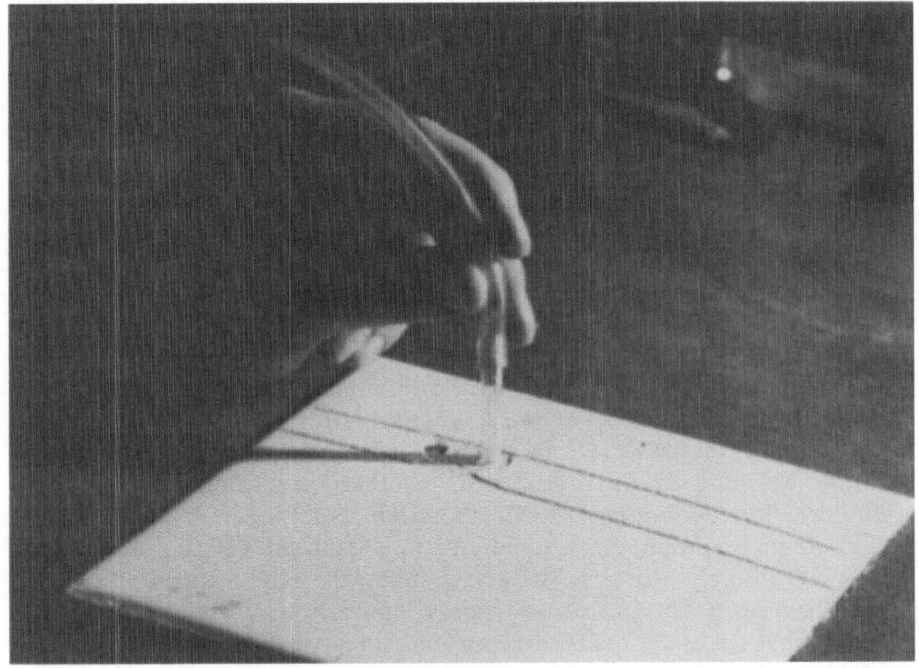

Fig. 5. Removal of adsorbent from thin-layer plates by vacuum.

Many naturally occurring esters are hydroxy esters, but the methyl esters of these hydroxy acids are generally not eluted with ease from polar columns or they have very long retention times. This is due to the polar hydroxy group. The difficulty may be overcome by acylation of the hydroxy group by reaction of the ester with acetic anhydride [19]. The acylated compound may then be readily eluted. The free hydroxy groups of such compounds may also be converted to the corresponding ether, generally the methyl ether by reaction with methyl iodide in the presence of silver oxide [20].

In both cases, the original compounds can be recovered by collection of gas chromatograph effluents and regeneration of the original hydroxy acids. Either manipulation can be used in conjunction with hydrogenation since the hydroxy esters may also be unsaturated. Another recently reported method that has achieved very wide use is that used for the formation of derivatives of hydroxy compounds from hexamethyldichlorosilazane (HMDS) and trimethylchlorosilane (TMCS), as shown below.

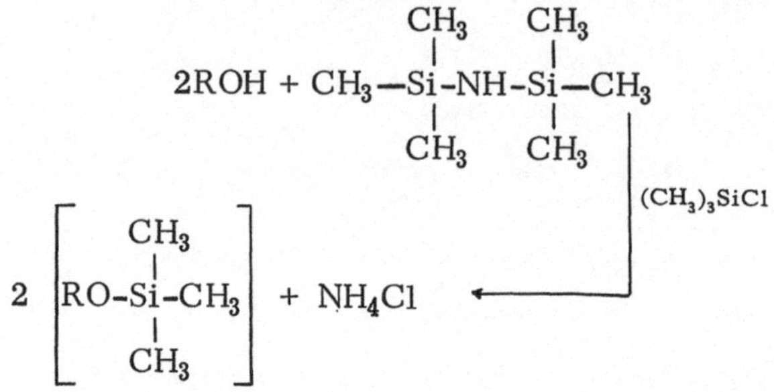

$$2ROH + CH_3-\underset{\underset{CH_3}{|}}{\overset{\overset{CH_3}{|}}{Si}}-NH-\underset{\underset{CH_3}{|}}{\overset{\overset{CH_3}{|}}{Si}}-CH_3$$

(CH₃)₃SiCl

$$2 \left[RO-\underset{\underset{CH_3}{|}}{\overset{\overset{CH_3}{|}}{Si}}-CH_3 \right] + NH_4Cl \longleftarrow$$

Either the trimethylsilyl reagent or dimethyl derivatives react readily with hydroxy acids to give the trimethyl silyl ether derivatives and ammonium chloride [21]. In some cases, where the presence of NH_3 can be harmful to the compound, another silylating reagent, bis-trimethylsilyl acetamide (BSA), as shown below, can react with hydroxy acids to form the trimethylsilyl (TMS) derivatives of hydroxy acids [22]. This material will also form the silyl ester derivatives of carboxylic acids, which are useful in certain cases. These compounds have greatly reduced retention times when compared with the hydroxy ester and acetyl and methyl ether derivatives of hydroxy esters. The original ester can be regenerated with facility upon hydrolysis.

"BSA"   N,O-Bis(trimethylsilyl)acetamide

$$R-\underset{\underset{OH}{|}}{CH}-(CH_2)_x-\overset{\overset{O}{\|}}{C}-OH$$

O – Si (CH₃)₃
|
CH₃– C = N – Si (CH₃)₃

$$R-\underset{\underset{Si(CH_3)_3}{\overset{|}{O}}}{CH}-(CH_2)_x-\overset{\overset{O}{\|}}{C}-O-Si(CH_3)_3$$

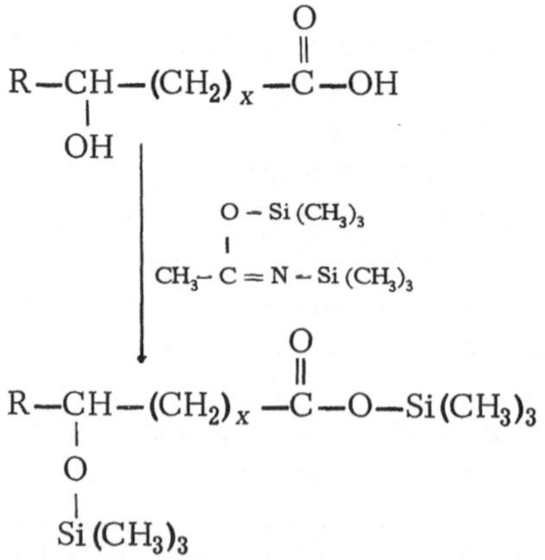

An elegant application of the TMS derivative method to the solution of a difficult separation problem may be seen in Fig. 6 which represents the analyses of mono- and diglycerides, polyglycerides, and polyglycerol esters [23].

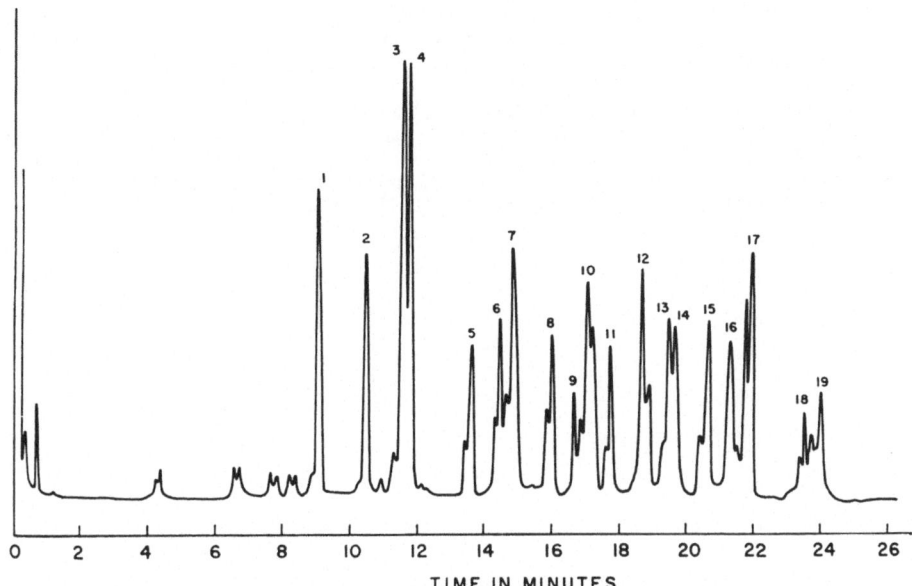

TIME IN MINUTES

Fig. 6. Separation of mono- and diglycerides, polyglycerols, and polyglycerol esters. Column: 3% JXR on 100:120 gas chrom. Q programmer from 125 to 325°C at 10°/min. Peaks 1 to 4 are monoglycerides: 1–myristate; 2–palmitate; 3–oleate; 4–stearate. Peaks 5 to 7 are monofatty acid esters of diglycerol: 5–palmitate; 6–oleate; 7–stearate. Peaks 8 to 10 are monofatty acid esters of triglycerol: 8–palmitate; 9–oleate; 10–stearate. Peaks 11 to 14 are diglycerides: 11–palmitate palmitate; 12–palmitate stearate; 13–oleate oleate; 14–stearate stearate. Peaks 15 to 17 are diglycerol difatty acid esters: 15–palmitate palmitate; 16–oleate oleate; 17–stearate stearate. Peaks 18 and 19 are triglycerol difatty acid esters: 18–oleate oleate; 19–stearate stearate. (Courtesy of Dr. M. Sahasrabudhe, Food and Drug Directorate, Ottawa, Canada.)

REFERENCES

1. D. J. Hanahan, J. Biol. Chem. 207:879 (1954).
2. D. J. Hanahan and I. L. Chaikoff, J. Biol. Chem. 168:233 (1947).
3. D. J. Hanahan and R. Vercamer, J. Am. Chem. Soc. 76:1804 (1954).
4. C. Entenman, J. Am. Oil Chemists, Soc. 38:534 (1961).
5. J. Folch, M. Lees, and S. G. H. Stanley, J. Biol. Chem. 226:497 (1957).
6. C. H. Lea and D. N. Rhodes, Biochem. J. 54:467 (1953).
7. J. Westley, J. J. Wren, and H. K. Mitchell, J. Biol. Chem. 229:131 (1957).
8. J. J. Biezenski, J. Lipid Res. 3:120 (1962).
9. M. A. Wells and J. C. Dittmer, Biochemistry 2:1259 (1963).
10. R. G. Sinclair, J. Biol. Chem. 174:343 (1948).
11. E. G. Perkins, Barber Colman Chromatogram 7:1 (1967).
12. E. U. A. Hashte and T. Nikkari, Acta Chem. Scand. 17:2565 (1963).
13. E. Stahl, Thin–Layer Chromatography, Academic Press, Inc., New York (1965).
14. H. Mangold, J. Am. Oil Chemists, Soc. 37:383 (1960).
15. J. W. Farquar, W. Insull, P. Resen, W. Stoffel, and E. H. Ahrens, Nutr. Revs. 17, Suppl. 8, Part II (1959).

16. R. A. Stein and N. Nicolaides, J. Lipid Res. 3:476 (1962).
17. A. L. Henne and W. L. Perilstein, J. Am. Chem. Soc. 65:2183 (1943).
18. R. G. Ackman, M. E. Retson, L. R. Gallay, and F. A. Vandenheuvel, Can. J. Chem. 39:1956 (1961).
19. J. S. Fritz and G. H. Schlenk, Anal. Chem. 31:1808 (1959).
20. T. Kishimoto and N. Radin, J. Lipid Res. 1:72 (1959).
21. C. Sweeley, R. Bentley, M. Makita, and W. Wells, J. Am. Chem. Soc. 85:2497 (1963).
22. J. F. Klebe, H. Finkbeiner, and D. M. White, J. Am. Chem. Soc. 88:3390 (1966).
23. M. Sahasrabudhe, J. Am. Oil Chemists' Soc. 44:376 (1967).

Index